T0136985

# Constraint Handling in Metaheuristics and Applications

Anand J. Kulkarni · Efrén Mezura-Montes ·
Yong Wang · Amir H. Gandomi ·
Ganesh Krishnasamy
Editors

# Constraint Handling in Metaheuristics and Applications

 Springer

*Editors*
Anand J. Kulkarni
Symbiosis Centre for Research
and Innovation
Symbiosis International University
Pune, Maharashtra, India

Yong Wang
School of Automation
Central South University
Changsha, China

Ganesh Krishnasamy
Monash University Malaysia
Selangor, Malaysia

Efrén Mezura-Montes
Artificial Intelligence Research Center
University of Veracruz
Xalapa, Veracruz, Mexico

Amir H. Gandomi
Faculty of Engineering and Information
Technology
University of Technology Sydney
Ultimo, NSW, Australia

ISBN 978-981-33-6712-8      ISBN 978-981-33-6710-4   (eBook)
https://doi.org/10.1007/978-981-33-6710-4

This Springer imprint is published by the registered company Springer Nature Singapore Pte Ltd.
The registered company address is: 152 Beach Road, #21-01/04 Gateway East, Singapore 189721,
Singapore

# Preface

Several nature-inspired metaheuristic methods have been developed in the past few years and can be categorized into Evolutionary Optimization and Swarm Intelligence methods. The notable examples of Evolutionary Optimization methods are Genetic Algorithms, Evolutionary Strategy, and Differential Evolution. The Swarm Intelligence Methods are Ant Colony Optimization, Particle Swarm Optimization, Artificial Bee Colony, Grey Wolf Optimizer, and krill herd. Besides biology, there are other sources of inspiration for these methods, such as art-inspired methods (e.g., Harmony Search and Interior Search) and socio-inspired methods (e.g., Cohort Intelligence and League Championship Algorithm). These methods can efficiently handle unconstrained problems; however, their performance is severely affected when applied to solve constrained problems. So far, several variations of penalty-based methods, feasibility-based methods, as well as repair-based approaches have been developed. In addition, several problem-specific heuristics have also been developed. The performance of the penalty-based methods is necessarily driven by the choice of the penalty parameter which necessitates a significant number of preliminary trials. The feasibility-based approaches are computationally quite intensive as iterative comparison among the available solution is required and then the solution closer to the feasible region is accepted. Furthermore, the repair approaches necessitate the function to be evaluated along with the constraints and then the solution is modified. This approach may become tedious with an increase in constraints. The development of strong, robust, and generalized constraint handling techniques is very much necessary to efficiently solve complex real-world problems. This edited book intends to provide a platform to discuss the state-of-the-art developments associated with generalized constraint handling approaches/techniques for the metaheuristics and/or the applications being addressed. The book also intends to discuss the core ideas, underlying principles, mathematical formulations, experimentations, solutions, and reviews and analysis of the different constraint handling approaches. It may provide guidelines to the potential researchers about the choice of such methods for solving a particular class of problems at hand. The contributions of the book may further help to explore new avenues leading toward multidisciplinary research discussions.

Every chapter submitted to the book has been critically evaluated by at least two expert reviewers. The critical suggestions by the reviewers helped and influenced the authors of the individual chapter to enrich the quality in terms of experimentation, performance evaluation, representation, etc. The book may serve as a valuable reference for the researchers working in the domain of metaheuristics for solving constrained problems.

The book is divided into two sections. The review of state-of-the-art constraint handling methods, testing, validation, etc., are discussed in Section I. The applications of the metaheuristics incorporated with different constraint handling techniques are discussed in Section II of the book.

## Section I: Constraint Handling Methods, Validation, and Reviews

Chapter "The Find-Fix-Finish-Exploit-Analyze (F3EA) Meta-Heuristic Algorithm with an Extended Constraint Handling Technique for Constrained Optimization" presents a constrained version of Find-Fix-Finish-Exploit-Analyze (F3EA) meta-heuristic algorithm. The epsilon (ε) constrained handling technique is employed to handle constraints within the body of F3EA metaheuristic algorithms and hence the method is referred to as εF3EA. The performance and efficiency of the approach are validated by solving nine well-known constraint benchmark functions and engineering design optimization problems. The algorithm is compared with eight other contemporary algorithms, which include SAFFa, COPSO, ISR, ATMES, SMES, ECHT-EP2, HCOEA, and αSimplex. The solutions highlighted that the εF3EA outperformed all of them. The chapter also underscores the prominent characteristics and steps of the F3EA, such as Find step introducing a new individual selection mechanism via imitating the military radar detection rationale; the Fix step exhibits how the reality of the target monitoring process can be transformed to a single variable constrained optimization problem to obtain a local search operator; the Finish step generates new solutions via a new adaptive mutation operator which is developed through the simulation of projectile motion using Physics equations; the Exploit step tries to take over opportunities presented by the generated potential solution and other members of the population; finally, in Analyze step, the population is updated.

Chapter "An Improved Cohort Intelligence with Panoptic Learning Behavior for Solving Constrained Problems" presents a modified version of Cohort Intelligence (CI) algorithm referred to as CI with Panoptic Learning (CI-PL). The novel PL approach makes every cohort candidate learn the most from the best candidate; however, it does not completely ignore the other candidates. The PL is assisted with a new sampling interval reduction method based on the standard deviation between the behaviors of the candidates. The chapter presents CI-PL solutions to a variety of well-known sets of unconstrained and constrained test problems as well as real-world problems. The approach produced competent and sufficiently robust results solving unconstrained, constrained test, and engineering problems. The analysis of the algorithmic parameters has been discussed using an illustration. It may also layout possible guidelines for parameter selection for other similar problems. The chapter also discusses associated strengths, weaknesses, and possible real-world problem areas where the algorithm can be applied.

Chapter "Nature-Inspired Metaheuristic Algorithms for Constraint Handling: Challenges, Issues, and Research Perspective" reviews the importance of meta-heuristic algorithms, their classification, various constraints handling techniques, applications, etc. The applications reviewed are associated with healthcare, data clustering, power system problem, prediction, etc. The metaheuristics are listed with associated strong application domains considering no-free-lunch approach. The constraint handling techniques such as penalty-based methods have been reviewed associated with the listed metaheuristics.

Chapter "Experimental Comparison of Constraint Handling Schemes in Particle Swarm Optimization" highlights the performance of metaheuristics degenerates when applied for solving constrained problems and further underscores the need for the development of constraint handling techniques. The chapter examines the performance of well-known Particle Swarm Optimization (PSO) algorithm on several benchmarks and constrained problems. In this work, six different penalty function-based approaches are utilized to incorporate the effects of constraints into the PSO algorithm, viz. variants of static penalty method, death penalty method, adaptive penalty, annealing-based penalty function, etc. These approaches are tested for their effectiveness by solving 12 constrained test problems. The results and associated analysis throw light on the suitability of a particular penalty function-based approaches for solving a certain type of constraints using PSO algorithm.

Similar to Chapter "Experimental Comparison of Constraint Handling Schemes in Particle Swarm Optimization", the necessity of specialized techniques as well as algorithmic modifications for the algorithm like PSO when dealing with constrained problems is also underscored in Chapter "Online Landscape Analysis for Guiding Constraint Handling in Particle Swarm Optimisation". Furthermore, it is highlighted that the problem type necessarily decides the choice of constraint handling technique. The pool of such techniques includes death penalty, weighted penalty, and variants of Deb's feasibility ranking. A landscape-aware approach is employed which exploits the rules derived from offline machine learning on a training set of problem instances. These rules drive the mechanism to prefer a specific one or switch between certain constraint handling techniques during PSO search.

Chapter "On the use of Gradient-Based Repair Method for Solving Constrained Multiobjective Optimization Problems—A Comparative Study" discusses the detailed analysis of the effect of repairing infeasible solutions using the gradient information for solving constrained multiobjective problems by employing multi-objective evolutionary algorithms. The gradient-based repair method is combined with a variety of classical constraint handling techniques, viz. constraint dominance principle, penalty function, C-MOEA/D, stochastic ranking, and -constrained and improved -constrained methods. The proposed repair approach exploits the gradient information derived from the constraint set to systematically guide the infeasible solution toward the feasible region. The approach is validated by solving 29 state-of-the-art problems with equality and inequality constraints. The performance of the gradient repair approach is characterized by the robustness of the solutions as well as well-diversified Pareto optimal solutions.

Chapter "MAP-Elites for Constrained Optimization" evaluates the applicability of MAP-Elites to locate the constrained search spaces by mapping them into feature spaces where each feature corresponds to a different constraint. The approach is validated by solving a large set of benchmark problems with a variety of dimensionalities. The approach is characterized by ease of adaptability to include customized evolutionary operators. Furthermore, it does not necessarily need explicit constraint handling techniques. Moreover, the approach preserves diversity, eases the definition of custom tolerance levels for each constraint and illuminates the search space as it provides additional information on the correlation between constraints and objective. It facilitates the interpretation of results through an intuitive visualization.

## Section II Applications of Constraint Handling Methods

Chapter "Optimization of Fresh Food Distribution Route Using Genetic Algorithm with the Best Selection Technique" addresses a methodology to resolve a capacitated model for the food supply chain. The model is a constrained mixed-integer nonlinear programming problem that minimizes the total cost under overall quality level constraints, due to the perishable nature of the products, and the other important such as demand, capacity, flow balance, other costs, etc. The associated practical application of the model is solved using an exact method based on the Branch-&-Bound technique and Genetic Algorithms with different selection methods. The selection methods chosen are tournament selection, stochastic sampling without Replacement and Boltzmann tournament selection. The performance of each selection method is established and validated using statistical analysis. It is important to note that the quality of the set of associated parameters is chosen using the Taguchi method.

The optimal selection of the cutting conditions associated with any of the material removal processes is quite important in the view of efficiency of the process and cost reduction. This is particularly important for the multi-pass face milling operations as it a highly complex problem both theoretically and practically. Chapter "Optimal Cutting Parameters Selection of Multi-Pass Face Milling Using Evolutionary Algorithms", presents various evolutionary optimization techniques such as Genetic Algorithm, hybrid Simulated Annealing Genetic Algorithm are employed to minimize the unit production cost of multi-pass face milling operations while considering several technological constraints. These constraints are associated with machining speed, feed rate, depth of cut, machining force, cutting power, surface roughness, etc. Similar to Chapter "Optimization of Fresh Food Distribution Route Using Genetic Algorithm with the Best Selection Technique", the algorithmic parameters are chosen using Taguchi method.

Chapter "Role of Constrained Optimization Technique in the Hybrid Cooling of High Heat Generating IC Chips Using PCM-Based Mini-channels" highlights the need of temperature control of the electronic components. This is because they are becoming very important in day-to-day practices and the shrinkage of their size has led to the reduction in the effective area available for the heat dissipation. The study emphasizes on the transient numerical simulations on seven asymmetric IC chips kept next to the Left-Right-Bottom-Top and Left-Right-Bottom mini-channels fabricated on the SMPS board using the three different phase change materials. Constrained Genetic Algorithm is used to identify the optimal temperature of the configuration. In addition, a sensitivity analysis is also carried out to critically study the effect of the constraints on the optimal temperature of the IC chips.

Chapter "Maximizing Downlink Channel Capacity of NOMA System Using Power Allocation Based on Channel Coefficients Using Particle Swarm Optimization and Back Propagation Neural Network" presents an attempt made to demonstrate the applicability of the machine learning approaches using Particle Swarm Optimization and Back Propagation Neural Network for the power allocation in Non Orthogonal Multiple Access (NOMA) downlink scenario. The Particle Swarm Optimization is employed to allocate the power such that total sum rate is maximized and Back Propagation Neural Network is used to make the results obtained using the earlier method more efficient constructing the relationship between input and target values. The work is practically important as NOMA is one of the entities of the 5G wireless communication especially when dealing with the channel capacity improvement.

Chapter "Rank Reduction and Diagonalization of Sensing Matrix for Millimeter Wave Hybrid Precoding Using Particle Swarm Optimization" highlights the need for estimation of the channel for mm-Wave wireless communication systems with hybrid precoding as the number of radio frequency chains is much smaller as compared to the number of antennas. The chapter demonstrates the methodology of using Particle Swarm Optimization to design the precoder and decoder of the Beam space channel model with the prior knowledge of associated Angle of arrival and Angle of departure. The algorithm yielded the results with high accuracy and speed for a large number of transmitter and receiver antennas.

Chapter "Comparative Analysis of Constraint Handling Techniques Based on Taguchi Design of Experiments" provides a detailed analysis of the effect of constraint handling techniques such as penalty functions, repair methods, and decoders have on a steady-state genetic algorithm. The performance of every approach is validated by solving several instances of the tourist trip design problem. Importantly, the chapter gives the mapping of the knapsack problem to solve the tourist trip design problem. The chapter in details describes the penalty functions, repair methods, and decoders highlighting their key advantages and limitations. It is worth mentioning that the differences among the constraint handling techniques are tested and validated using the Taguchi design of experiments.

Pune, India                                                    Anand J. Kulkarni
Xalapa, Mexico                                        Efrén Mezura-Montes
Changsha, China                                                  Yong Wang
Sydney, Australia                                          Amir H. Gandomi
Kuala Lumpur, Malaysia                           Ganesh Krishnasamy

# Acknowledgments

We are grateful to the reviewers of the book for their valuable time and efforts in critically reviewing the chapters. Their critical and constructive reviews certainly have helped in the enrichment of every chapter. The editors would like to thank Ms Kamiya Khattar and Divya Meiyazhagan Springer Nature India, for the editorial assistance and cooperation to produce this important scientific work. We hope that the readers will find this book useful and valuable to their research.

# Contents

**Rank Reduction and Diagonalization of Sensing Matrix
for Millimeter Wave Hybrid Precoding Using Particle Swarm
Optimization** ......................................................... 269
Mayank Lauwanshi and E. S. Gopi

**Comparative Analysis of Constraint Handling Techniques Based
on Taguchi Design of Experiments** ................................... 285
Misael Lopez-Sanchez, M. A. Cosío-León, and Anabel Martínez-Vargas

# About the Editors

**Dr. Anand J Kulkarni** holds a PhD in Distributed Optimization from Nanyang Technological University, Singapore, MS in Artificial Intelligence from University of Regina, Canada, Bachelor of Engineering from Shivaji University, India and Diploma from the Board of Technical Education, Mumbai. He worked as a Research Fellow on a Cross-border Supply-chain Disruption project at Odette School of Business, University of Windsor, Canada. Anand was Chair of the Mechanical Engineering Department at Symbiosis International (Deemed University) (SIU), Pune, India for three years. Currently, he is Associate Professor at the Symbiosis Center for Research and Innovation, SIU. His research interests include optimization algorithms, multi-objective optimization, continuous, discrete and combinatorial optimization, multi-agent systems, complex systems, probability collectives, swarm optimization, game theory, self-organizing systems and fault-tolerant systems. Anand pioneered socio-inspired optimization methodologies such as Cohort Intelligence, Ideology Algorithm, Expectation Algorithm, Socio Evolution & Learning Optimization algorithm. He is the founder and chairman of the Optimization and Agent Technology (OAT) Research Lab and has published over 50 research papers in peer-reviewed journals, chapters and conferences along with 4 authored and 5 edited books.

**Dr. Efrén Mezura-Montes** is a full-time researcher with the Artificial Intelligence Research Center at the University of Veracruz, MEXICO. His research interests are the design, analysis and application of bio-inspired algorithms to solve complex optimization problems. He has published over 150 papers in peer-reviewed journals and conferences. He also has one edited book and over 11 book chapters published by international publishing companies. From his work, Google Scholar reports over 6,900 citations. Dr Mezura-Montes is a member of the editorial board of the journals: "Swarm and Evolutionary Computation", "Complex & Intelligent Systems", "International Journal of Dynamics and Control", the "Journal of Optimization" and the "International Journal of Students' Research in Technology & Management". He is a former member of the editorial board of the journals "Computational Optimization and Applications" and "Soft Computing". He is also a reviewer for more than 20 international specialized journals, including the MIT Press Evolutionary Computation Journal, IEEE Transactions on Evolutionary Computation and the IEEE Transactions on Cybernetics. Dr Mezura-Montes is a member of the IEEE Computational Intelligence Society Evolutionary Computation Technical Committee and he is the founder of the IEEE Computational Intelligence Society task force on Nature-Inspired Constrained Optimization. He is a member of the IEEE Systems Man and Cybernetics Society Soft Computing Technical Committee. He is also regular member of the Machine Intelligence Research Labs (MIR Labs). Dr Mezura-Montes is a Level-2 member of the Mexican National Researchers System (SNI). Moreover, Dr Mezura-Montes is a regular member of the Mexican Academy of Sciences (AMC), a regular member of the Mexican Computing Academy (AMEXCOMP) and also a member of Technical Committee of the Mexican Science Council (CONACyT) Collaboration network on Applied Computational Intelligence.

**Dr. Amir H. Gandomi** is a Professor of Data Science and an ARC DECRA Fellow at the Faculty of Engineering & Information Technology, University of Technology Sydney. Prior to joining UTS, Prof. Gandomi was an Assistant Professor at Stevens Institute of Technology, USA and a distinguished research fellow in BEACON center, Michigan State University, USA. Prof. Gandomi has published over two hundred journal papers and seven books which collectively have been cited 20,000+ times (H-index = 66). He has been named as one of the most influential scientific mind and Highly Cited Researcher (top 1% publications and 0.1% researchers) for four consecutive years, 2017 to 2020. He also ranked 18th in GP bibliography among more than 12,000 researchers. He has served as associate editor, editor and guest editor in several prestigious journals such as AE of IEEE TBD and IEEE IoTJ. Prof Gandomi is active in delivering keynotes and invited talks. His research interests are global optimisation and (big) data analytics using machine learning and evolutionary computations in particular.

**Dr. Yong Wang** received the Ph.D. degree in control science and engineering from the Central South University, Changsha, China, in 2011. He is a Professor with the School of Automation, Central South University, Changsha, China. His current research interests include the theory, algorithm design, and interdisciplinary applications of computational intelligence. Dr. Wang is an Associate Editor for the IEEE Transactions on Evolutionary Computation and the Swarm and Evolutionary Computation. He was a recipient of Cheung Kong Young Scholar by the Ministry of Education, China, in 2018, and a Web of Science highly cited researcher in Computer Science in 2017 and 2018.

**Dr. Ganesh Krishnasamy** joined the School of Information Technology, Monash University Malaysia as a Lecturer in July 2019. He received the B.Eng. and M.Eng. degrees in electrical and electronic engineering from Universiti Kebangsaan Malaysia in 2004 and 2007. After working in the manufacturing industry for more than 4 years, he continued his doctoral studies at the University of Malaya where he completed his Ph.D. degree in Electrical Engineering. After obtaining his Ph.D., he worked at Sime Darby Plantation as a data scientist for about a year. His current research interests include the field of computer vision, machine learning, and optimization.

# List of Figures

**On the use of Gradient-Based Repair Method for Solving
Constrained Multiobjective Optimization Problems—A
Comparative Study**

## MAP-Elites for Constrained Optimization

## Optimization of Fresh Food Distribution Route Using Genetic Algorithm with the Best Selection Technique

## Optimal Cutting Parameters Selection of Multi-Pass Face Milling Using Evolutionary Algorithms

## Role of Constrained Optimization Technique in the Hybrid Cooling of High Heat Generating IC Chips Using PCM-Based Mini-channels

## Maximizing Downlink Channel Capacity of NOMA System Using Power Allocation Based on Channel Coefficients Using Particle Swarm Optimization and Back Propagation Neural Network

**Rank Reduction and Diagonalization of Sensing Matrix for
Millimeter Wave Hybrid Precoding Using Particle Swarm
Optimization**

**Comparative Analysis of Constraint Handling Techniques Based
on Taguchi Design of Experiments**

# List of Tables

**Optimal Cutting Parameters Selection of Multi-Pass Face Milling Using Evolutionary Algorithms**

**Role of Constrained Optimization Technique in the Hybrid Cooling of High Heat Generating IC Chips Using PCM-Based Mini-channels**

**Maximizing Downlink Channel Capacity of NOMA System Using Power Allocation Based on Channel Coefficients Using Particle Swarm Optimization and Back Propagation Neural Network**

**Rank Reduction and Diagonalization of Sensing Matrix for Millimeter Wave Hybrid Precoding Using Particle Swarm Optimization**

## Comparative Analysis of Constraint Handling Techniques Based on Taguchi Design of Experiments

# The Find-Fix-Finish-Exploit-Analyze (F3EA) Meta-Heuristic Algorithm with an Extended Constraint Handling Technique for Constrained Optimization

**Ali Husseinzadeh Kashan, Alireza Balavand, and Somayyeh Karimiyan**

**Abstract** As a novel population-based evolutionary algorithm, the Find-Fix-Finish-Exploit-Analyze (F3EA) meta-heuristic algorithm has been introduced for numerical optimization which develops new selection operators, a new parameter-free mutation operator, and a local search operator. The algorithm takes the surface of the objective function as the battleground and mimics the F3EA targeting process of object or installation selection for destruction in the warfare. It performs the main steps of Find-Fix-Finish-Exploit-Analyze (F3EA) in an iterative manner, wherein the Find step introduces a new individual selection mechanism via imitating the military radar detection rationale; in the Fix step, it is shown that how the reality of the target monitoring process can be transformed to a single variable constrained optimization problem to obtain a local search operator; In the finish step, new solutions are generated via a new adaptive mutation operator which is developed through the simulation of projectile motion using physics equations; in the exploit step it tries to take over opportunities presented by the generated potential solution and other members of the population; Finally, in analyze step, the population is updated. In this chapter, an extended epsilon constrained handling technique is used to handle constraints within the body of F3EA meta-heuristic algorithms that is called the $\varepsilon$F3EA. In order to evaluate the efficiency of the F3EA algorithm, nine constraint benchmark functions are used. In addition, the $\varepsilon$F3EA algorithm is compared with eight other algorithms, which included SAFFa, COPSO, ISR, ATMES, SMES, ECHT-EP2, HCOEA, and $\alpha$Simplex. Results indicate that the algorithm is the best of all on benchmark test problem instances.

A. H. Kashan (✉)
Faculty of Industrial and Systems Engineering, Tarbiat Modares University, Tehran, Iran
e-mail: a.kashan@modares.ac.ir

A. Balavand
Department of Industrial Engineering, Science and Research Branch, Islamic Azad University, Tehran, Iran
e-mail: a.balavand@srbiau.ac.ir

S. Karimiyan
Department of Civil Engineering, Islamshahr Branch, Islamic Azad University, Islamshahr, Iran
e-mail: s_karimiyan@iiau.ac.ir

© The Author(s), under exclusive license to Springer Nature Singapore Pte Ltd. 2021
A. J. Kulkarni et al. (eds.), *Constraint Handling in Metaheuristics and Applications*,
https://doi.org/10.1007/978-981-33-6710-4_1

1

**Keywords** Constrained optimization · Constraint-handling techniques · F3EA ·
Constraint violation

# 1 Introduction

Calculation of the optimum solutions for most of the optimization problems is a
hard and complex process. In practice, these solutions are achieved by heuristics
and meta-heuristic algorithms. Metaheuristic algorithms cover a set of approximate
optimization techniques that have been developed for the past three decades. Metaheuristic methods find acceptable solutions in a reasonable time for engineering and
science NP-hard problems. Unlike the exact optimization algorithm, metaheuristic
algorithms do not guarantee that obtained solutions are optimum solutions, but most
of the time, near the optimum solutions, are found by them. On the other hand, an
optimization algorithm provides satisfactory results, and the same algorithm may
have poor performance in other problems [1] especially in the constrained optimization problems. Most real-world optimization problems have constraints and usually,
their decision variables are more than classic functions that increase the dimension of
the problem. When constraints are added to the problem, the solution space has been
more limited, and feasible solution space will be changed. So, they are more complex
rather than classic optimization problems which are a challenge for meta-heuristic
algorithms in recent years [2–6]. Constrained optimization problems usually simulate real-world optimization problems. Different categories of algorithms have been
performed to solve the constrained optimization problem with different methods. A
constrained optimization problem with inequality constraints, equality constraints,
lower bound constraint, and upper bound constraint is shown as (1).

$$\min f(x)$$
$$s.t\, g_i(x) \leq 0 i = 1, \ldots, q$$
$$h_i(x) = 0 i = q + 1, \ldots, m$$
$$l_i \leq x_i \leq u_i i = 1, \ldots, n, \tag{1}$$

where $x = (x_1, x_2, \ldots, x_n)$ is a vector of the decision variables, $f(x)$ is an objective function, $g_i(x) \leq 0, h_i(x) = 0$ are $q$ inequality constraints and $m - q$ equality
constraints,. The constraints in optimization problems increase the complexity and no
longer a combination of some solutions is not feasible. There are a lot of constrained
optimization problems in the field of engineering design [7–11]. The handling
constraints in the constrained optimization problems is very important. Because
choosing the appropriate Bound Constraint Handling Method (BCHM) makes that
the optimization algorithms achieve the most efficiency in most of the problems.
So, different methods have been proposed in the field of BCHM [12]. There are
three methods with different approaches to handling constraints in the constrained
optimization problem [13]. In the following, these methods will be explained.

**Penalty functions**: One of these methods includes the penalty function method [14, 15]. The basic idea of the penalty function is to transfer constraints to the objective function with a penalty factor and transform the optimization problem into an unconstrained problem [15]. According to (2), in the penalty function method, the values of constraints violation $(v_{gi}, v_{hj}, v_{kl})$ are added to the objective functions and as many as the constraints violation are higher, the value of the objective function will be increased. It is important to mention that determining the penalty factor is an optimization problem and depends on the problem [15].

$$\min f(x)$$

s.t.

$$g_i(x) \geq g_0 \quad \forall i, \qquad v_{gi} = \max\left(1 - \frac{g_i(x)}{g_0}, 0\right)$$

$$h_j(x) \leq h_0 \quad \forall i, \qquad v_{hj} = \max\left(\frac{h_j(x)}{h_0} - 1, 0\right)$$

$$k_l(x) = k_0 \quad \forall, \qquad kv_{kl} = \left|\frac{k_l(x)}{k_0} - 1\right| \qquad (2)$$

The basic idea of the penalty function is shown in (3). According to this equation, the value of $v_{gi}$ is added to $f(x)$ when the $g_i(x) \geq g_0$, $v_{hj}$ is added to $f(x)$ when $h_j(x) \leq h_0$, and when $k_l(x) \geq k_0$, $v_{kl}$ will be added to $f(x)$. How adding the penalty function to the objective function is formulated based on (2). One of the critical challenges in the penalty function is how determining the appreciated value of $r$ [15]. The value of $r$ is changed for each constrained optimization problem and determined to appreciate the value of $r$ can have a great impact to reach the optimum value of decision variables.

$$v(x) = f(x) + \sum_{\forall i} r_i \times v_{gi} + \sum_{\forall j} r_j \times v_{hj} + \sum_{\forall l} r_l \times v_{kl} \qquad (3)$$

**Repair methods**: in this method, a generated feasible solution is replaced instead of an infeasible solution. After generating infeasible solutions, this solution has been repaired and replaced instead of an infeasible one.

**Feasibility preserving genetic operators**: in this method, the operators of the genetic algorithm always generate feasible individuals. According to this method, the mutation and crossover are done in a feasible area of solution space and as a result, generated solutions are feasible.

The first idea of handle constraint was proposed in the form of the penalty function method in the DE algorithm. There are different types of penalty functions [16]. The death penalty, quadratic penalty, and substitution penalty. In the death penalty, a constant large value is added to the infeasible individual that makes the value of fitness increase. While the square of constraints violations and fitness value of the repaired solution is the fitness of additive quadratic penalty method. The last method is the substitution quadratic penalty which is the combination of the death penalty and quadratic penalty. The feasibility is not computed for the feasible individual. Instead, the fitness values are the sum of the squared distance from the exceeded bound and a large value [12].

In this paper, an extended epsilon constrained handling technique is proposed to handle constraints within the body of F3EA meta-heuristic algorithm that is called $\varepsilon$F3EA. In the epsilon constraint method, the violation is defined as the sum of all the constraint violations $\emptyset(x)$ [17] based on the Eq. (4):

$$\emptyset(x) = \sum_{\forall i} v_{gi}{}^{p} + \sum_{\forall j} v_{hj}{}^{p} + \sum_{\forall l} v_{kl}{}^{p} \tag{4}$$

The $\varepsilon$F3EA includes several steps of generating random solutions, initialization of the $\varepsilon$ level, ranking the solutions, the fix step, the Find step, the Finish step, the Exploit step, the Analyze step, and checking stopping criteria. In generating random solutions step, random solutions are generated with uniform distribution and create the initial population. In the initialization of the $\varepsilon$ level step, the $\varepsilon$ value is defined. In the ranking of the solutions, the population is sorted based on constraint violations and function values. The Fix step, the Find step, the finish step, and the exploit step are implemented based on the body of the F3EA algorithm. In the analyze step, lexicographic order is used which will be explained. So, the proposed algorithm will be described in Sect. 3. In Sect. 4, the results of using the proposed method are presented on some real-world problems and will be compared with state of art methods, as well as the $\varepsilon$F3EA algorithm is examined, and the conclusion is presented in Sect. 5.

## 2   Background

In the $\varepsilon$ constraint handling method, the constraint violation is defined based on the sum of all violations of constraint in which if $violation > \varepsilon$, the solution is not feasible and the worth of this point is low. In this method, $\emptyset(x)$ precedes $f(x)$ which means the importance of feasibility is more than the objective function value of $f(x)$. The $\varepsilon$ levels have adjusted the value of precedence.

### 2.1   $\varepsilon$ Constrained Method

Assume that $f_1$, $f_2$ are the values of the objective function and $\emptyset_1$, $\emptyset_2$ are constraint violations based on solution $x_1$ and $x_2$. $\varepsilon$ levels comparison for $\varepsilon > 0$ in $(f_1, \emptyset_1)$, $(f_2, \emptyset_2)$ are defined as (5) and (6):

$$(f_1, \emptyset_1) <_\varepsilon (f_2, \emptyset_2) \leftrightarrow \begin{cases} f_1 < f_2, & \text{if } \emptyset_1, \emptyset_2 \le \varepsilon \\ f_1 < f_2, & \text{if } \emptyset_1 = \emptyset_2 \\ \emptyset_1 < \emptyset_2, & \text{otherwise} \end{cases} \tag{5}$$

$$(f_1, \emptyset_1) \le_\varepsilon (f_2, \emptyset_2) \leftrightarrow \begin{cases} f_1 \le f_2, & \text{if } \emptyset_1, \emptyset_2 \le \varepsilon \\ f_1 \le f_2, & \text{if } \emptyset_1 = \emptyset_2 \\ \emptyset_1 < \emptyset_2, & \text{otherwise} \end{cases} \tag{6}$$

The lexicographic order is used when $\varepsilon = 0$, $\varepsilon < 0$ and $\varepsilon \le 0$ in which constraint violation $\emptyset(x)$ precedes the objective value of $f(x)$. The ordinal comparisons between function values are used in the case of $\varepsilon = \infty$. In the constraint handling method, the constraints transfer into objective function and a constrained optimization problem converts into an unconstrained optimization problem using $\varepsilon$ level comparisons. So, constrained optimization problems can be solved by incorporating $\varepsilon$ level comparisons. Ordinary comparisons with $\varepsilon$ level comparisons are used for comparing until the value of $\varepsilon$ converges to 0. By reaching $\varepsilon$ to 0, the feasible solution has been reached. On the other hand, if the violation of a point is greater than $\varepsilon$, the point is infeasible and its worth is low.

The $\varepsilon$ constrained optimization method converts a constrained optimization problem into an unconstrained one with $\varepsilon$ level comparisons that use order relation for replacing. An optimization problem based on $\varepsilon$ a constrained optimization method is defined as (7):

$$minimize\ f(x)$$
$$subjectto\ \emptyset(x) \le \varepsilon \tag{7}$$

It is obvious that by converging $\varepsilon$ to 0, the optimal solution obtained as well as in the penalty function method the optimal solution obtained by increasing the penalty coefficient to infinity.

## 2.2 The F3EA Algorithm

In the field of population-based evolutionary algorithm, the Find-Fix-Finish-Exploit-Analyze (F3EA) meta-heuristic algorithm has been introduced for numerical optimization which develops new selection operators, new parameter-free mutation operator, and a local search operator [6]. The algorithm takes the surface of the objective function as the battleground and mimics the F3EA targeting process of object or installation selection for destruction in the warfare. This algorithm includes five main steps of the Find, Fix, Finish, Exploit, and Analyze step. The Find step mimics based on the object detection process follow by military radars. In the Fix step, the highest peak is found by a local heuristic method. In the Finish step, the intended target is destroyed. In the Exploit step contains information gathering from the target area. Finally, the Analyze step relates to the results of the attack on the target to identify more targeting opportunities [18]. In this chapter, an extended epsilon constrained handling technique is used to handle constraints by using F3EA meta-heuristic algorithm that is called $\varepsilon$F3EA. In the following, the structure of $\varepsilon$F3EA will be explained.

## 3 The F3EA Adapted to Constrained Optimization

The F3EA is a stochastic direct search method based on the population. F3EA has been successfully performed in some nonlinear, multimodal, and constrained engineering design problems. The results have been shown that F3EA is fast and has good convergence in these test functions [6]. F3EA has used the penalty function method to solve the constrained optimization problem so far. In this paper to develop the performance of F3EA for solving the constrained optimization problem, epsilon constrained handling with an automatic update of $\varepsilon$ is proposed. The algorithm of the $\varepsilon$F3EA is as follows:

**Step 1**: **generating random solutions** by which the initial population is randomly generated.

**Step 2**: **initialization of the $\varepsilon$ level** by which an initial $\varepsilon$ level is generated as $\varepsilon(0)$.

**Step 3**: **ranking the solutions**. by which the best members of the population are selected based on the best value of the constraint violations and objective function value, and the rank of individuals is defined by the $\varepsilon$ level comparison.

**Step 4**: **the Fix step**. In this step, the highest peak is found by $fminbnd$ function which is a function for optimization in the Matlab toolbox. In order to destroy the military tank, the projectile launch angle must be located in the highest peak position. So, the highest peak position is found by $fminbnd$ as a local optimum.

**Step 5**: **the Find step**. In this step, a new solution is generated as $p_i^t$ (solution $i$ at iteration $t$) which temporarily considers an artificial radar and other solutions ($p_j^t$) consider military facilities that may or may not be detected by the artificial radar. Then the distance between $p_i^t$ parent solution $i$ at iteration $t$ and child solution $j$ at iteration $t$ ($p_j^t$) is calculated as Eq. (8):

$$R_{ij}^t = \left| f\left(p_i^t\right) - f\left(p_j^t\right) \right| \tag{8}$$

This equation $f\left(p_i^t\right)$ shows the value of the objective function of $p_i^t$ and $f\left(p_i^t\right)$ shows the value of the objective function of $p_j^t$ that show in Eq. (9).

$$R_{maxij}^t = \left(f_{max}^t - f_{min}^t\right) \times u\left(p_i^t\right) \times \sqrt[4]{\frac{\left(1 - u\left(p_j^t\right)\right) \exp\left(-au\left(p_j^t\right)\right)}{u\left(p_i^t\right)}}, i \neq j. \tag{9}$$

where $R^t_{maxij}$ is the maximum range and based on the above equation, if $R^t_{ij} \leq R^t_{maxij}$, $p^t_j$ is detectable by $p^t_i$. Among detectable solutions $p^t_i$, one of the solutions will be selected in the Finish step.

**Step 6: the Finish step.** In this step, new solutions are generated via a new adaptive mutation operator which is developed through the simulation of projectile motion using physics equations. Assume that $p^t_i$ impersonates the position of a projectile launcher and $p^t_j$ shows the position of a target facility. $E^t_i$ is generated by $p^t_i$ and $p^t_j$ which is considered as the explosion position of the artificial projectile launched from $p^t_i$ toward $p^t_j$. How generating $E^t_i$ is shown in Eq. (10).

$$E^t_i = p^t_i + x\left(T^t_{ij}\right)\frac{\left(p^t_j - p^t_i\right)}{p^t_j - p^t_i} + z\left(T^t_{ij}\right)\frac{z^t_{ij}}{z^t_{ij}} \tag{10}$$

$x\left(T^t_{ij}\right)$ and $z\left(T^t_{ij}\right)$ are the projected explosion position on the $x$-axis and $z$-axis, respectively, which is defined as Eqs. (11) and (12):

$$x\left(T^t_{ij}\right) = w^t_{x_{ij}} T^t_{ij} + m^t_{ij}\left(v^t_{0_{ij}} \cos\alpha^t_{ij} - w^t_{x_{ij}}\right)\left(1 - e^{-T^t_{ij}/m^t_{ij}}\right) \tag{11}$$

$$z\left(T^t_{ij}\right) = w^t_{z_{ij}} T^t_{ij} - m^t_{ij} w^t_{z_{ij}}\left(1 - e^{-T^t_{ij}/m^t_{ij}}\right) \tag{12}$$

Based on these equations, $\alpha^t_{ij}$ is a lunch angel, $v^t_{0_{ij}}$ is the initial velocity of the projectile, the mass of projectile is $m^t_{ij}$, wind vector is $w^t_{ij}$, and $z^t_{ij}$ is perpendicular to the hyper-line connecting $p^t_i$ and $p^t_j$. In order to the further investigate the Finish step please refer to [6].

**Step 7: the Exploit step.** In the Exploit step, the information gathering from the target area is done in this step. This step utilizes the result of the Finish step because the result of the Finish step may have premature convergence. So, $c^t_i$ shows the number of changes made in $p^t_i$ and truncated geometric distribution [19] is used for simulating $c^t_i$ as Eq. (13):

$$c^t_i = \left\lceil \frac{\ln\left(1 - \left(1 - \left(1 - q^t_i\right)^n\right)rand(0, 1)\right)}{\ln\left(1 - q^t_i\right)} \right\rceil, c^t_i \in \{1, 2, \ldots, n\} \tag{13}$$

where $q^t_i < 1$, $q^t_i \neq 0$. By the above equation, $c^t_i$ the number of dimensions of $E^t_i$ is randomly selected which are assigned by $U^t_i$. $U^t_i$ is the output of the Exploit step.

**Step 8: Analyze step.** In the Analyze step, the new solution is analyzed that way a comparison between $f\left(U^t_i\right)$ and $f\left(p^t_i\right)$ is performed and if $f\left(U^t_i\right)$ is better than $f\left(p^t_i\right)$, $f\left(U^t_i\right)$ will be replaced. In this step, the comparison between solutions is based on two values of $\emptyset(x)$ and $f(x)$ value. Given that $\emptyset(x)$ precedes $f(x)$, So, at first, the value of $\emptyset(x)$ is checked and if the value of $\emptyset(x)$ is zero, then the current solution is compared based on $f(x)$ value. $f(U_i^t)$ is replaced instead of $G_{best}^t$, if in terms of violation and objective function value, $f(U_i^t)$ is better than $f(G_{best}^t)$. The process of replacing in this step is defined as (14) and (15):

$$(f_i, \emptyset_i) <_\varepsilon (G_{best}, \emptyset_{best}) \leftrightarrow \begin{cases} f_i < G_{best}; if \ \emptyset_i, \emptyset_{best} \leq \varepsilon \\ f_i < G_{best}; if \ \emptyset_i = \emptyset_{best} \\ \emptyset_i < \emptyset_{best}; otherwise \end{cases} \quad (14)$$

$$(f_i, \emptyset_i) \leq_\varepsilon (G_{best}, \emptyset_{best}) \leftrightarrow \begin{cases} f_i \leq G_{best}; if \ \emptyset_1, \emptyset_2 \leq \varepsilon \\ f_i \leq G_{best}; if \ \emptyset_1 = \emptyset_2 \\ \emptyset_i < \emptyset_{best}, ; otherwise \end{cases} \quad (15)$$

In the following, the pseudocode of $\varepsilon$F3EA will be explained based on Fig. 1.

In order to create an initial population of the F3EA, a set of random variables is generated between the lower and upper bounds. Given that feasibility always precedes objective function value, at first the constraint violation of the current population is checked and then the solution is compared based on objective function value. If the constraint violation is smaller than the epsilon value and the objective function value is smaller than the global best, the current solution is replaced as shown in (15), the global best solution is updated and the algorithm enters the main loop.

```
1- generate a set of initial random population and set initial parameters
2- if constraint violation<epsilon threshold
3-   if objective function value<global best value
4-     update global best
5- calculate R_{ij}^f, R_{ij}^{cv}
6- while FE<FEmax
7-   perform the fix step
8-     if constraint violation<epsilon threshold
9-       if objective function value<global best value
10-        update global best
11-  perform the find step
12-    calculate R_{max}^f, R_{max}^{cv} and p_i, p_j
13-  for offspring=1:N0
14-    perform the finish step
15-      calculate E_i^f, E_i^{cv}
16-    perform the exploit step
17-      perform crossover methods and generate offspring
18-      compare offspring with their parent in analyze step
19-        if constraint violation<epsilon threshold
20-          if objective function value<global best value
21-            update global best
22-  end for
23-  update epsilon threshold
25- check stopping criteria
26-   if stopping criteria has not been met
27-     return to line 7
28-     save results
28- else
29-     break
```

**Fig. 1** The pseudocode of $\varepsilon$F3EA

In the Fix step, the new solution is generated based on $fminbnd$ [20] function that this function is one of the classic optimization methods available in the Matlab optimization toolbox. After generating a new solution in the Fix step, the feasibility of the solution is checked and this solution is replaced global best if the better solution is found based on (15). The values of $R^{cv}_{maxij}$ and $R^{ofv}_{maxij}$ are calculated in the Find step. In original paper of F3EA $R^{cv}_{maxij}$ is calculated for all population based on Eq. (9), but in $\varepsilon$F3EA approach the constraint violation ($cv$) and objective function value ($ofv$) are calculated for all solutions in population, therefore, the values of $R^{cv}_{maxij}$ and $R^{ofv}_{maxij}$ are calculated as follow:

$$
\begin{cases}
R^{ofv}_{maxij} = \left(f^t_{max} - f^t_{min}\right) \times u\left(p^{tobv}_i\right) \times \sqrt[4]{\dfrac{\left(1 - u\left(p^{tobv}_j\right)\right) \exp\left(-au\left(p^{tobv}_j\right)\right)}{u\left(p^{tobv}_i\right)}}, i \neq j. if cv < ep \\[4mm]
R^{cv}_{maxij} = \left(cv^t_{max} - cv^t_{min}\right) \times u\left(p^{tcv}_i\right) \times \sqrt[4]{\dfrac{\left(1 - u\left(p^{tcv}_j\right)\right) \exp\left(-au\left(p^{tcv}_j\right)\right)}{u\left(p^{tcv}_i\right)}}, i \neq j. if cv > ep
\end{cases}
\tag{16}
$$

According to Eq. (16), $cv^t_{max}$ and $cv^t_{min}$ are the maximum value of constraint violation of population and the minimum value of constraint violation of population, respectively. $p^{tobv}_i$ is the value of the objective function of population $i$th and $p^{tcv}_i$ is the value of the constraint violation of population $i$th. In this approach, if constraint violation ($cv$) of population $i$th is smaller than epsilon threshold ($ep$), the above section of Eq. (16) is performed, otherwise, the bottom section is performed. Finding detectable solutions is done in the Find step, that way if $R^{tofv}_{ij} \leq R^{tofv}_{maxij}$, $p^t_i$ is detectable by $p^t_i$. This process is also performed based on $R^{cv}_{maxij}$. Mutation is done in the finish step. In this step to increase the exploitation power of $\varepsilon$F3EA, the offspring of each selected solution from the find step are generated. This idea is taken from LCA [3]. In order to identify the number of offspring, the following Eq. (17) is used.

$$
N0 = 2 \times 0.1 \times N
\tag{17}
$$

As mentioned before in the $\varepsilon$F3EA algorithm each solution is investigated by two criteria of objective function value and constraint violation. This method has been also used in the Finish step. That means all of the solutions are once compared based on constraint violation and then are compared based on objective function value. Thus, this process is one of the differences between $\varepsilon$F3EA and F3EA. Generating a new solution in the Finish step depends on $p^t_i, p^t_j$. Assume that $p^t_i$ impersonates the position of the projectile launcher and $p^t_j$ simulates the position of a target facility. $E^t_i$ is the position of explosion that has been thrown from $p^t_i$ toward $p^t_j$. Let $\left[p^t_i, u\left(p^{tobv}_i\right)\right]$, $\left[p^t_i, u\left(p^{tcv}_i\right)\right]$, $\left[p^t_j, u\left(p^{tobv}_j\right)\right]$ and $\left[p^t_j, u\left(p^{tcv}_j\right)\right]$. In order to simulate the launch, it is needed to the inclination ($\beta_{ij}$), launch angle ($\beta_{ij}$), the projectile initial velocity ($v_{0ij}$), and distance between $p^t_i$ and $p^t_j$. All of the Finish stage has been explained in [6].

In the Finish step, the mutation is done based on Eq. (10) with the difference that each parent has the number of $N0$ offspring. After generating $E^t_i$ in each iteration, the crossover method is done on $E_i^t$ in the exploit step. There are two kinds of crossover in $\varepsilon$F3EA that one of them is related to Eq. (13) and another crossover will be explained in the following:

- To the number of population, a permutation vector was generated as $(p)$.
- The first three members of $p$ were chosen as $x^1, x^2, x^3$ in each iteration of the current population.
- $x_{ap}^t$ was created based on Eq. (18):

$$x_{ap}^{t+1} = \frac{\left(x^{1,t}.x^{2,t}\right)}{x^{3,t}} \tag{18}$$

$x^{1,t}$, $x^{2,t}$, and $x^{3,t}$ represent the first, second, and third members of the vector $p$ in the iteration t, respectively. Equation (18) improved the quality and diversity of the solutions produced. In the exploit step for each parent, the $N0$ offspring are generated by crossover methods and all of the offspring are compared with their parents. All of the generated solutions are compared based on Eq. (15) in analyze step which means at first, the constraint violation, and then the objective function value is checked. If a solution is better than the global best, the global best is updated. Finally, the value of epsilon becomes smaller in each iteration that the value of epsilon is calculated as Eq. (19):

$$\varepsilon = \varepsilon \times \left(1 - \frac{FE}{Fmax}\right)^\delta \tag{19}$$

According to Eq. (19), $FE$ is the number of function evaluations, $Fmax$ is the maximum number of function evaluations and $\delta$ is a constant number and has been considered 50. This process makes the value of epsilon is converged to zero that means that in the initial iterations the solution with positive constraint violation is accepted and as many as the algorithm approach the final iterations, the constraint violation of all solution should be zero and the solution are compared based on objective function value.

If the number of functions evaluation $(FE)$ exceeds the maximum number of functions evaluation $(FE_{max})$, the algorithm is stopped.

## 4 Experiments

In this section, the $\varepsilon$F3EA performs to evaluate the benchmark constrained optimization problems. These benchmark functions include 13 well-studied problems (g01–g09) of the CEC 2006 test suite [21], where we compare $\varepsilon$F3EA with the result of a study that has been written by Husseinzadeh Kashan [3]. In that study, the league champion algorithm (LCA) [22] has been compared with eight state of the art algorithms. G01 to g09 has some features of the number of decision variables ($n$), estimates the ratio between the feasible region and the entire search space ($\rho$), the number of linear inequalities, the number of nonlinear inequalities, the number of linear equalities, and the number of nonlinear equalities. In this study, the $\varepsilon$F3EA will be compared with eight other algorithms of SAFFa [23], COPSO [24], ISR [25], ATMES [26], SMES [27], ECHT-EP2 [28], HCOEA [29], and $\alpha$Simplex [30].

In order to solve the benchmark functions, it is necessary to explain that with 225,000 number of function evaluations, $\varepsilon$F3EA almost finishes its search, since all individuals converge to the global optimum with acceptable precision.

## 4.1  G01 Benchmark Function

According to (20), the objective function is a Quadratic function with 13 decision variables. Estimating the ratio between the feasible region and the entire search space ($\rho$) is 0.0111. The number of linear inequalities constraints is nine, this function does not have the linear inequalities constraints, in nonlinear equalities constraints, and nonlinear equalities constraints. Constraints of g1, g2, g3, g7, g8, and g9 are active.

$$Minimize\ f(X) = 5\sum_{d=1}^{4} x_d - 5\sum_{d=1}^{4} x_d^2 - \sum_{d=5}^{13} x_d$$

subject to:

$$g_1(X) = 2x_1 + 2x_2 + x_{10} + x_{11} - 10 \le 0$$
$$g_2(X) = 2x_1 + 2x_3 + x_{10} + x_{12} - 10 \le 0$$
$$g_3(X) = 2x_2 + 2x_3 + x_{11} + x_{12} - 10 \le 0$$
$$g_4(X) = -8x_1 + x_{10} \le 0$$
$$g_5(X) = -8x_2 + x_{11} \le 0$$
$$g_6(X) = -8x_3 + x_{12} \le 0$$
$$g_7(X) = -2x_4 - x_5 + x_{10} \le 0$$
$$g_8(X) = -2x_6 - x_7 + x_{11} \le 0$$
$$g_9(X) = -2x_8 - x_9 + x_{12} \le 0 \tag{20}$$

In Table 1, BV is the best value, AV is the average value, WV is the worst value and STD shows the standard deviation value. The results of Table 1 show that there is no significant difference between BV, AV, WV, and STD. The best result is related to the $\varepsilon$F3EA, SMES, COPSO, and SAFFa algorithms.

**Table 1** Comparative results of $\varepsilon$F3EA with other algorithms in g01 problem

| Parameters | SAFFa | COPSO | ISR | ATMES | SMES | ECHT-EP2 | HCOEA | αSimplex | εF3EA |
|---|---|---|---|---|---|---|---|---|---|
| BV | −15.0000 | −15.000000 | −15.000 | −15.000 | −15.0000 | −15.0000 | −15.000 | −14.9999998 | −15 |
| AV | −15.0000 | −15.000000 | −15.000 | −15.000 | −15.0000 | −15.0000 | −15.000 | −14.9999998 | −15 |
| WV | −15.0000 | −15.000000 | −15.000 | −15.000 | −15.0000 | −15.0000 | −15.000 | −14.9999998 | −15 |
| STD | 0 | 0 | 5.8E−14 | 1.6E−14 | 0 | 0.00E+00 | 4.3E−7 | 6.4E−06 | 0 |

## 4.2   G02 Benchmark Function

This function is nonlinear which has 20 variables based on (21). The value of $\rho$ is 99.8474. The number of linear inequalities constraints, linear inequalities constraints, nonlinear equalities constraints, and nonlinear equalities constraints are 1, 1, 0, 0, respectively. The constraint g1 is close to being active.

$$Minimize\ f(X) = -\left| \left( \sum_{d=1}^{n} \cos^4(x_d) - 2\prod_{d=1}^{n} \cos^2(x_d) \right) \middle/ \sqrt{\sum_{d=1}^{n} dx_d^2} \right|$$

subject to:

$$g_1(X) = 0.75 - \prod_{d=1}^{n} x_d \leq 0$$

$$g_2(X) = \sum_{d=1}^{n} x_d - 0.75n \leq 0 \tag{21}$$

where $n = 20$ and $0 \leq x_d \leq 1$. The best known objective value is $f(X^*) = -0.80361910412559$.

The results in Table 2 show that the COPSO, ISR, $\alpha$Simplex, and ECHT-EP2 have had the best value among the algorithms. Whatever the value of AV, WV, and STD is less, it means that the performance of the algorithm is better. Accordingly, in terms of the AV, WV, and STD, the $\varepsilon$F3EA has had the best values among eight other algorithms.

## 4.3   G03 Benchmark Function

The g03 function is Polynomial with 10 decision variables in which the value of $\rho$ is zero. It has only one equality nonlinear constraints. The equation of g03 is as (22):

$$Minimize\ f(X) = -(\sqrt{n})^n \prod_{d=1}^{n} x_d$$

subject to:

$$h(X) = \sum_{d=1}^{n} x_d^2 - 1 = 0 \tag{22}$$

where $n = 10$ and $0 \leq x_d \leq 1$ $(d = 1, \ldots, n)$. The global objective value is $f(X^*) = -1$.

**Table 2** Comparative results of $\varepsilon$F3EA with other algorithms in g02 problem

| Parameters | SAFFa | COPSO | ISR | ATMES | SMES | ECHT-EP2 | HCOEA | αSimplex | εF3EA |
|---|---|---|---|---|---|---|---|---|---|
| BV | -0.80297 | -0.803619 | -0.803619 | -0.803388 | -0.803601 | -0.8036191 | -0.803241 | -0.8036191 | -0.8036069 |
| AV | -0.79010 | -0.801320 | -0.782715 | -0.790148 | -0.785238 | -0.7998220 | -0.801258 | -0.7841868 | -0.8035734 |
| WV | -0.76043 | -0.786566 | -0.723591 | -0.756986 | -0.751322 | -0.7851820 | -0.792363 | -0.7542585 | -0.803489 |
| STD | 1.2E-02 | 4.59E-03 | 2.2E-02 | 1.3E-2 | 1.67E-2 | 6.29E-3 | 3.8E-03 | 1.3E-02 | 2.6324E-05 |

Given that the g03 function has less complexity compared to other benchmarks, the difference values between functions are not significant based on the results of Table 3. But in terms of the STD, the $\varepsilon$F3EA has had the best performance compared to other results in this table and it means that in all iterations, the $\varepsilon$F3EA has reached the optimal solution.

## 4.4 G04 Benchmark Function

The g04 function is quadratic with five decision variables. Estimating the ratio between the feasible region and the entire search space ($\rho$) is 52.1230. There are only six nonlinear equalities constraints. Constraints g1 and g6 are active which its equation is based on the (23):

$$Minimize\ f(X) = 5.3578547x_3^2 + 08356891x_1x_5 + 37.293239x_1 - 40792.141$$
subject to:

$$g_1(X) = 85.334407 + 0.0056858x_2x_5 + 0.0006262x_1x_4 - 0.0022053x_3x_5 - 92 \le 0$$
$$g_2(X) = -85.334407 - 0.0056858x_2x_5 - 0.0006262x_1x_4 + 0.0022053x_3x_5 \le 0$$
$$g_3(X) = 80.51249 + 0.0071317x_2x_5 + 0.0029955x_1x_2 + 0.0021813x_3^2 - 110 \le 0$$
$$g_4(X) = -80.51249 - 0.0071317x_2x_5 - 0.0029955x_1x_2 - 0.0021813x_3^2 + 90 \le 0$$
$$g_5(X) = 9.300961 + 0.0047026x_3x_5 + 0.0012547x_1x_3 + 0.0019085x_3x_4 - 25 \le 0$$
$$g_6(X) = -9.300961 - 0.0047026x_3x_5 - 0.0012547x_1x_3 - 0.0019085x_3x_4 + 20 \le 0 \qquad (23)$$

With boundary conditions $78 \le x_1 \le 102$, $33 \le x_2 \le 45$, and $27 \le x_d \le 45 (d = 3, 4, 5)$. The optimum objective value is $f(X^*) = -30665.539$.

Table 4 shows the superior performance of all algorithms in the g04 function. The ATMES and $\varepsilon$F3EA have had the best performance compared to other algorithms.

## 4.5 G05 Benchmark Function

The G05 benchmark is a cubic function in which $\rho = 0$. According to (24), this function has two linear inequalities constraints and three nonlinear equalities constraints. The equation of g05 has defined as follows:

$$Minimize\ f(X) = 3x_1 + 0.000001x_1^3 + 2x_2 + (0.000002/3)x_2^3$$

subject to:

$$g_1(X) = -x_4 + x_3 - 0.55 \le 0$$
$$g_2(X) = -x_3 + x_4 - 0.55 \le 0$$

**Table 3** Comparative results of $\varepsilon$F3EA with other algorithms in g03 problem

| Parameters | SAFFa | COPSO | ISR | ATMES | SMES | ECHT-EP2 | HCOEA | $\alpha$Simplex | $\varepsilon$F3EA |
|---|---|---|---|---|---|---|---|---|---|
| BV | −1.00000 | −1.000005 | −1.001 | −1.000 | −1.000 | −1.0005 | −1.000 | −1.0005001 | −1.0000 |
| AV | −0.99990 | −1.000005 | −1.001 | −1.000 | −1.000 | −1.0005 | −1.000 | −1.0005001 | −1.0000 |
| WV | −0.99970 | −1.000003 | −1.001 | −1.000 | −1.000 | −1.0005 | −1.000 | −1.0005001 | −1.0000 |
| STD | 7.5E−05 | 3.16E−07 | 8.2E−09 | 5.9E−5 | 2.09E−4 | 0.00E+00 | 1.3E−12 | 8.5E−14 | 0 |

**Table 4** Comparative results of $\varepsilon$F3EA with other algorithms in g04 problem

| Parameters | SAFFa | COPSO | ISR | ATMES | SMES | ECHT-EP2 | HCOEA | αSimplex | εF3EA |
|---|---|---|---|---|---|---|---|---|---|
| BV | −30665.50 | −30665.538 | −30665.539 | −30665.539 | −30665.539 | −30665.538 | −30665.539 | −30665.538 | −30665.539 |
| AV | −30665.20 | −30665.538 | −30665.539 | −30665.539 | −30665.539 | −30665.5387 | −30665.539 | −30665.538 | −30665.539 |
| WV | −30663.30 | −30665.538 | −30665.539 | −30665.539 | −30665.539 | −30665.538 | −30665.539 | −30665.538 | −30665.539 |
| STD | 4.85E−01 | 0 | 1.1E−11 | 7.4E−12 | 0 | 0.00E+00 | 5.4E−7 | 4.2E−11 | 1.0042E−11 |

$$h_3(X) = 1000 \sin(-x_3 - 0.25) + 1000 \sin(-x_4 - 0.25) + 894.8 - x_1 = 0$$
$$h_4(X) = 1000 \sin(x_3 - 0.25) + 1000 \sin(x_3 - x_4 - 0.25) + 894.8 - x_2 = 0$$
$$h_5(X) = 1000 \sin(x_4 - 0.25) + 1000 \sin(x_4 - x_3 - 0.25) + 1294.8 = 0 \tag{24}$$

With boundary conditions $0 \leq x_1 \leq 1200$, $0 \leq x_2 \leq 1200$, $-0.55 \leq x_3 \leq 0.55$, and $-0.55 \leq x_4 \leq 0.55$. The best known objective value is $f(X^*) = 5126.49671$.

In Table 5, the comparative results of all the algorithms have been shown for solving the g05 problem. These results show that all algorithms have had good results that except the SAFFa, the COPSO, and the SMES, other algorithms have reached the optimal solution.

### 4.6   G06 Benchmark Function

The g06 is a kind of cubic function with two decision variables and two nonlinear inequalities constraints. Estimating the ratio between the feasible region and the entire search space is 0.0066. in this function, all constraints are active. Its equation is based on (25):

$$Minimize \ f(X) = (x_1 - 10)^3 + (x_2 - 20)^3$$

subject to:

$$g_2(X) = (x_1 - 6)^2 + (x_2 - 5)^2 - 82.81 \leq 0 \tag{25}$$

With boundary conditions $13 \leq x_1 \leq 100$, and $0 \leq x_2 \leq 100$. The optimum objective value is $f(X^*) = -6961.8139$.

Given that the g05 is two dimensions function and it has only two constraints, the comparative results are not a significant difference according to the results of Table 7. COPSO, $\alpha$Simplex, ECHT-EP2, and $\varepsilon$F3EA have had the best values. The values of AV and WV are similar to each other in four functions. But in terms of STD, the COPSO is outperformed.

### 4.7   G07 Benchmark Function

According to (26), it is obvious that g07 has 10 decision variables and is quadratic. The value of $\rho$ is 0.0003. It has three linear inequalities constraints and five nonlinear inequalities constraints. The constraints g1, g2, g3, g4, g5, and g6 are active.

$$Minimize \ f(X) = x_1^2 + x_2^2 + x_1 x_2 - 14x_1 - 16x_2 + (x_3 - 10)^2 + 4(x_4 - 5)^2 +$$

**Table 5** Comparative results of $\varepsilon$F3EA with other algorithms in g05 problem

| Parameters | SAFFa | COPSO | ISR | ATMES | SMES | ECHT-EP3 | HCOEA | αSimplex | εF3EA |
|---|---|---|---|---|---|---|---|---|---|
| BV | 5126.989 | 5126.498096 | 5126.497 | 5126.497 | 5126.599 | 5126.4967 | 5126.498 | 5126.4967 | 5126.4967 |
| AV | 5432.08 | 5126.498096 | 5126.497 | 5127.648 | 5174.492 | 5126.4967 | 5126.498 | 5126.4967 | 5126.4967 |
| WV | 6089.43 | 5126.498096 | 5126.497 | 5135.256 | 5304.167 | 5126.4967 | 5126.498 | 5126.4967 | 5126.4967 |
| STD | 3.89E+03 | 0 | 7.2E−13 | 1.80E+00 | 5.00E+01 | 0.00E+00 | 1.7E−7 | 3.5E−11 | 4.13E−13 |

**Table 6** Comparative results of εF3EA with other algorithms in g06 problem

| Parameters | SAFFa | COPSO | ISR | ATMES | SMES | ECHT-EP2 | HCOEA | αSimplex | εF3EA |
|---|---|---|---|---|---|---|---|---|---|
| BV | −6961.800 | −6961.8139 | −6961.814 | −6961.814 | −6961.814 | −6961.8139 | −6961.814 | −6961.8139 | −6961.8139 |
| AV | −6961.800 | −6961.8139 | −6961.814 | −6961.814 | −6961.284 | −6961.8139 | −6961.814 | −6961.8139 | −6961.8139 |
| WV | −6961.800 | −6961.8139 | −6961.814 | −6961.814 | −6952.482 | −6961.8139 | −6961.814 | −6961.8139 | −6961.8139 |
| STD | 0 | 0 | 1.9E−12 | 4.6E−12 | 1.85E+0 | 0.00E+00 | 8.5E−12 | 1.85E−10 | 1.85E−12 |

$$(x_5 - 3)^2 + 2(x_6 - 1)^2 + 5x_7^2$$
$$+7(x_8 - 11)^2 + 2(x_9 - 10)^2 + (x_{10} - 7)^2 + 45$$

subject to:

$$g_1(X) = -105 + 4x_1 + 5x_2 - 3x_7 + 9x_8 \leq 0$$
$$g_2(X) = 10x_1 - 8x_2 - 17x_7 + 2x_8 \leq 0$$
$$g_3(X) = -8x_1 + 2x_2 + 5x_9 - 2x_{10} - 12 \leq 0$$
$$g_4(X) = 3(x1 - 2)^2 + 4(x_2 - 3)^2 + 2x_3^2 - 7x_4 - 120 \leq 0$$
$$g_5(X) = 5x_1^2 + 8x_2 + (x_3 - 6)^2 - 2x^4 - 40 \leq 0$$
$$g_6(X) = x_1^2 + 2(x_2 - 2)^2 - 2x_1x_2 + 14x_5 - 6x_6 \leq 0$$
$$g_7(X) = 0.5(x_1 - 8)^2 + 2(x_2 - 4)^2 + 3x_5^2 - x_6 - 30 \leq 0$$
$$g_8(X) = -3x_1 + 6x_2 + 12(x_9 - 8)^2 - 7x_{10} \leq 0 \qquad (26)$$

where $-10 \leq x_d \leq 10 \, (d = 1, \ldots, 10)$. The global objective value is $f(X^*) = 24.3062$.

According to the results of Table 7, the best performance is related to $\varepsilon$F3EA. Because the standard deviation of $\varepsilon$F3EA is smaller than other algorithms that means that the best values have been generated close to the best solution.

## 4.8 G08 Benchmark Function

This benchmark function is nonlinear and it has only two nonlinear inequalities constraints. The number of variables is two and the value of $\rho$ is equal to 0.8560. The g08 equation is as (27):

$$Minimize \, f(X) = -\frac{\sin^3(2\pi x_1)\sin(2\pi x_2)}{x_1^3(x_1 + x_2)}$$

subject to:

$$g_1(X) = x_1^2 - x_2 + 1 \leq 0$$
$$g_2(X) = 1 - x_1 + (x_2 - 4)^2 \leq 0 \qquad (27)$$

where $0 \leq x_d \leq 10 (d = 1, 2)$. The global objective value is $f(X^*) = -0.095825$.

Given that the low number of constraints and decision variables, the g08 is a simple function. So, according to Table 8, there are no significant differences between the results of algorithms, and all algorithms have found the optimal solution.

**Table 7** Comparative results of εF3EA with other algorithms in g07 problem

| Parameters | SAFFa | COPSO | ISR | ATMES | SMES | ECHT-EP2 | HCOEA | αSimplex | εF3EA |
|------------|-------|-------|-----|-------|------|----------|-------|----------|-------|
| BV | 24.48 | 24.3062 | 24.306 | 24.306 | 24.327 | 24.3062 | 24.306 | 24.3062 | 24.3062 |
| AV | 26.58 | 24.3062 | 24.306 | 24.316 | 24.475 | 24.3063 | 24.307 | 24.3062 | 24.3062 |
| WV | 28.40 | 24.3062 | 24.306 | 24.359 | 24.843 | 24.3063 | 24.309 | 24.3068 | 24.3062 |
| STD | 1.14E+00 | 3.34E−06 | 6.3E−05 | 1.1E−2 | 1.32E−1 | 3.19E−05 | 7.1E−04 | 1.3E−04 | 3.32716E−13 |

**Table 8** Comparative results of $\varepsilon$F3EA with other algorithms in g08 problem

| Parameters | SAFFa | COPSO | ISR | ATMES | SMES | ECHT-EP2 | HCOEA | αSimplex | εF3EA |
|---|---|---|---|---|---|---|---|---|---|
| BV | −0.095825 | −0.095825 | −0.095825 | −0.095825 | −0.095825 | −0.095825 | −0.095825 | −0.095825 | −0.095825 |
| AV | −0.095825 | −0.095825 | −0.095825 | −0.095825 | −0.095825 | −0.095825 | −0.095825 | −0.095825 | −0.095825 |
| WV | −0.095825 | −0.095825 | −0.095825 | −0.095825 | −0.095825 | −0.095825 | −0.095825 | −0.095825 | −0.095825 |
| STD | 0 | 0 | 2.7E−17 | 2.8E−17 | 0 | 0.00E+00 | 2.4E−14 | 3.8E−13 | 2.27598E−17 |

### 4.9 g09 Benchmark Function

The g09 function is Polynomial and has seven decision variables. The value of $\rho$ is 0.5121 and it has only two nonlinear inequalities constraints. The constraint g1 and g4 are active which its equation is as (28):

$$\text{Minimize } f(X) = (x_1 - 10)^2 + 5(x_2 - 12)^2 + x_3^4 + 3(x_4 - 11)^2 + 10x_5^6$$
$$+7x_6^2 + x_7^4 - 4x_6x_7 - 10x_6 - 8x_7$$

subject to:

$$g_2(X) = -282 + 7x_1 + 3x_2 + 10x_3^2 + x_4 - x_5 \leq 0$$
$$g_3(X) = -196 + 23x_1 + x_2^2 + 6x_6^2 - 8x_7 \leq 0$$
$$g_4(X) = 4x_1^2 + x_2^2 - 3x_1x_2 + 2x_3^2 + 5x_6 - 11x_7 \leq 0 \tag{28}$$

where $-10 \leq x_d \leq 10$ $(d = 1, \ldots, 7)$. The global objective value is $f(X^*) = 680.630057$.

The results of Table 9 show that most algorithms have found the optimal solution and SAFFa has had the worst performance. In terms of STD, the $\varepsilon$F3EA has had the best result among all algorithms.

## 5 Conclusion

In this paper, an extended epsilon constrained handling version of F3EA algorithm was proposed to handle constraints that was called $\varepsilon$F3EA. The F3EA algorithm is classified into the population-based algorithm which simulates battleground and mimics the F3EA targeting process of object or installation selection for destruction in the warfare. The $\varepsilon$F3EA algorithm was used for solving constraint optimization problems with $\varepsilon$ constraint handling techniques. In this way, this algorithm was divided into nine steps of generating random solutions, initialization of the $\varepsilon$ level, ranking the solutions, the Fix step, the Find step, the Finish step, the Exploit step, the Analyze step, and checking stopping criteria. For solving the constraint optimization problem, The lexicographic order was used when $\varepsilon > 0$ in which constraint violation precedes the objective value. That way if $\emptyset(x) \leq \varepsilon$ then the objective value was compared and if the better solution was found, it was replaced. In each iteration, the $\varepsilon$ value was decreased at a constant rate. With this process, in the last iterations, the solutions without constraint violation were accepted and then in terms of objective function values, the solutions are accepted. In order to evaluate the efficiency of the $\varepsilon$F3EA algorithm, nine constraint benchmark functions were used. Functions categories included three quadratic functions, two nonlinear functions, two polynomial functions, and two cubic functions. In addition, the $\varepsilon$F3EA algorithm was compared

**Table 9** Comparative results of $\varepsilon$F3EA with other algorithms in g09 problem

| Parameters | SAFFa | COPSO | ISR | ATMES | SMES | ECHT-EP2 | HCOEA | $\alpha$Simplex | $\varepsilon$F3EA |
|---|---|---|---|---|---|---|---|---|---|
| BV | 680.64 | 680.630057 | 680.630057 | 680.630 | 680.632 | 680.630057 | 680.630 | 680.630057 | 680.630057 |
| AV | 680.72 | 680.630057 | 680.630057 | 680.639 | 680.643 | 680.630057 | 680.630 | 680.630057 | 680.630057 |
| WV | 680.87 | 680.630057 | 680.630057 | 680.673 | 680.719 | 680.630057 | 680.630 | 680.630057 | 680.630057 |
| STD | 5.92E−2 | 0 | 3.2E−13 | 1.0E−2 | 1.55E−2 | 2.61E−08 | 9.4E−8 | 2.9E−10 | 4.45339E−13 |

with eight other algorithms, which included SAFFa, COPSO, ISR, ATMES, SMES, ECHT-EP2, HCOEA, and $\alpha$Simplex. According to the results, the $\varepsilon$F3EA showed the fast convergence toward optimum solutions and it provided the appropriate results on all benchmark functions.

# References

1. Wolpert, D.H., Macready, W.G.: No free lunch theorems for optimization. IEEE Trans. Evol. Comput. **1**(1), 67–82 (1997)
2. Askarzadeh, A.: A novel metaheuristic method for solving constrained engineering optimization problems: crow search algorithm. Comput. Struct. **169**, 1–12 (2016)
3. Husseinzadeh Kashan, A.: An efficient algorithm for constrained global optimization and application to mechanical engineering design: League championship algorithm (LCA). Comput.-Aided Des. **43**(12), 1769–1792 (2011)
4. Sadollah, A., et al.: Mine blast algorithm: a new population based algorithm for solving constrained engineering optimization problems. Appl. Soft Comput. **13**(5), 2592–2612 (2013)
5. Eskandar, H., et al.: Water cycle algorithm–a novel metaheuristic optimization method for solving constrained engineering optimization problems. Comput. Struct. **110**, 151–166 (2012)
6. Husseinzadeh Kashan, A., Tavakkoli-Moghaddam, R., Gen, M.: Find-Fix-Finish-Exploit-Analyze (F3EA) meta-heuristic algorithm: an effective algorithm with new evolutionary operators for global optimization. Comput. Ind. Eng. **128**, 192–218 (2019)
7. Yildiz, A.R.: A comparative study of population-based optimization algorithms for turning operations. Inf. Sci. **210**, 81–88 (2012)
8. Yildiz, A.R.: Comparison of evolutionary-based optimization algorithms for structural design optimization. Eng. Appl. Artif. Intell. **26**(1), 327–333 (2013)
9. Karagöz, S., Yıldız, A.R.: A comparison of recent metaheuristic algorithms for crashworthiness optimisation of vehicle thin-walled tubes considering sheet metal forming effects. Int. J. Veh. Des. **73**(1–3), 179–188 (2017)
10. Jalili, S., Husseinzadeh Kashan A., Hosseinzadeh, Y.: League championship algorithms for optimum design of pin-jointed structures. J. Comput. Civ. Eng. **31**(2), 04016048 (2017)
11. Husseinzadeh Kashan, A., Karimi, B., Noktehdan, A.: A novel discrete particle swarm optimization algorithm for the manufacturing cell formation problem. Int. J. Adv. Manuf. Technol. **73**(9–12), 1543–1556 (2014)
12. Biedrzycki, R., Arabas, J., Jagodziński, D.: Bound constraints handling in differential evolution: an experimental study. Swarm Evol. Comput. **50**, 100453 (2019)
13. Coello, C.A.C., Zacatenco, C.S.P.: List of references on constraint-handling techniques used with evolutionary algorithms. Power **80**(10), 1286–1292 (2010)
14. Michalewicz, Z.: A survey of constraint handling techniques in evolutionary computation methods. Evol. Prog. **4**, 135–155 (1995)
15. Coello, C.A.C.: Theoretical and numerical constraint-handling techniques used with evolutionary algorithms: a survey of the state of the art. Comput. Methods Appl. Mech. Eng. **191**(11–12), 1245–1287 (2002)
16. Jagodziński, D., Arabas, J.: A differential evolution strategy. In: 2017 IEEE Congress on Evolutionary Computation (CEC). IEEE (2017)
17. Takahama, T., Sakai, S.: Solving constrained optimization problems by the $\varepsilon$ constrained particle swarm optimizer with adaptive velocity limit control. In: 2006 IEEE Conference on Cybernetics and Intelligent Systems. IEEE (2006)
18. Husseinzadeh Kashan, A., Tavakkoli-Moghaddam, R., Gen, M.: A warfare inspired optimization algorithm: the Find-Fix-Finish-Exploit-Analyze (F3EA) metaheuristic algorithm. In: Proceedings of the Tenth International Conference on Management Science and Engineering Management. Springer (2017)

19. Husseinzadeh Kashan, A., Karimi, B., Jolai, F.: Effective hybrid genetic algorithm for minimizing makespan on a single-batch-processing machine with non-identical job sizes. Int. J. Prod. Res. **44**(12), 2337–2360 (2006)

20. Forsythe, G.E.: Computer methods for mathematical computations. Prentice-Hall Ser. Autom. Comput. **259** (1977)

21. Liang, J., et al.: Problem definitions and evaluation criteria for the CEC 2006 special session on constrained real-parameter optimization. J. Appl. Mech. **41**(8), 8–31 (2006)

22. Husseinzadeh Kashan, A.: League Championship Algorithm (LCA): An algorithm for global optimization inspired by sport championships. Appl. Soft Comput. **16**, 171–200 (2014)

23. Farmani, R., Wright, J.A.: Self-adaptive fitness formulation for constrained optimization. IEEE Trans. Evol. Comput. **7**(5), 445–455 (2003)

24. Aguirre, A.H., et al.: COPSO: Constrained optimization via PSO algorithm. In: Center for Research in Mathematics (CIMAT). Technical report no. I-07-04/22-02-2007 (2007), p. 77

25. Runarsson, T.P., Yao, X.: Search biases in constrained evolutionary optimization. IEEE Trans. Syst., Man, Cybern., Part C (Appl. Rev.) **35**(2), 233–243 (2005)

26. Wang, Y., et al.: An adaptive tradeoff model for constrained evolutionary optimization. IEEE Trans. Evol. Comput. **12**(1), 80–92 (2008)

27. Mezura-Montes, E., Coello, C.A.C.: A simple multimembered evolution strategy to solve constrained optimization problems. IEEE Trans. Evol. Comput. **9**(1), 1–17 (2005)

28. Mallipeddi, R., Suganthan, P.N.: Ensemble of constraint handling techniques. IEEE Trans. Evol. Comput. **14**(4), 561–579 (2010)

29. Wang, Y., et al.: Multiobjective optimization and hybrid evolutionary algorithm to solve constrained optimization problems. IEEE Trans. Syst., Man, Cybern. Part B (Cybern.) **37**(3), 560–575 (2007)

30. Takahama, T., Sakai, S.: Constrained optimization by applying the/spl alpha/constrained method to the nonlinear simplex method with mutations. IEEE Trans. Evol. Comput. **9**(5), 437–451 (2005)

# An Improved Cohort Intelligence with Panoptic Learning Behavior for Solving Constrained Problems

Ganesh Krishnasamy, Anand J. Kulkarni, and Apoorva S. Shastri

**Abstract** In this paper, we present a new optimization algorithm referred to as Cohort Intelligence with Panoptic learning (CI-PL). This proposed algorithm is a modified version of Cohort Intelligence (CI), where Panoptic learning (PL) is incorporated into CI which makes every cohort candidate learn the most from the best candidate but at same time it does not completely ignore the other candidates. The PL is assisted with a new sampling interval reduction method based on the standard deviation between the behaviors of the cohort candidates. A variety of well-known set of unconstrained and constrained test problems have been successfully solved by using the proposed algorithm. The CI-PL approach produced competent and sufficiently robust results solving unconstrained, constrained, and engineering problems. The associated strengths, weaknesses, and possible real-world extensions are also discussed.

**Keywords** Cohort intelligence · Panoptic learning · Nature-inspired optimization · Unconstrained · Constrained test problems

G. Krishnasamy
Monash University Malaysia, Jalan Lagoon Selatan Bandar Sunway, 47500 Subang Jaya, Selangor, Malaysia
e-mail: ganesh.krishnasamy@monash.edu

A. J. Kulkarni (✉)
Symbiosis Center for Research and Innovation, Symbiosis International (Deemed University), Pune 412115, Maharashtra, India
e-mail: anand.kulkarni@sitpune.edu.in

A. S. Shastri
Symbiosis Institute of Technology, Symbiosis International (Deemed University), Pune 412115, Maharashtra, India
e-mail: apoorva.shastri@sitpune.edu.in

# 1 Introduction

Several nature-inspired optimization algorithms are in past few years. Some notable examples are genetic algorithm (GA) [1–4], particle swarm optimization (PSO) [5, 4], ant colony optimization (ACO) [6, 7], simulated annealing (SA) and Bee Algorithm (BA) [8, 9]. These algorithms usually perform well for solving unconstrained problems. However, their performance degenerates when applied to constrained problems.

An Artificial Intelligence (AI)-based socio-inspired optimization methodology referred to as Cohort Intelligence (CI) was proposed by Kulkarni et al. [10]. Recently, CI has been applied for solving several optimization applications such as clustering [11], problems from combinatorial domain such as the 0–1 Knapsack Problem [12], traveling salesman problem [13], cyclic bottleneck problem, and large-sized combinatorial problems such as sea-cargo mix problem and selection of cross-border shipper problem [14]. Recently, CI was applied for solving mechanical engineering problems such as heat exchanger design [15], discrete and mixed variable engineering problems [16] and cup forming design problems [14]. Recently several variations of CI were proposed by Patankar and Kulkarni [17]. Shastri and Kulkarni [18] proposed Multi-Cohort Intelligence (Multi-CI) algorithm having intra and inter group learning mechanism. In addition, CI with Cognitive Computing (CICC) was applied for solving steganography problems by Sarmah and Kulkarni [19, 20].

The original CI discussed in above literature is based on roulette approach. It has tendency of being trapped in a local minima and exhibited slow convergence when solving high dimension/features problems [10, 12, 14, 21]. In this paper, a new learning approach referred to as Panoptic Learning (PL) is introduced to replace the roulette wheel selection approach. The PL approach is inspired from the natural cohort learning behavior to learn from every candidate in the cohort partially in every learning attempt as opposed to roulette wheel approach which makes every candidate learn from a single candidate in a particular learning attempt. The approach of PL makes every candidate learn the most from the best candidate but at the same time it does not completely ignore the other candidates. The PL-based approach is better suited to imitate the cohort learning behavior than roulette wheel-based approach.

In addition, we present a new sampling interval reduction technique based on the standard deviation between the behaviors of candidates to replace the neighborhood reduction method as implemented in the original CI [10, 22]. A standard deviation approach computes the probability of the behavior selection based on all the behaviors in the cohort, i.e., every candidate observes all the behaviors in the cohort and devises its own. This helps every candidate to avoid trapping into local minima as compared to the original CI approach in which only one behavior was followed in every learning attempt. Overall, the proposed algorithm is able to simulate the cohort learning behavior more realistically, in which yields a faster convergence and improved solution than original CI. Furthermore, our proposed method is simple yet effective and does not require a tedious iterative optimization procedure compared to other methods. The CI-PL was validated by solving four unconstrained test problems. A penalty function approach was incorporated in CI-PL and successfully tested on constrained test problems.

This manuscript is organized as follows: Sect. 2 describes the framework and formulation of the CI-PL. In Sect. 3, the solution of unconstrained, constrained, and engineering problems using CI-PL algorithm is presented. Also, a brief discussion on the results, the evident features, advantages, and some limitations of the CI-PL methodology are presented in this section. The conclusions along with a note on future directions are presented in Sect. 4 of the paper.

## 2    Cohort Intelligence with Panoptic Learning (CI-PL)

Consider an unconstrained problem as follows:

$$\text{Minimize } f(\mathbf{x}) = f(x_1, \ldots, x_i, \ldots, x_N)$$
$$\text{Subject to } \Psi_i^{lower} \leq x_i \leq \Psi_i^{upper}, \quad i = 1, \ldots, N \tag{1}$$

In the context of CI-PL the objective function $f(\mathbf{x})$ is considered as the behavior of an individual candidate in each cohort with set of qualities $\mathbf{x}^c = (x_1^c, \ldots, x_i^c, \ldots, x_N^c)$. The algorithm is described below.

The number of candidates $C$ are chosen, learning attempt counter $n = 1$, sampling interval $\Psi_i$ for each quality $x_i$ are initialized. In a cohort, the behavior $f(\mathbf{x}^c)$ is computed for every individual candidate $c$ $(c = 1, 2, \ldots, C)$. The individual tries to improve itself by modifying its qualities. The qualities obtained will be closer to the qualities of candidate whose behavior is best in cohort in the current learning attempt. However, such qualities will not be solely derived from best candidate and will also be influenced by the qualities of rest of the candidates in the cohort. Furthermore, the shrinking of sampling interval is based on the concept of standard deviation $\delta_i$ associated with a particular quality $x_i^c$ of all the candidates in the cohort, i.e., $\forall i \, \delta_i = \sqrt{\frac{1}{C} \sum_{c=1}^{C} \left( x_i^c - \mu_i^c \right)}$ where $\mu_i^c = \sum_{c=1}^{C} \frac{x_i^c}{C}, i = 1, \ldots, N$. The values of $C$, $t$, standard deviation expansion factor $\gamma$, and convergence parameter $\varepsilon$ are selected depending on preliminary trials of the algorithm.

The algorithm steps are discussed below (refer Fig. 1 for flowchart).

***Step 1***: The selection probability of the function/behavior $f^*(\mathbf{x}^c)$ is as follows:

$$p^c = \frac{1/f^*(x^c)}{\sum_{c=1}^{C} 1/f^*(x^c)}, \quad (c = 1, \ldots, C) \tag{2}$$

***Step 2***: The candidates generate a set of qualities $\mathbf{x}^{\hat{c}} = \left( x_1^{\hat{c}}, \ldots, x_i^{\hat{c}}, \ldots, x_N^{\hat{c}} \right)$, $\hat{c} \in (1, \ldots, C)$, which they follow in the next learning attempt. It is calculated as follows:

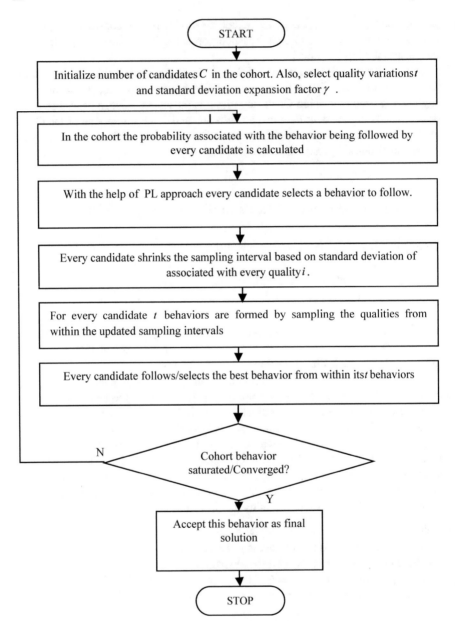

**Fig. 1** CI-PL flowchart

$$\mathbf{x}^{\hat{c}} = \begin{bmatrix} p^1 \cdots p^{c-2} & p^{c-1} & p^{c+1} & p^{c+2} \cdots p^c \end{bmatrix} \times \begin{bmatrix} x_1^1 & \cdots\cdots & x_N^1 \\ \vdots & \ddots & \vdots \\ x_1^{c-2} & \cdots\cdots & x_N^{c-2} \\ x_1^{c-1} & \cdots\cdots & x_N^{c-1} \\ x_1^{c+1} & \cdots\cdots & x_N^{c+1} \\ x_1^{c+2} & \cdots\cdots & x_N^{c+2} \\ \vdots & \ddots & \vdots \\ x_1^c & \cdots\cdots & x_N^c \end{bmatrix} \tag{3}$$

The notation $\hat{c}$ indicates that the candidate selected is not known in advance.

**Step 3**: Every candidate $c$ shrinks interval $\Psi_i^{\hat{c}}$ associated with every variable $x_i^{\hat{c}}$ to its current solution neighborhood. The shrinking of the interval $\Psi_i^{\hat{c}}$ is based on the standard deviation as follows:

$$\Psi_i^{\hat{c}} \in \left[ x_i^{\hat{c}} - (\delta_i \times \gamma), \ x_i^{\hat{c}} + (\delta_i \times \gamma) \right], \quad i = 1, \dots, N \tag{4}$$

where $\gamma = $ standard deviation expansion factor.

**Step 4**: Each candidate $c$ then samples $t$ qualities from within the updated interval $\Psi_i^{\hat{c}}$ and generates a set of $t$ updated behaviors, i.e., $\mathbf{F}^{c,t} = \left\{ f(\mathbf{x}^c)^1, \dots, f(\mathbf{x}^c)^j, \dots, f(\mathbf{x}^c)^t \right\}$, and selects the best behavior $f^*(\mathbf{x}^c)$ from within. Then the cohort with updated behaviors is as follows: $\mathbf{F}^C = \left\{ f^*(\mathbf{x}^1), \dots, f^*(\mathbf{x}^c), \dots, f^*(\mathbf{x}^C) \right\}$.

**Step 5**: The above procedure continues till the behavior of every candidate in the cohort becomes almost the same for significant number of learning attempts, and accepted as final solution.

## 3 Solution to Test Problems

The CI-PL algorithm was coded in MATLAB 7.14 (R2012a) on a Windows platform using Intel Atom CPU N450, 1.67 GHz processor, and 2 GB RAM. Every unconstrained and constrained test problems was solved 20 times. The required parameters for the implementation of CI-PL for unconstrained and constrained problems are shown in Tables 2 and 10, respectively. These parameters were derived empirically over numerous experiments.

## 3.1 Optimization of Unconstrained Test Problems

The CI-PL algorithm performance was evaluated by solving four unconstrained benchmark problems such as Ackley, Rosenbrock, Sphere, and Griewank. The results

**Table 1** Performance comparison of CI-PL with other optimization algorithms

| Problem | RHPSO | CPSO | LDWPSO | SQP | CI | CI-PL |
|---|---|---|---|---|---|---|
| | Best<br>Mean<br>Worst | Best<br>Mean<br>Worst | Best<br>Mean<br>Worst | Best<br>Mean<br>Worst | Best<br>Mean<br>Worst | Best<br>Mean<br>Worst |
| Sphere | 1.5000E−323<br>3.5078E−245<br>5.0380E−248 | 1.4356E−81<br>3.4213E−12<br>1.7103E−10 | 1.5387E−06<br>1.2102E−04<br>1.1486E−03 | 3.5657E−28<br>2.5749E−27<br>8.8173E−27 | 2.0000E−15<br>2.4900E−06<br>1.7780E−05 | **2.1081E−35**<br>**5.5536E−29**<br>**2.7768E−28** |
| Rosenbrock | 1.5606E−08<br>1.2061E−07<br>3.0398E−07 | 1.1856E−08<br>9.3949E−03<br>9.0066E−02 | 2.8453E_03<br>3.1101E+00<br>1.1050E+01 | 7.5595E−12<br>1.4352E+00<br>3.9866E+00 | 0.0000E+00<br>0.0000E+00<br>0.0000E+00 | **4.1213E−15**<br>**1.1893E−04**<br>**5.9315E−04** |
| Ackley | 0.0000E+00<br>0.0000E+00<br>0.0000E+00 | 8.8178E−16<br>1.5952E−08<br>6.3330E−07 | 1.3078E−04<br>5.9934E−03<br>2.5325E−02 | 1.5245E+01<br>1.9090E+01<br>1.9959E+01 | 1.2322E−07<br>2.0911E−07<br>2.6499E−07 | **8.8817E−16**<br>**8.8817E−16**<br>**8.8817E−16** |
| Griewank | 0.0000E+00<br>0.0000E+00<br>0.0000E+00 | 0.0000E+00<br>2.1287E−10<br>6.4174E−09 | 1.6949E−02<br>1.7072E−01<br>7.2835E−01 | 2.8879E−09<br>3.5357E−01<br>3.6312E+00 | 7.3960E−03<br>1.7100E−02<br>4.9183E−02 | **4.5253E−13**<br>**4.8000E−03**<br>**7.4000E−03** |

were compared with various optimization algorithms such as Robust Hybrid PSO (RHPSO) [23], Sequential Quadratic Programming (SQP) [24], Chaos-PSO (CPSO) [25], Linearly Decreasing Weight PSO (LDWPSO) [25] and original CI [10].

From Tables 1 and 2, it is observed that the overall performance of CI-PL is comparable with other optimization techniques. The best solution obtained by CI-PL is substantially better than CI in both the cases, i.e., unimodal Sphere function and multimodal functions such as Ackley, Rosenbrock and Griewank. From Table 2, it can be observed that the function evaluations (FE) were improved significantly, whereas the computational time was marginally improved in case of CI-PL. The convergence plot for sphere function is presented in Fig. 2f along with associated variable/quality convergence plots in Fig. 2a–e. These plots exhibited the candidates self-supervised learning behavior and convergence on best solution.

## 3.2 Optimization of Constrained Test Problems

The CI-PL with penalty function method was compared with several other constrained optimization techniques such as a self-adaptive penalty approach [26], a dynamic penalty scheme [27], GA with penalty function approach [2], variation of feasibility-based rule [28], HPSO [29], PSO [30], homomorphous mapping [31], cultural algorithm [32], cultural differential evolution (CDE) [33], gradient repair method [34] integrated into PSO [35], evolution strategies (ES) [36], differential evolution (DE) [37], genetics adaptive search (GeneAS) [38], Lagrange multiplier method [39], branch and bound technique and geometric programming approach [40], modified differential evolution (COMDE) [41], and cooperative coevolutionary differential evolution with improved augmented Lagrangian (CCiLF) [42].

**Table 2** Comparison of CI-PL and CI

| Problem | Original CI | | | | | CI-PL | | | | |
|---|---|---|---|---|---|---|---|---|---|---|
| | Best Mean Worst | C, t, r | FE | SD | Time (s) | Best Mean Worst | C, t, r | FE | SD | Time (s) |
| Sphere | 2.0000E−15 2.4900E−06 1.7780E−05 | 5, 15, 0.80 | 18750 | 4.5800E−03 | 1.55 | 2.1081E−35 5.5536E−29 2.7768E−28 | 5, 25, 25 | 625 | 4.9672E−29 | 1.34 |
| Rosenbrock | 0.0000E+00 0.0000E+00 0.0000E+00 | 5, 15, 0.80 | 9750 | 0.0000E+00 | 5.20 | 4.1213E−15 1.1893E−04 5.9315E−04 | 5, 25 25 | 500 | 4.7422E−05 | 0.9746 |
| Ackley | 1.2322E−07 2.0911E−07 2.6499E−07 | 5, 15, 0.85 | 11250 | 4.3200E−08 | 1.50 | 8.8818E−16 8.8818E−16 8.8818E−16 | 5, 25, 25 | 3375 | 0.0000E+00 | 2.98 |
| Griewank | 7.3960E−03 1.7100E−02 4.9183E−02 | 5, 15, 0.997 | 18750 | 8.8300E−03 | 2.00 | 4.5253E−13 4.8000E−03 7.4000E−03 | 5, 25, 15 | 1500 | 1.4148E−03 | 1.50 |

Using CI-PL algorithm incorporated with penalty function approach we have solved the constrained test problems [33]. There are four maximization problems: G02, G03, G12, G08, and twelve minimization problems G01, G04 to G07, G09 to G11, G13 to G15, and G24. For the linear objective function (LOF) problems, CI-PL achieved the best known solution for G24 and achieved a feasible solution for G10. The best result obtained for G24 was $f(\mathbf{x}) = -5.5080$ with the global optimum x = {2.3295, 3.1785} and $g_i(\mathbf{x}) = \{0.0000, 0.0000\}$. In case of G10, the best result obtained was feasible with $f(\mathbf{x}) = 7057.70$. Its global optimum was x = {604.2, 1451.2, 5020.2, 183.7, 299.5, 217.0, 284.0, 399.4} and the values of $g_i(\mathbf{x})$ are = {−0.0028, 0.0000, −0.0008, −347.3246, −917.6246, −288.9541} for the constraints. The closeness to the best reported solution was 0.18%. Regarding the problems with nonlinear objective function (NLOF), CI-PL found feasible solutions in all the problems and showed robustness in all of those problems. The details of the solutions achieved by CI-PL for the NLOF problems are provided here. In problem G04, the best solution obtained was $f(\mathbf{x}) = -30665.5389$ with x = {78.0000, 33.0000, 29.9953, 45.0000, 36.7758} and the values of $g_i(\mathbf{x})$ are = {−92.0000, 0.0000, −8.8405, −11.1595, 0.0000, −5.0000} for the constraints. In problem G05, the best result obtained was $f(\mathbf{x}) = 5126.498$ with x = {678.3, 1027.8, 0.1, − 0.4} and the values of $g_i(\mathbf{x})$ are = {−0.0343, −1.0657, 0.0001, 0.0001, 0.0001} for the constraints. In problem G08, which is a maximization problem, the best result obtained was $f(\mathbf{x}) = 0.095825$ with x = {1.2280, 4.2454} and the values of $g_i(\mathbf{x})$ are = −1.7375, −0.1678} for the constraints. In problem G09, the best result obtained was $f(\mathbf{x}) = 680.6463$ with x = {2.3138, 1.9583, −0.4678, 4.3512, −0.6181, 1.0377, 1.5712} and the values of $g_i(\mathbf{x})$ are = {0.0000, −252.7696, −145.0568, 0.0001} for the constraints. In the problem G12, which is also a maximization problem, the best result obtained was $f(\mathbf{x}) = 1.0000$ with x = {5.0000, 5.0000, 5.0000} and the values of $g_i(\mathbf{x})$ is = {−0.0625} for the constraints. In problem G15, the best result obtained was $f(\mathbf{x}) = 961.7236$ with x = {3.4360, 0.2238, 3.6254} and the values of $g_i(\mathbf{x})$ are = {0.0001, 0.0000} for the constraints (Table 3).

In problem G01, the best solution obtained was $f(\mathbf{x}) = -14.9988$ with x = {1.0000, 1.0000, 1.0000, 1.0000, 1.0000, 1.0000, 1.0000, 1.0000, 1.0000, 3.0015, 3.0021, 3.0008, 1.0000} and the values of $g_i(\mathbf{x})$ are = {−0.0035, −0.0050, −0.0045, −4.9994, −4.9995, −5.0029, −0.0011, −0.0000, −0.0017} for the constraints. In problem G01, the best result obtained was $f(\mathbf{x}) = -14.9988$ with x = {1.0000, 1.0000, 1.0000, 1.0000, 1.0000, 1.0000, 1.0000, 1.0000, 1.0000, 3.0015, 3.0021, 3.0008, 1.0000} and the values of $g_i(\mathbf{x})$ are = {−0.0035, −0.0050, −0.0045, −4.9994, −4.9995, −5.0029, −0.0011, −0.0000, −0.0017} for the constraints. In problem G02, which is a maximization problem, the best result obtained was $f(\mathbf{x}) = 0.799$ with x = {3.1141, 3.1038, 3.1327, 3.0538, 3.0696, 2.9923, 2.9595, 2.9302, 0.5249, 0.4608, 0.4896, 0.3409, 0.5213, 0.4746, 0.4718, 0.4453, 0.4264, 0.5483, 0.4561, 0.4550} and the values of $g_i(\mathbf{x})$ are = {0.0000, −120.0288} for the constraints. In problem G03, which is a maximization problem, the best result obtained was $f(\mathbf{x}) = 1.0322$ with x = {0.4501, 0.4501, 0.4501, 0.4500, 0.4501} and the values of $g_i(\mathbf{x})$ are = {0.0127} for the constraints. In problem G06, the best result obtained was $f(\mathbf{x}) = -6959.0878$ with x = {14.0962, 0.8454} and the

**Table 3** Statistical solutions of the constrained test problems

| Problem | | G01 Best Avg. Worst | G02 Best Avg. Worst | G03 Best Avg. Worst | G04 Best Avg. Worst | G05 Best Avg. Worst | G06 Best Avg. Worst | G07 Best Avg. Worst | G08 Best Avg. Worst | G09 Best Avg. Worst | G10 Best Avg. Worst | G11 Best Avg. Worst | G12 Best Avg. Worst | G13 Best Avg. Worst |
|---|---|---|---|---|---|---|---|---|---|---|---|---|---|---|
| Best known solutions | | -15 | 0.803619 | 1 | -30665.539 | 5126.498 | -6961.8138 | 24.306 | 0.095825 | 680.630057 | 7049.3307 | 0.75 | 1 | 0.0539498 |
| Methods | Koziel and Michalewicz [31] | -14.7184 / -14.6478 / -14.5732 | 0.79486 / 0.78722 / 0.78279 | 0.9997 / 0.9989 / 0.9960 | -30664.5 / -30655.3 / -30645.9 | - | - 6952.1 / - 6342.6 / -5473.9 | 24.620 / 24.826 / 24.069 | 0.095825 / 0.089156 / 0.029143 | 680.91 / 681.16 / 683.18 | 7321.1. / 7498.6 / 7685.8 | 0.75 / 0.75 / 0.75 | - | - |
| | Runarsson and Yao [46] | -15.000 / -15.000 / -15.000 | 0.803515 / 0.781975 / 0.726288 | 1.000 / 1.000 / 1.000 | -30665.539 / -30665.539 / -30665.539 | 5126.497 / 5128.881 / 5142.472 | -6961.814 / -6875.940 / -6350.262 | 24.307 / 24.374 / 24.642 | 0.095825 / 0.095825 / 0.095825 | 680.630 / 680.656 / 680.763 | 7054.316 / 7559.192 / 8835.655 | 0.750 / 0.750 / 0.750 | 1.00000 / 1.00000 / 1.00000 | 0.053957 / 0.067543 / 0.216915 |
| | Hamida and Schoenauer [44] | - | - | 1 / 0.99989 / N.A. | -30665.5 / -30665.5 / N.A. | 5126.5 / 5141.65 / N.A. | - 6961.81 / -6961.81 / N.A. | 24.3323 / 24.6636 / N.A. | 0.09582 / 0.09582 / N.A. | 680.630 / 680.641 / N.A. | - | 0.75 / 0.75 / N.A. | - | - |
| | Montes et al. [37] | - | - | 1.001038 / 1.000989 / 1.000579 | -30665.53906 / -30665.53906 / -30665.53906 | 5126.59960 / 5174.49230 / 5304.16699 | -6961.8139 / -6961.2839 / -6961.4819 | 24.32671 / 24.47492 / 24.84282 | 0.095826 / 0.095826 / 0.095826 | 680.631592 / 680.643410 / 680.719299 | - | 0.749090 / 0.749358 / 0.749830 | - | - |
| | Hedar and Fukushima [45] | 14.999105 / 14.993316 / 14.979977 | 0.7549125 / 0.3717081 / 0.213110 | 1.000001 / 0.999187 / 0.991518 | -30665.5380 / -30665.4665 / -30664.6880 | 5126.4981 / 5126.4981 / 5126.4981 | -6961.8138 / -6961.8138 / -6961.8138 | 24.31057 / 24.37952 / 24.64439 | 0.095825 / 0.095825 / 0.095825 | 680.63008 / 680.63642 / 680.69832 | 7059.8635 / 7059.3210 / 9398.6492 | 0.749999 / 0.749999 / 0.749999 | 1.00000 / 1.00000 / 1.00000 | 0.539498 / 0.2977204 / 0.4388511 |
| | Becerra and Coello Coello [33] | 15.000000 / 14.999996 / 14.999993 | 0.803619 / 0.724886 / 0.590098 | 0.995143 / 0.788635 / 0.639920 | -30,665.5386 / -30,665.5386 / -30,665.5386 | 5126.5709 / 5207.4106 / 5327.3904 | -6961.8138 / -6961.8138 / -6961.8138 | 24.30620 / 24.30621 / 24.30621 | 0.095825 / 0.095825 / 0.095825 | 680.630057 / 680.630057 / 680.630057 | 7049.2480 / 7049.2482 / 7049.2484 | 0.749900 / 0.757995 / 0.796455 | 1.00000 / 1.00000 / 1.00000 | 0.056180 / 0.288324 / 0.392100 |
| | Lampinen | - | - | - | - | 5126.484 / 5126.484 / 5126.484 | -6961.814 / -6961.814 / -6961.814 | - | 0.095825 / 0.095825 / 0.095825 | 680.630 / 680.630 / 680.630 | - | 0.74900 / 0.74900 / 0.74900 | - | - |
| | Chootinan and Chen [34] | -15.00000 / -15.00000 / -15.00000 | 0.801119 / 0.785476 / 0.745329 | 0.99998 / N.A. / 0.99979 | -30,665.5386 / -30,665.5386 / -30,665.5386 | 5126.4981 / 5126.4981 / 5126.4981 | -6961.8139 / -6961.8139 / -6961.8139 | - | 0.095825 / 0.095825 / 0.095825 | 680.6303 / 680.6381 / 680.6538 | 7049.2607 / 7049.5659 / 7051.6857 | 0.7500 / 0.7500 / 0.7500 | - | - |
| | Farmani and Wright [27] | -15.00000 / -14.99930 / -14.99800 | 0.79989 / 0.77512 / 0.74398 | 0.99978 / 0.99930 / 0.99830 | -30,665.5000 / -30,665.2000 / -30,663.3000 | 5126.9890 / 5432.08 / N.A. | -6961.8000 / -6961.8000 / -6961.8000 | 24.59 / 27.83 / 32.69 | 0.095825 / 0.095825 / 0.095825 | 680.6400 / 680.7200 / 680.8700 | 7070.23 / 7760.54 / 8568.81 | 0.7500 / 0.7500 / 0.7500 | - | - |

(continued)

**Table 3** (continued)

| Problem | | G01 | G02 | G03 | G04 | G05 | G06 | G07 | G08 | G09 | G10 | G11 | G12 | G13 |
|---|---|---|---|---|---|---|---|---|---|---|---|---|---|---|
| Methods | Deb [2] | −15.00000 | – | – | −30614.814 | – | – | 24.372 | – | 680.659424 | 7065.742 | – | – | 0.053950 |
| | | −15.00000 | | | −30196.404 | | | 24.409 | | 681.525635 | 8274.830 | | | 0.241289 |
| | | −13.00000 | | | −29606.596 | | | 25.075 | | 687.188599 | 10925.165 | | | 0.507761 |
| | Dong et al. [43] | – | – | – | −30664.7 | – | – | – | – | – | 7114.84 | – | – | 0.05451 |
| | | | | | −30656.1 | | | | | | 7222.389 | | | 0.05451 |
| | | | | | −30662.8 | | | | | | 7225.979 | | | 0.05451 |
| | He and Wang [29] | – | – | – | −30665.539 | – | – | – | 0.095825 | – | – | – | – | – |
| | | | | | −30665.539 | | | | 0.095825 | | | | | |
| | | | | | −30665.539 | | | | 0.095825 | | | | | |
| | Hu Eberhart [30] | −15.00000 | – | 1 | −30665.5 | – | −6961.7 | 24.4420 | 0.09583 | 680.657 | 7131.01 | 0.7500 | 1.00000 | – |
| | | −15.00000 | | 1 | −30665.5 | | −6960.7 | 26.7188 | 0.09583 | 680.876 | 7594.65 | 0.7500 | 1.00000 | |
| | | −15.00000 | | 1 | −30665.5 | | −6956.8 | 31.1843 | 0.09583 | 681.675 | 8823.56 | 0.7500 | 1.00000 | |
| | Ray et al. [4] | – | – | – | −30651.662 | – | −6737.0479 | – | – | – | – | – | 1.00000 | – |
| | | | | | −30647.105 | | −6852.5630 | | | | | | 1.00000 | |
| | | | | | −30619.047 | | −6744.0864 | | | | | | 1.00000 | |
| | CCiALF | −15 | −0.803617 | 1.0005 | −30665.539 | 5126.4967 | −6961.814 | 24.3062 | −0.09582 | 680.6300 | 7049.248 | 0.7498 | 1.00000 | 0.05394 |
| | | −15 | −0.793087 | 1.0005 | −30665.539 | 5126.4970 | −6961.814 | 24.3062 | −0.09582 | 680.6300 | 7049.248 | 0.7498 | 1.00000 | 0.05394 |
| | | −15 | −0.761706 | 1.0005 | −30665.539 | 5126.4970 | −6961.814 | 24.3062 | −0.09582 | 680.6300 | 7049.248 | 0.7498 | 1.00000 | 0.05394 |
| | COMDE | −15 | −0.803601 | 1.001 | −30665.539 | 5126.497 | −6961.814 | 24.3062 | −0.09582 | 680.6300 | 7049.248 | 0.7499 | 1.00000 | 0.05394 |
| | | −15 | −0.803518 | 1.001 | −30665.539 | 5127.542 | −6961.814 | 24.3062 | −0.09582 | 680.6300 | 7077.682 | 0.7499 | 1.00000 | 0.05394 |
| | | N.A. | N.A. | N.A. | N.A. | N.A. | N.A. | N.A. | N.A. | N.A. | N.A. | N.A. | N.A. | N.A. |
| | **CI-PL** | **−14.99880** | **0.7991** | **1.0322** | **−30665.538** | **5126.498** | **−6959.08786** | **24.4145** | **0.095825** | **680.6463** | **7057.70** | **0.7490** | **1.00000** | **0.0565** |
| | | **−14.99600** | **0.7875** | **1.0322** | **−30665.533** | **5126.760** | **−6903.17034** | **24.6753** | **0.095825** | **680.7735** | **7462.90** | **0.7490** | **1.00000** | **0.0937** |
| | | **−14.99080** | **0.7832** | **1.0322** | **−30665.525** | **5127.104** | **−6637.47132** | **25.0526** | **0.095825** | **680.9530** | **8897.40** | **0.7490** | **1.00000** | **0.3034** |

(continued)

**Table 3** (continued)

| Problem | | G14 | G15 | G24 |
|---|---|---|---|---|
| | | Best<br>Avg.<br>Worst | Best<br>Avg.<br>Worst | Best<br>Avg.<br>Worst |
| Best known solutions | | −47.764884 | 961.715022 | −5.5080 |
| Methods | Montes and Lopez-Ramirez (DE) [36] | 47.765<br>−47.765<br>NA | 961.715<br>961.715<br>NA | −5.508<br>−5.508<br>NA |
| | Montes et al. (ES) [37] | 47.765<br>47.765<br>NA | 961.715<br>961.949<br>NA | −5.508<br>−5.508<br>NA |
| | Montes et al. | −47.764888<br>−47.722542<br>−47.036510 | 961.715022<br>961.715022<br>961.715022 | −5.5080<br>−5.5080<br>−5.5080 |
| | CI-PL | **−47.2080**<br>**−46.2879**<br>**−44.5988** | **961.7236**<br>**962.2494**<br>**962.8885** | **−5.5080**<br>**−5.5080**<br>**−5.5080** |

G. Krishnasamy et al.

**Table 4** Variable and constraint value comparison

| Design variables | Best solutions found | | | | | | | | | |
|---|---|---|---|---|---|---|---|---|---|---|
| | Coello Coello and Becerra [32] | Arora [47] | Coello Coello [26] | Coello Coello and Montes [48] | He and Wang [49] | He and Wang [29] | Kulkarni and Tai [50] | Behrooz et al. [42] | Mohamed and Sabry [41] | CI-PL |
| $x_1$ | 0.05000 | 0.05339 | 0.05148 | 0.05198 | 0.05170 | 0.051728 | 0.05060 | 0.05168 | NA | **0.052216** |
| $x_2$ | 0.31739 | 0.39918 | 0.35166 | 0.36396 | 0.35712 | 0.357644 | 0.32781 | 0.35671 | NA | **0.369507** |
| $x_3$ | 14.03179 | 9.18540 | 11.63220 | 10.89052 | 11.26508 | 11.244543 | 14.05670 | 11.28896 | NA | **10.577393** |
| $g_1(\mathbf{X})$ | 0.00000 | 0.00001 | −0.00330 | −0.000013 | −0.00000 | −0.000845 | −0.05290 | 2.22e−16 | −4.041e−12 | **−4.0763e−010** |
| $g_2(\mathbf{X})$ | −0.00007 | −0.00001 | −0.00010 | −0.000021 | 0.00000 | −1.2600e−05 | −0.00740 | 1.11e−16 | −1.454e12 | **−4.5099e−005** |
| $g_3(\mathbf{X})$ | −3.96796 | −4.12383 | −4.02630 | −4.061338 | −4.05460 | −4.051300 | −3.70440 | 4.05 | −4.053785 | **−4.0781** |
| $g_4(\mathbf{X})$ | −0.75507 | −0.69828 | −0.73120 | −0.722698 | −0.7274 | −0.727090 | −0.74769 | 7.28e−01 | −0.727729 | **−0.7189** |
| $f(\mathbf{X})$ | 0.01272 | 0.01273 | 0.01270 | 0.01268 | 0.01266 | 0.0126747 | 0.01350 | 0.012665 | 0.012665 | **0.0126713** |

**Table 5** Statistical solutions of various algorithms

| Methods | Best | Mean | Worst | Std. |
|---|---|---|---|---|
| Coello Coello and Becerra [32] | 0.0127210 | 0.0135681 | 0.015116 | 8.4152e−004 |
| Arora [47] | 0.0127303 | N.A. | N.A. | N.A. |
| Coello Coello [26] | 0.0127048 | 0.0127690 | 0.012822 | 3.9390e−005 |
| Coello Coello and Montes [48] | 0.0126810 | 0.0127420 | 0.012973 | 5.9000e−005 |
| He and Wang [49] | 0.0126652 | 0.0127072 | 0.0127191 | 1.5824e−005 |
| He and Wang [29] | 0.0126747 | 0.0127300 | 0.012924 | 5.1985e−004 |
| Kulkarni and Tai [50] | 0.01350 | 0.02607 | 0.05270 | N.A. |
| Behrooz et al. [42] | 0.012665 | 0.012665 | 0.012665 | 9.87e−08 |
| Mohamed and Sabry [41] | 0.012665 | 0.012667 | 0.01271 | 3.09e−09 |
| **CI-PL** | **0.0126713** | **0.012758** | **0.0128457** | **2.8526e−005** |

values of $g_i(\mathbf{x})$ are = $\{-0.0024, -0.0001\}$ for the constraints. In problem G07, the best result obtained was $f(\mathbf{x}) = 24.4145$ with x = $\{2.1405, 2.4070, 8.7451, 5.0264, 1.0057, 1.5052, 1.2874, 9.7996, 8.2563, 8.3588\}$ and the values of $g_i(\mathbf{x})$ are = $\{-0.0689, -0.1365, -0.2537, -0.7639, -0.3521, -6.2287, -0.3427, -49.7028\}$ for the constraints. In problem G11, the best result obtained was $f(\mathbf{x}) = 0.7490$ with x = $\{-0.7064, 0.5000\}$ and the values of $g_i(\mathbf{x})$ are = $\{0.001\}$ for the constraints. In problem G14, the best result obtained was $f(\mathbf{x}) = -47.2080$ with x = $\{= 0.8398, 0.2898, 0.2421, 0.3404, 0.2922, 0.0932, 0.0874, 0.1166, 0.2397, 0.1348\}$ and the values of $g_i(\mathbf{x})$ are = $\{0.0491, 0.0041, 0.0296\}$ for the constraints.

For problems G05, G08, G12, and G24, CI-PL yielded solutions that are equal to the best-known solutions. The best results obtained by CI-PL algorithm for problems G01, G02, G03, G06, G07, G11, and G14 are very close to the best known solutions which are 15.0000, 0.8036, 1.0000, −6961.8138, 24.30, 0.7500, and −47.764884, respectively. The closeness of the obtained solutions is around 0.001% to the best known solutions. However, for G13 problem, the difference between best results obtained by CI-PL and best reported solution is quite high, which is around 4.7%. This may be due to the moderated dimensionality, nonlinear equality constraints and small feasible region. The best solutions obtained Koziel and Michalewicz [31] are comparable with solutions obtained by CI-PL, though the algorithm in [31] was unable to tackle problems with both nonlinear objective function and more than 6 active constraints function as in case of G01 and G07. The best solutions obtained in G01 and G07 were −14.7184 and 24.620, respectively, against −14.9980 and 24.4145 obtained by CI-PL. The solutions obtained by CI-PL and Runarsson and Yao [46] were comparable, except for the solution obtained for G03 by Runarsson and Yao [46] was an optimal solution whereas the solution obtained by CI-PL was not an optimal but feasible. Although the best solutions obtained by algorithm in [31] and CI-PL in case of G02, G06, G10 and G13 were near about optimal but the average solutions of obtained by the both algorithms showed a degraded performance. In [45], the optimal solution was achieved for G06 but in case of G02, the solution obtained was not optimal. The results obtained by Hedar and Fukushima [45] in G01, G02, G03, G07, and G10 were not optimal and hence it appears that the algorithm was unable to handle problems with moderated dimensionality ($n >$ 5). The optimal solutions were obtained by Becerra and Coello Coello [33] for G01, G06, G07, and G10. Therefore, the algorithm in [33] appears to be able to handle problems with active constraints ($a > 6$) except the problems with small feasible region (($\mathbf{x}\%$) $\approx 0$) as shown through its performances for the case of G03, G05, G11, and G13. Though a few selected problems were solved by Chootinan and Chen [34], the solutions obtained in case of G01, G02, G03, G06, G10 were better than the solutions obtained by CI-PL. The unavailability of the solutions to all test problems hinders a detailed analysis of algorithm by Chootinan and Chen [34]. Therefore, a comprehensive comparison cannot be carried out between [34] and CI-PL. The best solutions obtained by Farmani and Wright [27] were comparable with the best solutions of CI-PL. However, the standard deviation between the test solutions of [27] is quite high. Therefore, this algorithm does not display the robustness and consistency as shown by CI-PL. The solutions obtained by Deb [2] also show a high standard deviation, therefore its less robust than CI-PL. Furthermore, the solutions obtained by Dong et al. [43], He and Wang [29], Hu and Eberhart [30], Ray et al. [4], Montes and Lopez-Ramirez [36], and Montes et al. [37] are comparable with the solutions obtained by CI-PL. For NEC problems, CI-PL yielded feasible solutions in all of them and the best solutions were achieved in two problems (G03 and G11). CI-PL found best known solution in one of seven problems with a moderated high dimensionality (MD). We can conclude based on the results that the performance of CI-PL deteriorated marginally for higher dimensionality problems. For the problems with more than 6 active constraints, CI-PL was still able to achieve feasible solution. Finally, for problems with very small feasible region, CI-PL was able to find two best known solutions in them.

## 3.3 Optimization of the Engineering Problems

Three well-studied engineering design problems were used to evaluate the performances of our proposed algorithm. Then, we compared the performances of CI-PL with various well-known optimization algorithms. For each problem, 20 independent CI-PL runs were carried out.

### 3.3.1 Spring Design Problem

The mathematical formulation of this problem is:

$$\text{Minimize } fcost(\{x\}) = (N + 2) \times d \times D \tag{5}$$

Subjected to

$$g_1\{x\} = 1 - \frac{N \times d^3}{71785 \times D^4} \le 0 \tag{6}$$

$$g_2\{x\} = \frac{4 \times d^2 - D \times d}{12566 \times (d \times D^3 - D^4)} + \frac{1}{5108 \times D^2} - 1 \le 0 \tag{7}$$

$$g_3\{x\} = 1 - \frac{140.45 \times D}{N \times d^2} \le 0 \tag{8}$$

$$g_4\{x\} = \frac{d + D}{1.5} - 1 \le 0 \tag{9}$$

$$0.05 \le D \le 2, 0.25 \le d \le 1.3, 2 \le N \le 15$$

Table 4 presents the performance comparison between CI-PL with different optimization algorithms for the spring design problem. Table 5 shows the statistical simulation results.

### 3.3.2 Welded Beam Design Problem

The mathematical formulation of this problem is:

$$\text{Minimize:} f(\vec{y}) = 1.10471 y_2 y_1^2 + 0.04811 y_3 y_4 (14 - y_2) \tag{10}$$

$$\text{Subject to:} g_1(\vec{y}) = \tau(\vec{y}) - 13000 \le 0 \tag{11}$$

$$g_2(\vec{y}) = \sigma(\vec{y}) - 30000 \le 0 \tag{12}$$

$$g_3(\vec{y}) = y_1 - y_4 \le 0 \tag{13}$$

$$g_4(\vec{y}) = 1.10471 y_1^2 + 0.04811 y_3 y_4 (14 + y_2) - 5 \le 0 \tag{14}$$

$$g_5(\vec{y}) = 0.125 - y_1 \le 0 \tag{15}$$

$$g_6(\vec{y}) = \delta(\vec{y}) - 0.25 \le 0 \tag{16}$$

$$g_7(\vec{y}) = 6000 - Pc(\vec{y}) \le 0 \tag{17}$$

where $0.1 \le y_1 \le 2, 0.1 \le y_2 \le 10, 0.1 \le y_3 \le 10, 0.1 \le y_4 \le 2$.

**Table 6** Variables and constraint value comparison

| Design variables | Best solutions found | | | | | | | | | | CI-PL |
|---|---|---|---|---|---|---|---|---|---|---|---|
| | Coello Coello [26] | Coello Coello and Montes [48] | Coello Coello and Becerra [32] | He and Wang [49] | He and Wang [29] | Deb [2] | Siddall [51] | Ragsdell and Phillips [40] | Behrooz et al. [42] | Mohamed and Sabry [41] | |
| $x_1$ | 0.208800 | 0.205986 | 0.205700 | 0.202369 | 0.20573 | 0.2489 | 0.2444 | 0.2455 | 0.2057 | NA | **0.2057** |
| $x_2$ | 3.420500 | 3.471328 | 3.470500 | 3.544214 | 3.47048 | 6.1730 | 6.2189 | 6.1960 | 3.4704 | NA | **3.4708** |
| $x_3$ | 8.997500 | 9.020224 | 9.036600 | 9.048210 | 9.03662 | 8.1789 | 8.2915 | 8.2730 | 9.0366 | NA | **9.0369** |
| $x_4$ | 0.210000 | 0.206480 | 0.205700 | 0.205723 | 0.20573 | −0.2533 | 0.2444 | 0.2455 | 0.2057 | NA | **0.2057** |
| $g_1(X)$ | −0.337812 | −0.074092 | −0.00047 | −12.8397 | 0.00010 | −5758.6037 | 5743.50202 | −5743.8265 | 7.24e−10 | −1.8190e12 | −0.7654 |
| $g_2(X)$ | −353.9026 | −0.266227 | −0.00156 | −1.24746 | −0.02656 | −255.57690 | 4.015209 | −4.715097 | 2.00e−08 | 0.000000 | −1.1976 |
| $g_3(X)$ | −0.00120 | −0.000495 | 0.000000 | −0.00149 | 0.00000 | −0.004400 | 0.000000 | 0.000000 | 3.32e−13 | 0.000000 | −5.7441e−006 |
| $g_4(X)$ | −3.411865 | −3.430043 | −3.43298 | −3.42934 | −3.43298 | −2.982866 | 3.022561 | −3.020289 | 3.4329 | −3.432983 | −3.4329 |
| $g_5(X)$ | −0.08380 | −0.080986 | −0.08073 | −0.07938 | −0.08070 | −0.123900 | 0.119400 | −0.120500 | 0.0807 | −0.080729 | −0.8070 |
| $g_6(X)$ | −0.235649 | −0.235514 | −0.23554 | −0.23553 | −0.23554 | −0.234160 | 0.234243 | −0.234208 | 0.2355 | 0.235540 | −0.2355 |
| $g_7(X)$ | −363.2323 | −58.66644 | −0.00077 | −11.6813 | −0.02980 | −4465.27092 | 3490.46941 | −3604.2750 | 1.88e−08 | 3.638e12 | −0.1629 |
| $f(X)$ | 1.748309 | 1.728226 | 1.724852 | 1.72802 | 1.72485 | 2.43311600 | 2.38154338 | 2.3859373 | 1.7248 | 1.724852 | **1.724865** |

**Table 7** Statistical solutions of various algorithms

| Methods | Best | Mean | Worst | Std. |
|---------|------|------|-------|------|
| Coello Coello [26] | 1.748309 | 1.771973 | 1.785835 | 0.011220 |
| Coello Coello and Montes [48] | 1.728226 | 1.792654 | 1.993408 | 0.074713 |
| Coello Coello and Becerra [32] | 1.724852 | 1.971809 | 3.179709 | 0.443131 |
| He and Wang [49] | 1.728024 | 1.748831 | 1.782143 | 0.012926 |
| He and Wang [29] | 1.724852 | 1.749040 | 1.814295 | 0.040049 |
| Deb [2] | 2.38145 | 2.38263 | 2.38355 | N.A. |
| Siddall [51] | 2.3815 | N.A. | N.A. | N.A. |
| Ragsdell and Phillips [40] | 2.3859 | N.A. | N.A. | N.A. |
| Behrooz et al. [42] | 1.724852 | 1.724852 | 1.724854 | 5.11E−07 |
| Mohamed and Sabry [41] | 1.724852 | 1.724852 | 1.724854 | 1.60E−12 |
| **CI-PL** | **1.724865** | **1.724878** | **1.724916** | **8.5554e−006** |

Table 6 presents the performance comparison between CI-PL with different optimization algorithms for the Welded Beam Design Problem. The statistical simulation results are summarized in Table 7. From Table 6, we can observe that our algorithm is able to achieve a comparable best solution compared with other algorithms. In addition, CI-PL achieved a very small standard deviation for its solution as shown in Table 7.

### 3.3.3 Pressure Vessel Design Problem

The mathematical formulation of this problem is:

Minimize:

$$f(\vec{y}) = 0.6624y_1y_3y_4 + 1.7781y_2y_3^2 + 3.1661y_1^2y_4 + 19.84y_1^2y_3 \tag{18}$$

Subject to:

$$g_1(\vec{y}) = -y_1 + 0.0193y_3 \leq 0 \tag{19}$$

$$g_2(\vec{y}) = -y_2 + 0.00954y_3 \leq 0 \tag{20}$$

$$g_3(\vec{y}) = -\pi y_3^2 y_4^2 - \frac{4}{3}\pi y_3^2 + 1296000 \leq 0 \tag{21}$$

$$g_4(\vec{y}) = y_4 - 240 \leq 0$$

**Table 8** Performance comparison

| Design variables | Best solutions found | | | | | | | | | |
|---|---|---|---|---|---|---|---|---|---|---|
| | Sandgren [52] | Kannan and Kramer [39] | Deb [2] | Coello Coello [26] | Coello Coello and Montes [48] | He and Wang [49] | He and Wang [29] | Behrooz et al. [42] | Mohamed and Sabry [41] | CI-PL |
| $x_1$ | 1.1250 | 1.1250 | 0.9375 | 0.8125 | 0.8125 | 0.8125 | 0.81250 | 0.8125 | NA | **0.8125** |
| $x_2$ | 0.6250 | 0.6250 | 0.5000 | 0.4375 | 0.4375 | 0.4375 | 0.43750 | 0.4375 | NA | **0.4375** |
| $x_3$ | 47.7000 | 58.2910 | 48.3290 | 40.3239 | 42.0974 | 42.0913 | 42.09840 | 42.0984 | NA | **41.9284** |
| $x_3$ | 117.7010 | 43.6900 | 112.6790 | 200.0000 | 176.6540 | 176.7465 | 176.6366 | 176.6365 | NA | **178.7556** |
| $g_1(\mathbf{X})$ | −0.204390 | 0.000016 | −0.004750 | −0.034324 | −0.000020 | −0.000139 | $-8.8 \times 10^{-7}$ | 0.0000 | 0.0000 | **−0.0033** |
| $g_2(\mathbf{X})$ | −0.169942 | −0.068904 | −0.038941 | −0.052847 | −0.035891 | −0.035949 | −0.03590 | 0.0359 | −0.035880 | **−0.0375** |
| $g_3(\mathbf{X})$ | 54.226012 | −21.22010 | −3652.8768 | −27.10584 | 27.886075 | −116.38270 | 3.12270 | 0.000 | −8.6147e09 | **0.0000** |
| $g_4(\mathbf{X})$ | −122.29900 | −196.31000 | −127.32100 | −40.00000 | −63.34595 | −63.253500 | 63.36340 | 63.36 | −63.363404 | **−61.2444** |
| $f(\mathbf{X})$ | 8129.8000 | 7198.0428 | 6410.3811 | 6288.7445 | 6059.9463 | 6061.0777 | 6059.7143 | 6059.7143 | 6053.7143 | **6080.5357** |

**Table 9** Statistical solutions of various algorithms

| Methods | Best | Mean | Worst | Std. |
|---|---|---|---|---|
| Sandgren [52] | 8129.8000 | N.A. | N.A. | N.A. |
| Kannan and Kramer [39] | 7198.0428 | N.A. | N.A. | N.A. |
| Deb [38] | 6410.3811 | N.A. | N.A. | N.A. |
| Coello Coello [26] | 6288.7445 | 6293.8432 | 6308.1497 | 7.4133 |
| Coello Coello and Montes [48] | 6059.9463 | 6177.2533 | 6469.3220 | 130.9297 |
| He and Wang [49] | 6061.0777 | 6147.1332 | 6363.8041 | 86.4545 |
| He and Wang [29] | 6059.7143 | 6099.9323 | 6288.6770 | 86.2022 |
| Behrooz et al. [42] | 6059.7143 | 6059.7143 | 6059.7143 | 1.01E−11 |
| Mohamed and Sabry [41] | 6059.7143 | 6059.7143 | 6059.7143 | 3.62E−10 |
| **CI-PL** | **6080.5357** | **6089.5990** | **6093.5516** | **2.07046** |

where $1 \times 0.0625 \leq y_1 \leq 99 \times 0.0625$, $1 \times 0.0625 \leq y_2 \leq 99 \times 0.0625$, $10 \leq y_3 \leq 200$, and $10 \leq y_4 \leq 200$.

As stated in [1] the variables $y_1$ and $y_2$ are discrete values which are integer multiples of 0.0625 inch. Hence, the upper and lower bounds of the ranges of $y_1$ and $y_2$ are multiplied by 0.0625 as shown.

Table 8 presents the performance comparison between CI-PL with different optimization algorithms for the pressure vessel Design Problem and their statistical simulation results are shown in Table 9. From Tables 8 and 9, it can be seen that the best solution found by CI-PL is comparable with other algorithms.

## 3.4 Discussion

The CI-PL varies from the original CI [10] in learning approach as well as the reduction of the sampling interval. The PL approach helps to generate a set of qualities that it may follow in the subsequent learning attempt. In the original CI approach, the chances of any candidate to select the better set of qualities will increase according to the associated probability stake $p^c$, which is directly proportional to the quality of the behavior $f(\mathbf{x}^c)$. However, in the CI-PL approach, the behavior $f(\mathbf{x}^{c[?]})$ which every candidate may follow tends to be closer to the best candidate in the cohort. In other words, it may totally prevent any candidate from selecting worse qualities as may be the case in roulette wheel approach. Therefore, it may attain a better set of qualities and hence improved behavior. This may further make the CI-PL algorithm to reach the optimal solution in comparatively fewer number of learning attempts than the original CI. The reduction of sampling interval in this paper is based on the standard deviation between the qualities of the candidates. It is important to mention here that all the candidates tend to arrive at the optimum solution simultaneously. As they approach the optimum solution, their behaviors and qualities start getting similar. In the initial

**Table 10** The summary of CI-PL solutions details

| Problem | Solutions Best Avg. Worst | Standard deviation | Avg. no. of function evaluations | Avg. comp. time (s) | Closeness to the best reported solution (%) | Variations t | Expansion factor $\gamma$ |
|---------|---------------------------|--------------------|----------------------------------|---------------------|---------------------------------------------|--------------|---------------------------|
| G01 | −14.99880 −14.99600 −14.99080 | 0.00128 | 13600 | 13.363 | 8.00e−005 | 40 | 25.0 |
| G02 | 0.7991 0.7875 0.7832 | 0.00264 | 11600 | 12.279 | 5.62e−003 | 40 | 20.0 |
| G03 | 1.0322 1.0322 1.0322 | 0.00000 | 4250 | 0.849 | 0.0322 | 25 | 15.0 |
| G04 | −30665.538 −30665.533 −30665.525 | 0.00242 | 8250 | 6.839 | 3.26e−008 | 25 | 23.7 |
| G05 | 5126.498 5126.760 5127.104 | 0.11171 | 3375 | 2.279 | 0 | 25 | 15.0 |
| G06 | −6959.08786 −6903.17034 −6637.47132 | 59.51767 | 15800 | 1.252 | 3.91e−004 | 100 | 30.0 |
| G07 | 24.4145 24.6753 25.0526 | 0.10659 | 9375 | 8.902 | 4.46e−003 | 25 | 20.0 |
| G08 | 0.095825 0.095825 0.095825 | 0.000000 | 1500 | 0.895 | 0 | 25 | 17.5 |
| G09 | 680.6463 680.7735 680.9530 | 0.04841 | 8125 | 5.208 | 2.38e−005 | 25 | 17.5 |
| G10 | 7057.70 7462.90 8897.40 | 313.4418 | 8375 | 3.987 | 1.18e−003 | 25 | 21 |
| G11 | 0.7490 0.7490 0.7490 | 0.000000 | 2800 | 1.395 | 1.33e−003 | 35 | 15.0 |
| G12 | 1.00000 1.00000 1.00000 | 0.00000 | 625 | 1.239 | 0 | 25 | 15.0 |
| G13 | 0.0565 0.0937 0.3034 | 0.0233 | 1875 | 1.686 | 0.0472 | 25 | 15.0 |

(continued)

**Table 10** (continued)

| Problem | Solutions Best Avg. Worst | Standard deviation | Avg. no. of function evaluations | Avg. comp. time (s) | Closeness to the best reported solution (%) | Variations t | Expansion factor $\gamma$ |
|---|---|---|---|---|---|---|---|
| G14 | −47.2080 −46.2879 −44.5988 | 0.39807 | 10000 | 16.628 | 9.25e−003 | 25 | 25.0 |
| G15 | 961.7236 962.2494 962.8885 | 0.20784 | 3875 | 5.574 | 8.91e−006 | 25 | 25.0 |
| G24 | −5.5080 −5.5080 −5.5080 | 0.00000 | 4000 | 1.566 | 0 | 25 | 35.0 |
| Spring design problem | 0.0126713 0.0127580 0.0128457 | 0.00002 | 3500 | 1.225 | 4.73e−004 | 25 | 20.0 |
| Welded beam design problem | 1.724865 1.724878 1.724916 | 8.5554e−006 | 5000 | 2.701 | 7.53e−006 | 25 | 25.0 |
| Pressure vessel design problem | 6080.5357 6089.5990 6093.5516 | 2.07046 | 7500 | 2.588 | 3.43e−006 | 50 | 20.0 |

learning attempts when the candidates' behaviors are marginally different, thus, the standard deviation among the candidates could be large signifying a wider sampling interval to search the improved qualities. As their behaviors start getting similar, the standard deviation reduces and it results in a smaller sampling interval to search for an improved solution. This method ensures a faster convergence. From Tables 1 and 2, the solutions to the test problems was marginally improved when compared with other algorithms for the benchmark unconstrained test problems such as Sphere function, Ackley function, Rosenbrock function, and the Griewank function. A smaller standard deviation of CI-PL compared with CI highlights the improved robustness of the algorithm. In addition, the computational cost, i.e., function evaluations (FE) and computational time was marginally improved. As an illustration, the convergence of the candidates is presented in Fig. 2a–e for sphere function along with the solution convergence plot in Fig. 2f. The reduction of sampling interval for all candidates for one quality is shown in Fig. 2g. In Fig. 2h the standard deviation between the qualities of all candidates in cohort is shown. In the early learning attempts, the standard deviation was large enough because the candidates had different initialization points. Another point worth mentioning about this approach is that for each learning attempt, although a candidate learns the most from the best candidate, it also learns in some proportion from the other candidates of cohort. This approach ensures that the

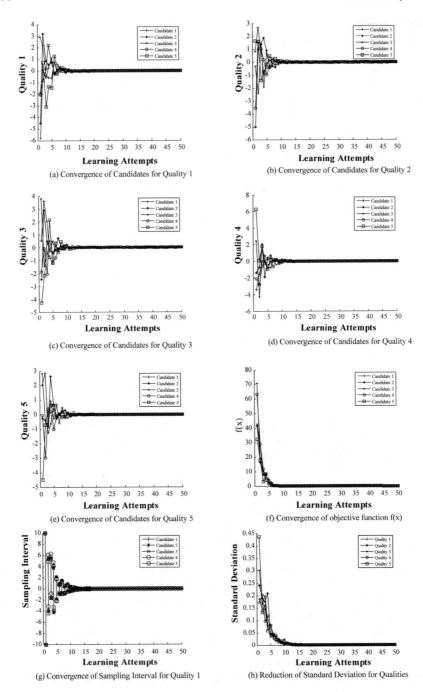

**Fig. 2** Convergence of qualities, objective function, sampling interval and reduction of standard deviation for 5 variable (qualities) sphere function

cohort may not get stuck in local minima. Furthermore, in the original CI approach, the sampling interval associated with every quality was expanded to the original one when no change in the candidates' solution was observed for a considerable number of learning attempts. In CI-PL, such approach is not required. This was because PL makes every candidate to follow the qualities that are closer to the qualities of the candidate whose behavior is best in cohort in the current learning attempt.

CI approach with penalty function approach was also applied to several constrained test problems [50, 53] incorporated into it. It is evident from the results presented in Table 3 through Table 10, that the CI methodology with panoptic learning approach is capable of successfully solving problems with equality and inequality of constraints. The results also exhibited competitive performance when compared with the existing algorithms. Referring to Table 10, it is clear that the standard deviation highlighted the robustness with reasonable number of function evaluations and CPU time as well as closeness to the best reported solutions so far.

The parameters: number of candidates $C$, numbers of variations in the behavior $t$, and standard deviation expansion factor $\gamma$ were chosen using preliminary trials of the algorithm. It is worth mentioning that during the development of the CI algorithm with PL approach several other variations were tried without much success. For example, in the first case, every candidate learn only from the best candidate for the current learning attempt. While, in the second case, all candidates of cohort learn from the other candidate whose behavior was similar with them. In the third case, the candidates learn from another candidate whose behavior was most different from them. In first case, it was observed that the cohort got stuck in the local minima if the cohort only followed a particular candidate for a large number of learning attempts who was heading toward the local minima. In the second case, groups were formed in cohort where some candidates converged to one solution and others converged to some other solutions. Both the solutions obtained by the groups of cohort were not optimum and hence optimum solution was hardly ever found using this case. In the third case, the candidates of cohort oscillated about the optimum solution without ever reaching the solution.

## 4  Conclusions and Future Directions

The paper has validated the CI-PL methodology by solving well-known uncon-strained, constrained, and engineering problems. The PL-based approach resembles the natural cohort learning behavior and is better suited to imitate the cohort learning behavior than roulette wheel-based approach [10]. The ability of proposed CI-PL algorithm has been demonstrated by solving several unconstrained, constrained, and engineering problems. Comparable if not better results exhibited its ability to solve the constrained problems efficiently with acceptable computational cost and robust-ness. The algorithm proposed here exhibited few limitations also. In addition to the advantages, few limitations are also observed. The computational performance was dependent on several parameters. A self-adaptive scheme can be developed for the estimation of these parameters. To make the CI-PL more generalized and to further improve its performances a new method can be developed in which it would weigh the individual qualities of every candidate based on their merit and not just the whole behavior. In addition, the constraint handling technique may be further improved/developed using a multi-criteria optimization approach [4, 54].

# References

1. Changdar, C., Mahapatra, G.S., Kumar Pal, R.: An improved genetic algorithm based approach to solve constrained knapsack problem in fuzzy environment. Expert Syst. Appl. **42**(4), 2276–2286 (2015)
2. Deb, K.: An efficient constraint handling method for genetic algorithms. Comput. Methods Appl. Mech. Eng. **186**, 311–338 (2000)
3. Li, X., Parrott, L.: An improved genetic algorithm for spatial optimization of multi-objective and multi-site land use allocation. Comput. Environ. Urban Syst. **59**, 184–194 (2016). ISSN 0198-9715
4. Ray, T., Tai, K., Seow, K.: Multiobjective design optimization by an evolutionary algorithm. Eng. Optim. **33**(4), 399–424 (2001)
5. Chen, Z., Xiong, R., Cao, J.: Particle swarm optimization-based optimal power management of plug-in hybrid electric vehicles considering uncertain driving conditions. Energy **96**(1), 197–208 (2016)
6. Dorigo, M., Birattari, M., Stitzle, T.: Ant colony optimization: arificial ants as a computational intelligence technique. IEEE Comput. Intell. Mag. 28–39 (2006)
7. Wu, W., Tian, Y., Jin, T.: A label based ant colony algorithm for heterogeneous vehicle routing with mixed backhaul. Appl. Soft Comput. **47**, 224–234 (2016)
8. Kavousi-Fard, A., Niknam, T., Fotuhi-Firuzabad, M.: A novel stochastic framework based on cloud theory and modified bat algorithm to solve the distribution feeder reconfiguration. IEEE Trans. Smart Grid **7**(2), 740–750 (2016)
9. Pham, D., Ghanbarzadeh, A., Koc, E., Otri, S., Rahim, S., Zaidi, M.: The Bees Algorithm. Technical Note, Manufacturing Engineering Centre, Cardiff University, UK (2005)
10. Kulkarni, A., Durugkar, I., Kumar, M.: Cohort intelligence: a self-supervised learning behavior. In: Proceedings of the IEEE Conference on Systems, Man and Cybernetics, pp. 1396–1400 (2013)
11. Krishnasamy, G., Kulkarni, A., Paramesran, R.: A hybrid approach for data clustering based on modified cohort intelligence and k-means. Expert Syst. Appl. 6009–6016 (2013)
12. Kulkarni, A.J., Shabir, H.: Solving 0-1 Knapsack problem using cohort intelligence algorithm. Int. J. Mach. Learn. Cybern. **7**(3), 427–441 (2016)
13. Kulkarni, A.J., Krishnasamy, G., Abraham, A.: Cohort intelligence: a socio-inspired optimization method. In: Intelligent Systems Reference Library, vol. 114. Springer (2017). https://doi.org/10.1007/978-3-319-44254-9. (ISBN: 978-3-319-44254-9)
14. Kulkarni, O., Kulkarni, N., Kulkarni, A.J., Kakandikar, G.: Constrained Cohort intelligence using static and dynamic penalty function approach for mechanical components design. Int. J. Parallel Emerg. Distrib. Syst. (In Press) (2016). https://doi.org/10.1080/17445760.2016.1242728)
15. Dhavle, S.V., Kulkarni, A.J., Shastri, A., Kale, I.R.: Design and economic optimization of shell-and-tube heat exchanger using cohort intelligence algorithm. Neural Comput. Appl. (In Press) (2017)
16. Kale, I.R., Kulkarni, A.J.: Cohort intelligence algorithm for discrete and mixed variable engineering problems. Int. J. Parallel Emerg. Distrib. Syst. (In Press) (2017)

17. Patankar, N.S., Kulkarni, A.J.: Variations of cohort intelligence. Soft. Comput. **22**(6), 1731–1747 (2018)
18. Shastri, A.S., Kulkarni, A.J.: Multi-cohort intelligence algorithm: an intra-and inter-group learning behaviour based socio-inspired optimisation methodology. Int. J. Parallel Emerg. Distrib. Syst. **33**(6), 675–715 (2018)
19. Sarmah, D., Kulkarni, A.J.: JPEG based steganography methods using cohort intelligence with cognitive computing and modified multi random start local search optimization algorithms. Inf. Sci. **430–431**, 378–396 (2018)
20. Sarmah, D., Kulkarni, A.J.: Image steganography capacity improvement using cohort intelligence and modified multi random start local search methods. Arab. J. Sci. Eng. (In Press) (2017)
21. Krishnasamy, G., Kulkarni, A.J., Raveendran, P.: A hybrid approach for data clustering based on modified cohort intelligence and K-means. Expert Syst. Appl. 6009–6016 (2014)
22. Kulkarni, A.J., Baki, M.F., Chaouch, B.A.: Application of the Cohort-intelligence optimization method to three selected combinatorial optimization problems. Eur. J. Oper. Res. **250**(2), 427–447 (2016)
23. Xu, W., Geng, Z., Zhu, Q., Gu, X.: A piecewise linear chaotic map and sequential quadratic programming based robust hybrid particle swarm optimization. Inf. Sci. 85–102 (2013)
24. Schittkowski, K.: NLQPL: a FORTRAN-subroutine solving constrained nonlinear programming problems. Ann. Oper. Res. 485–500 (1985)
25. Liu, L., Zhong, W., Qian, F.: An improved chaos-particle swarm optimization algorithm. J. East China Univ. Sci. Technol. (Natl. Sci. Ed.) 267–272 (2010)
26. Coello Coello, C.: Use of self-adaptive penalty approach for engineering optimization problems. Comput. Ind. 113–127 (2000)
27. Farmani, R., Wright, J.: Self-adaptive fitness formulation for constrained optimization. EEE Trans. Evol. Comput. 445–455 (2003)
28. Lampinen, J.: A constraint handling approach for the differential evolution algorithm. IEEE Congr. Evol. Comput. 1468–1473 (2002)
29. He, Q., Wang, L.: A hybrid particle swarm optimization with a feasibility-based rule for constrained optimization. Appl. Math. Comput. 1407–1422 (2007)
30. Hu, X., Eberhart, R.: Solving constrained nonlinear optimization problems with particle swarm optimization. In: 6th World Multi-Conference on Systemics, Cybernetics and Informatics (2002)
31. Koziel, S., Michalewicz, Z.: Evolutionary algorithms, homomorphous mapping, and constrained parameter optimization. Evol. Comput. 19–44 (1999)
32. Coello Coello, C., Becerra, R.: Efficient evolutionary optimization through the use of a cultural algorithm. Eng. Optim. 219–236 (2004)
33. Becerra, R., Coello Coello, C.: Cultured differential evolution for constrained optimization. Comput. Methods Appl. Mech. Eng. 4303–4322 (2006)
34. Chootinan, P., Chen, A.: Constraint handling in genetic algorithms using a gradient-based repair method. Comput. Oper. Res. 2263–2281 (2006)
35. Zahara, E., Hu, C.: Solving constrained optimization problems with hybrid particle swarm optimization. Eng. Optim. 1031–1049 (2008)
36. Montes, E., Lopez-Ramirez, B.: Comparing bio-inspired algorithms in constrained optimization problems. IEEE Congr. Evol. Comput. 662–669 (2007). Singapore
37. Montes, E., Varela, M., Caemen, R., Ramon, G.: Differential evolution in constrained numerical optimization: a empirical study. Inf. Sci. 4223–4262 (2010)
38. Deb, K.: GeneAS: a robust optimal design technique for mechanical component design. In: Dasgupta, D., Michalewicz, Z. (eds.) Evolutionary Algorithms in Engineering Applications, pp. 497–514. Springer (1997)
39. Kannan, B., Kramer, S.: An augmented lagrange multiplier based method for mixed integer discrete continuous optimization and its applications to mechanical design. ASME J. Mech. Des. 405–411 (1994)

40. Ragsdell, K., Phillips, D.: Optimal design of a class of welded structures using geometric programming. SME J. Eng. Ind. Ser. B 1021–1025 (1976)
41. Mohamed, A.W., Sabry, H.Z.: Constrained optimization based on modified differential evolution algorithm. Inf. Sci. **194**, 171–208 (2012)
42. Behrooz, G., Li, X., Ozlen, M.: Cooperative coevolutionary differential evolution with improved augmented Lagrangian to solve constrained optimisation problems. Inf. Sci. **369**(C), 441–456 (2016)
43. Dong, Y., Tang, J., Xu, B., Wang, D.: An application of swarm optimization to nonlinear programming. Comput. Math. Appl. 1655–1668 (2005)
44. Hamida, S., Schoenauer, M.: ASCHEA: new results using adaptive segregational constraint handling. IEEE Congr. Evol. Comput. 884–889 (2002)
45. Hedar, A.R., Fukushima, M.: Derivative-free simulated annealing method for constrained continuous globale optimization. J. Glob. Optim. 521–549 (2006)
46. Runarsson, T., Yao, X.: Stochastic ranking for constrained evolutionary optimization. IEEE Trans. Evol. Comput. 284–294 (2000)
47. Arora, J.: Introduction to Optimum Design. Elsevier Academic Press (2004)
48. Coello Coello, C., Montes, E.: Constraint-handling in genetic algorithms through the use of dominance-based tournament selection. Adv. Eng. Inf. 193–203 (2002)
49. He, Q., Wang, L.: An effective co-evolutionary particle swarm optimization for constrained engineering design problems. Eng. Appl. Artif. Intell. 89–99 (2006)
50. Kulkarni, A., Tai, K.: Solving constrained optimization problems using probability collectives and a penalty function approach. Int. J. Comput. Intell. Appl. **10**(4), 445–470 (2011)
51. Siddall, J.: Analytical Design-Making in Engineering Design. Prentice-Hall (1972)
52. Sandgren, E.: Nonlinear integer and discrete programming in mechanical design. In: ASME Design Technology Conference, pp. 95–105 (1988)
53. Vanderplaat, G.: Numerical Optimization Techniques for Engineering Design. Mcgraw-Hill (1984)
54. Metkar, S., Kulkarni, A.: Boundary searching genetic algorithm: a multi-objective approach for constrained problems. In: Satapathy, S., Biswal, B., Udgata, S. (ed.) In Advances in Intelligent and Soft Computing, pp. 269–276. Springer (2013)

# Nature-Inspired Metaheuristic Algorithms for Constraint Handling: Challenges, Issues, and Research Perspective

**Surabhi Kaul and Yogesh Kumar**

**Abstract** Because of getting the efficient and accurate results in the field of optimization problem solving, researchers are taking much attentiveness in heuristic and metaheuristic approaches. Through utilizing the experimentation strategy Heuristic algorithms help in generating the accurate results to every problem. The response time of metaheuristic algorithms are much higher as compared with others. Various nature-based metaheuristic algorithms are easily accessible. And because of their effective applications and high power they are being widely used in various literatures like in a field of their applications, analysis, comparison, and algorithms. But still knowing its wide sights it is also called as "black box" because some time metaheuristic algorithms perform better and sometime results are too low on are given optimization problems. Metaheuristics are said to be most efficient for solving constraints in optimization problems. Metaheuristic algorithms can be categorized over various classes for separating them between different searching patterns and describe how the algorithm copy a specific phenomenal performance in the search area, diverse classification explored. The main focus of this chapter is to get the light over various constraints handling techniques, Importance of metaheuristic algorithms in constraint handling, and metaheuristic classification approach with proper flow diagram. In this paper, we have highlighted various interesting metaheuristic Algorithms and their application areas in different field. This chapter targets to review all metaheuristics applications in different fields like healthcare, data clustering, power system problem, optimization problem, and prediction process. Further taxonomy about metaheuristic algorithms is also the part of the chapter. One of the sections in this chapter gives the Comparison of different optimization algorithms in different research fields.

**Keywords** Metaheuristics algorithms · Constraint handling · Crow search algorithm · Whale optimization · Grasshopper optimization

S. Kaul
CSE, Chandigarh University, Gharuan, Mohali, Punjab, India
e-mail: surabhi93kaul@gmail.com

Y. Kumar (✉)
CSE, Chandigarh Group of Colleges, Landran, Mohali, Punjab, India
e-mail: Yogesh.arora10744@gmail.com

© The Author(s), under exclusive license to Springer Nature Singapore Pte Ltd. 2021
A. J. Kulkarni et al. (eds.), *Constraint Handling in Metaheuristics and Applications*,
https://doi.org/10.1007/978-981-33-6710-4_3

# 1   Introduction

In many ways' optimization algorithms [7] can be categorized. Out of which one way is to study the nature of the algorithm which can be called as stochastic or deterministic algorithm. The deterministic type of algorithms trails, difficult procedures and uses various consequences for getting the value of variables and functions. Whereas stochastic algorithms are purely unplanned in their nature. Other different optimization algorithms can be studied by mixing the stochastic and deterministic algorithm together. The mixer of both algorithms can be called as hybrid algorithm. The deterministic can be classified as classical methods, whereas on the other hands stochastic algorithms can be classified as heuristic and metaheuristic. These heuristic and metaheuristic algorithms are taking much attention of researcher nowadays in their fields. The word "heuristic," is of Greek origin, which means to discover or to find [13]. The definition of this term, given by author [4], is as follows- heuristic means various schemes used for problem solving processes in machine and man, readily accessible information to control. The trial and error method used for solving roots of quadratic equations can be one of the examples for heuristics. Other examples related to heuristics may include drawing a picture to visualize the problem, examining a concrete example, instead of the abstract one and many others.

In terms of mathematics, the "heuristic" means the procedures for getting the solutions which can be practically better and adequately best as compared to classical methods which are too slow or not so feasible. The best way to understand this can be explained with the help of example, in place of listing down all promising solution to some particular problem and checking the solutions one by one, heuristic approaches make guesses based over practice to slim down the users' solutions. The word "meta" is also Greek origin, whose meaning is "beyond." So the word "metaheuristic" [15], stands for a set of procedures theoretically categorized above the heuristics in the way that they guide their design. Many researchers use these both terms "heuristic" and "metaheuristic" interchangeably. It is noted that metaheuristic algorithms are of higher level as compared to heuristic algorithms. Hence help in providing us with better and efficient results. Ideally talking, the most appropriate solution for any assumed optimization problem can be solved by brute force, where all the possible users' solutions in the entire solution space are counted and are further monitored by monotonous trial and error process. But this monotonous strategy is not much successful when there are much bigger number of decision variables. Some conformist methods can also be used for solving various optimization problems that may include augmented Lagrangian and the conjugate gradient methods. But these methods, in terms of iterative modification approaches come up with various limitations, especially when there exist several local optima in the solution space. The gradient methods become unstable and complicated when there is a multiple peak of objective or cost function. To evade such problems, metaheuristic algorithms again its importance. The metaheuristic algorithms can be characterized into four main classes which are evolutionary algorithms, physics-based algorithms, swarm-based algorithms, and human-based algorithms. Unlike gradient methods derives the solutions based on previous solution or from some derivative information, metaheuristic algorithms produce the solution by merging multiple good solutions that are stored in memory to get the best result for the problem.

The two main elements [14] that set up the foundation for flow of algorithms in any metaheuristic algorithms are exploitation and exploration. The word exploration stands for global search in a general manner. As related to algorithm it means exploring each solution chosen by the user for providing the result. Whereas word exploitation stands for local search which in algorithm means fine study over the explored solution. This stage is as important as previous one because in this stage preferred solution to handle the problems are identified at this stage. While designing metaheuristic algorithms, it is very important to take care among both the phase's, i.e., exploitation and exploration.

Different algorithms [21] are used these days for solving optimization problems. Among all, nature inspired optimization algorithms are currently being used and are in trend these days for solving various optimization problems. In our chapter, mainly nature inspired Metaheuristic algorithms are to be discussed. Moreover, the use of recent natural inspired algorithms for achieving more efficient and effective results is also the part of this novel. The nature inspired optimization algorithms [24] work more efficiently for handling the constraints of optimization problems and providing us with much feasible solutions. Constraints can make an easy problem hard and hard problems even harder. Surprisingly, in the past only little research efforts have been devoted to the development of efficient and effective constraint-handling techniques in contrast to the energy invested in the development of new methods for unconstrained optimization. Different constraints handling techniques are also the part of this chapter and our main work is to handle different constraint-handling techniques for solving various optimization problems using metaheuristic algorithms.

# 2 Motivation

Metaheuristics is a framework that consists of group of algorithmic structures whose purpose is to find the nearest optimal solutions with the help of integration of cleverness and various search processes. As metaheuristics algorithms are much easier and flexible they are used for solving various optimization problems. In the beginning, for solving various optimization problems, genetic algorithms were used. But these days various new and trending algorithms are being developed which are either inspired from man-made process or from natural phenomena. Various Metaheuristic algorithms which were used earlier are: artificial bee colony, particle swarm optimization, ant colony optimization, and many more. Various metaheuristic algorithms which have gained their importance these days are: elephant herding optimization, crow search algorithm, lion optimization algorithm, grasshopper optimization, and many more. Moreover, enthusiasm to do work in the respective field is given by

its numerous wonderful applications such as in the field of solving various optimization problems like in health care [51], power system and many others. In this
novel work, researchers are putting their best efforts for understanding the concept of
constraints handing using metaheuristic techniques [9]. Comparison among various
latest metaheuristic algorithms in different fields is also the part of this chapter.

## 3 Background Study

In this section, we explicate the introduction to the Constraint handling techniques,
Importance of metaheuristic algorithms in constraint handling and metaheuristic
classification approach with proper flow diagram. Various interesting metaheuristic
Algorithms and their application areas in different field are also presented in this
section.

### 3.1 Constraint Handling Techniques

An enormous bulk of problems in the field of engineering and science is framed
as the part of optimization problem which further consists of a set of constraints
categorized as inequality and equality sets. These problems are very hard to solve
not because of their high nonlinearity problems, but also due to very challenging
search domains bounded by various constraints [18]. The choice for choosing the
appropriate optimization algorithm and the various ways for handling the constraint
plays are very important role. For every problem, it is not possible to achieve efficient
algorithm. Even sometimes for a given problem, efficient algorithm works in different
ways for handling various constraints which may lead accuracy and efficiency in
results.

Various different constraints handling techniques [8] are being listed out till now,
which can be ranged from traditional methods to refined adaptive methods and
further more stochastic ranking. There are lots of methods that help in converting the
constrained optimization problem into an unconstrained by revising its objectives to
make it easily approachable. The main advantage of this conversion from constrained
to unconstrained help in lowering the search domain and problem becomes much
smooth to be solved [5]. In addition to this conversion many more parameters are
being included for handling the problem and getting the proper set of results. In
many cases, this conversion works astonishingly well if appropriate values are used,
and further the transformed unconstrained problem can be resolved excellently with
the help of different optimization methods. In this section, several methods [23] and
their function for handling constraints are being discussed. Some of them are listed
as feasibility method, $\epsilon$ -constrained method, static penalty method, dynamic penalty
method, and stochastic ranking [19].

a.    Static Penalty and Dynamic Penalty Methods

## Static Penalty

The approach [19] proposed by researchers defines levels of violation of the constraints (and penalty factors associated with them):

$$\text{fitness}(\vec{x}) = f(\vec{x}) + \sum_{i=1}^{m} \left( R_{k,i} \times \max[0, g_i(\vec{x})]^2 \right) \tag{1}$$

where $R_{k,i}$ are the penalty coefficients used, m is total the number of constraints, f(x) is the unpenalized objective function, and $k = 1, 2, \dots, 1$, where 1 is the number of levels of violation defined by the user.

## Criticism to Static Penalty

- It may not be a good idea to keep the same penalty factors, along the entire evolutionary process.
- Penalty factors are, in general, problem-dependent.
- The approach is simple, although in some cases the user may need to set up a high number of penalty factors.

## Dynamic Penalty

Within this category, we will consider any penalty [43] function in which the current generation number is involved in the computation of the corresponding penalty factors (normally the penalty factors are defined in such a way that they increase over time, i.e., generations).

An example of this sort of approach is the following: The approaches have been evaluated at generation t using:

$$\text{fitness}(\vec{x}) = f(\vec{x}) + (C \times t)^{\alpha} \times SVC(\beta, \vec{x}) \tag{2}$$

where $C$, $\alpha$, and $\beta$ are constants defined by the user ($C = 0.5$, $\alpha = 1$ or 2, and $\beta = 1$ or 2).

$SVC(\beta, x)$ is defined as:

$$SVC(\beta, \vec{x}) = \sum_{i=1}^{n} D_i^{\beta}(\vec{x}) + \sum_{j=1}^{p} D_j(\vec{x}) \tag{3}$$

And

$$D_i(\vec{x}) = \begin{cases} 0 & g_i(\vec{x}) \leq 0 \\ |g_i(\vec{x})| & \text{otherwise} \end{cases} \quad 1 \leq i \leq n$$

$$D_j(\vec{x}) = \begin{cases} 0 & -\varepsilon \leq h_j(\vec{x}) \leq \varepsilon \\ |h_i(\vec{x})| & \text{otherwise} \end{cases} \quad 1 \leq j \leq p \tag{4}$$

This dynamic function increases the penalty as we progress through generations.

**Criticism to Dynamic Penalty**

- Some researchers have argued that dynamic penalties work better than static penalties.
- In fact, many EC researchers consider dynamic penalty as a good choice when dealing with an arbitrary constrained optimization problem.
- Note, however, that it is difficult to derive good dynamic penalty functions in practice as it is difficult to produce good penalty factors for static functions.

b.    Barrier Function Method

Lagrangian multipliers help in solving the equality constraints [25] but to deal with inequality constraints was very difficult and challenging task. One way to overcome such difficulty was to use barrier function and the logarithmic barrier functions together. The formula to generate it is written as below.

$$L(x) = -\mu \sum_{j=1}^{N} \log\left[-\psi_j(x)\right] \tag{5}$$

Here the range of $\mu > 0$ can vary during iterations ($t$).

c.    Feasibility Criteria

In this technique author had categorized three feasible selection mechanisms [33], which are discussed as below:

- Among all the feasible solution and infeasible solution one of each is being selected.
- When two feasible solutions are compared with each other, solutions which have lower objective value will be taken into account for getting the right solution.
- When two infeasible solutions are compared, the one with lower degree of constraint violation will be chosen.

The degree of the violation of constraints can be stated by the penalty term as given below:

$$P(x) = \sum_{i=1}^{M} |\phi_i(x)| + \sum_{j=1}^{N} \max\left\{0, \psi_j(x)\right\}^2. \tag{6}$$

Such rules can be considered as fitness ranking and preference of low constraint violation.

d.   Stochastic Ranking

Stochastic ranking [34] is another technique for constraint handling which gains its importance very rapidly. In this technique, control parameter is pre-defined and is bonded to the range between $0 < p_f < 0$ by the user to check the balance among feasibility and infeasibility solution. In this method no penalty parameter is used. On the basis of relative value and the sum of constraint violations the choice between solutions is being performed. With the help of bubble sort ranking of solution can be done.

Steps involved in this method are, first, to make a uniformly distributed variable denoted by "$u$" and compare that variable with pre-defined constant denoted by "$p_f$." If after comparison we get $u < p_f$ (when solution are feasible), interchange if $f(x_j) > f(x_i)$. But if both the solution is infeasible, interchange if $P(x_j) > P(x_i)$.

The ranking is done using the constant "$p_s$," which is given below.

$$p_s(x) = p_o p_f + p_v(1 - p_f),\qquad (7)$$

"$p_o$" stands for number of chances that a user can win, which on further depends on objective value. Whereas "$p_v$" is the number of chances to win the particular solution, which depends upon the violation of the constraints. "$p_s$" is the ranking constant used for getting the appropriate results for various equations.

Advantages:

- The motivation of stochastic ranking comes from the need for balancing objective and penalty functions directly and explicitly in optimization.
- Ranking is achieved by a bubble-sort-like procedure using stochastic ranking
- The $\epsilon$-Constrained Approach

Another technique for handling constraints was developed known as $\epsilon$-constrained method [19]. This method contains two main steps: limit relaxation and lexicographical ordering. Two solutions, i.e., xi and xj are compared and ranked with f (xi) its objective values. The equation can be written as:

$$\{f(x_i), P(x_i)\} \leq \varepsilon\{f(x_i), P(x_i)\},\qquad (8)$$

Advantages:

The $\varepsilon$ constrained method provides a transformation method, that help in converting algorithms for unconstrained problems with algorithms for constrained problems using the $\varepsilon$ level comparison.

## 3.2   Objective of the Research

In this section, objectives have been discussed that are the part of this chapter.

- To study and analysis of the different metaheuristic algorithms for optimizing the research problems.
- To apply the metaheuristic classification approach to various constraint handling techniques.
- To explore and apply the new generation metaheuristic optimization methods in various research areas.

## 3.3   The Importance of Metaheuristic Algorithms in Constraint Handing

For solving various problems related to optimization, over the last decades, many algorithms were trending. Algorithms [17] used for solving the diverse constrained engineering optimization problems are mainly based on two methods either numerical linear and nonlinear programming. But when talking about the constraints related to multiple or sharp peaks at that time it becomes difficult for gradient search to achieve the goals. Such type of drawbacks in gradient search has motivated researcher to rely more over metaheuristic algorithms which are based on the replications and copying the performance of the natural behavior of various humans for solving hard and complex optimization problems. The main idea used in metaheuristic algorithms is that they put together rules and randomness to emulate natural phenomena [30]. Many nature inspired metaheuristic algorithms [37] used these days have proved their excellence and efficiency. The algorithms have proved best for handling constraints of optimization problems and achieving the feasible solution. Or else, several solution efforts may be fruitless and restrictions may be violated.

Metaheuristic is used more oftenly for solving the problem related to metaheuristic computing that mainly uses various heuristic rules for solving various computational problems. A Metaheuristic is a model that uses various heuristic methods for solving complex and hard problems. In simple words, metaheuristic looks like an algorithmic framework which are basically used for evaluating efficient results for various complex or hard optimization problems. Some properties [41] that can be used to define metaheuristic are as follows:

- They help in guiding us the strategies for exploration procedure.
- They are not a problem specific.
- They are not deterministic in nature, but can be counted as approximate.
- Techniques used in this field may range from local search processes to complex learning processes.
- The main goal is to get an efficient optimal solution by exploring the search space.

Metaheuristic basically uses high-level approaches for exploring search spaces with the help of diverse methods. Metaheuristic may consist of various assorted classifications which are being constructed over exploitation and exploration, the two metaphors used for search procedures.

The classification based on meta heuristics is based on two things, i.e., population-based meta heuristics and other is trajectory-based meta heuristic. The population-based metaheuristics algorithm produces the population, which is improved from the previous population with the help of various advanced search iteration. The new advance population is the best solution that we get from either whole population or from a portion of it [48]. As compared with a trajectory-based algorithm, it starts with the single solution and at each instance new best solution is achieved and is compared to its efficiency. The results of trajectory-based algorithm are more exploitation as compared to population based whose results are more exploration in nature. The metaheuristics algorithm can be classified as local search, where minor modification can lead to much efficient results than the previous one.

Researchers [35] have distributed meta heuristics into two parts, i.e., nature inspired and non-nature inspired. Nature inspired metaheuristics algorithms are further classified into following types which are: physics based, swarm intelligence, bio inspired and many more that are characterized on the basis of their behavior, social, emotions, etc.

The taxonomy [44] used for metaheuristic mainly uses methods for objective function usage, use of memory and memory less and neighborhood structures. However the classification of,meta heuristics can be performed either by nature-inspired and non-nature-inspired or Population-based and trajectory based.

Lots of surveys in the past have been done on metaheuristic. The main purpose of these surveys is to get the information related to its applications, its algorithms, its analysis and its comparisons. These Metaheuristic algorithms are said to be a black box at some time they are good at solving the optimization problems in one domain and not in another domain [6]. A comprehensive survey has been done in 2019 to get an idea for metaheuristics based upon research (till now 1222 has been done) over past 33 years. The main ways to achieve the size of study directed in this particular area. There are a few more terms that can be used for in comparison with meta heuristics are evolutionary computing, local minima versus global minima, swarm intelligence, neighborhood search, and few others [19].

# 4 Study of Novel Metaheuristic Algorithms

In this section, we discussed about the related work, which is best presenting the findings of metaheuristic algorithms in the different fields like health care, data clustering, power system problem, optimization problem, and prediction process. Further taxonomy [50] about metaheuristic algorithms is also the part of the section.

## 4.1   Related Work

Best findings about the various roles and the impact of metaheuristic algorithms [2] are discussed below.

### 4.1.1   Metaheuristic Algorithms in Field of Health Care

Beloufa et al. [10] proposed that the disease diabetes can be diagnosed with the help of various techniques used by artificial intelligences. In order to enhance performance of ABC algorithm a mutation operator is added. Further, to improve the variety of ABC without any compromise with solution quality, another operator of a genetic algorithm is used, if the current best solution cannot be modernized. The main function of this updated ABC algorithm is to work for creating and amend automatically in association operations and regulation directly from data. With the help of UCI machine learning diabetes dataset can be used. Capability of the given method is based on the specificity standards, cataloging rate and sensitivity with the help of 10-fold cross-validation technique. Classification rate is 84.21% and at the time when compared with the prior research it looked more capable for the same problem.

Subanya et al. [45] stated that the motive of this paper was to establish the finest element subset with better classification correctness by using a metaheuristic algorithm in cardiovascular disease diagnosis. In order to obtain the best characteristics in disease detection, artificial bee colony (ABC) algorithm is used. Support Vector Machine (SVM) classification can be used to calculate the suitability of ABC. The performance of the concerned algorithm is validated against the Cleveland Heart disease dataset which received from the VCI machine learning repository. Artificial bee colony (ABC) algorithm and Support Vector Machine (SVM) in comparison to Feature selection doing better.

Dash, S. et al. [12] explained about the main purpose of the maps is to choose primary values of population and amend inclusion coefficient so that enlarge populations and get a better exploration procedure to attain overall optima ignoring the local optima. In choosing most appropriate features from Naive Bayesian stochastic algorithm and search space algorithm, various estimation techniques for learning algorithm is applied over various features like stability, generalization and accuracy. The performance of other chaotic models has been lowered in comparison to chaos-based logistic model.

### 4.1.2   Data Clustering Using Metaheuristic Algorithms

Karaboga et al. [29] explained that Artificial Bee Colony (ABC) algorithm is newly established optimization algorithms. There is another important device is Clustering analysis, which can be used in different applications and also used to recognize the similar groups of entities according to their elements. The working of the artificial bee Colony algorithm and Particle Swarm Optimization (PSO) compared with each other. Artificial Bee Colony algorithm is mainly used for data clustering. There are different data sets by UCI Machine Learning Repository which is used to express the outcome of different techniques. ABC algorithm is one of the most adaptive algorithms which are used in the field multivariate data clustering.

Chander et al. [11] presented the adaptive directive operative fractional lion algorithm. All the solutions which are produced by the proposed algorithm depend upon the Adaptive Directive Fractional Lion Algorithm. The new MKWLI function can be used for assessing the accurate value. The dynamic directive operative searching algorithm is used for upgrading the performance of the Female Lion algorithm. Therefore, to figure out the accurate cluster core reiteration the most effective and suggested algorithm is the Adaptive Directive Operative Fractional Lion Algorithm.

### 4.1.3 Impact of Metaheuristic Algorithms for Predicting System

Kumar et al. [31] stated online repurchase intentions of purchaser and for this, artificial bee colony algorithm and other different techniques of machine learning are selected to achieve this purpose. Artificial Bee Colony Algorithm is used to know about the consumer repurchase reasons by way of feature selection of customer and shopping complex peculiarity for the prediction model.

Jaafari et al. [26] described the expansion and affirmation of different hybrid models and adaptive neuro-fuzzy inference system (ANFIS) which is used for clear prediction of wildfire possibilities. After inspecting a set of descriptive alterable elements like (elevation, temperature, slope, land use, aspect, rainwater, soil order, wind speed consequence, human settlements distance to roads) discover spatial database discovered with the help of 32 fire events. On the basis of the power of spatial organization among all sets and according to the possibility of wildfire each set of variables allocated weights by using the frequency ratio model. The purpose of the weights is to use for providing guidance to different mixture models such as ANFIS-FA and ANFIS-GA.

### 4.1.4 Role of Metaheuristic Algorithms for Solving Optimization Problems

Jayabarathi, T et al. [27] described that the inventor of metaheuristic algorithm was Xin-She Yang, who made these algorithms in 2010 and further altered and used for various optimization problems in engineering field. The author has explained in this novel about the study of BA, few sample real-world optimization, its elements and instructions for the upcoming study.

Majhi et al. [32] explained some efforts in the explorative activities in a CSA through space transform search (STS) procedure in their work. The name of the suggested algorithm is STS-CSA. An STS-CSA algorithm is used to merge space transformation search technique and determine a result of current search space and at the same time is used to alter search space for discovering the way out of global optimum solution. STS-CSA has been assessed through standard IEEE CEC 2017 benchmark operations for resolving the optimization problems performance. For the checking of efficiency of the suggested algorithm real-world engineering problems are corroborated to resolve practical problems. Further author had described that the study shows complication measure, statistical measure, and convergence investigation provides that suggested method is effective and consistent for resolving optimization problems.

Shirke et al. [42] stated about the Crow Search algorithm that crows are very clever as they hide their food in different hiding places and used according to their requirements. The outcomes of the CSA's prove that there is need of capacity for resolving engineering-related optimization problems. In order to resolve different engineering design problems CSA is efficient for solving these problems.

Delalic et al. [22] stated the application from the behavior of elephants, which is known as elephant herding algorithm. This algorithm is rapidly gaining its importance these days. For testing the efficiency, the EHO algorithm is applied to some instance of TSPLIB and the result boosts up with more efficient and accurate performance as compared with any other algorithm.

Johari et al. [28] presented a variety of applications in which several optimization problems are being solved by Firefly algorithm. Optimization is a technique which is used to formulate best results with the help of minimum and maximum parameters is included in the problem. The performance of the firefly's algorithm is used in different optimization problem like multi-objective, chaotic, and discrete. The Fireflies algorithm mainly used in the field of Engineering and Computer Science for resolving optimization problems. For obtaining superior results some other updated procedures also used for better results. Fireflies algorithm gives better and efficient performance and results as compared to other metaheuristic algorithms.

Feng et al. [16] described that in a study or research by using different techniques there are different elements involved such as converging speed, searching ability, and exploitation ability. While testing the performance of different algorithm techniques in different sets, the study shows that result of EGOA founds better than metaheuristic algorithms. EGOA algorithm is used in different engineering problems. Further elaborating EGOA in bin packing problem used EGOA algorithm, at the same time other metaheuristic algorithms also used for the same problem, but after checking the results it was found that the results coming from EGOA are much better from other algorithms.

### 4.1.5 Effectively Handling the Power System Problems Using Metaheuristic Methods

Sambariya et al. [38] presented the working of the bat algorithm (BA) the working of this algorithm is based on fuzzy logic-based power system stabilizer which is used to improved small-signal stability. In bat algorithm technique suggested optimization is measured by objective function which is based on square minimization which helps to guarantee the nonlinear model's stability. This BA FPSS is rapidly used in IEEE ten machine thirty-nine-bus system models and after obtaining the results it is compared with a robust fuzzy controller. For checking the excellence performance between BAFPSS and FPSS the test is conducted in four different models in which fault locations are different.

Sayed et al. [25] stated that to resolve the multifarious engineering problems, metaheuristic algorithms are rapidly used as optimization algorithms. On the basis of the behavior of the crows, CSA is used in present days. This model considers exciting features and when it applies to make multi-model its exploration tactics founds high difficulties. The study of upgrading version of CSA is representing which is used to resolve energy optimization problems. In the updated form of CSA some changes are made in his old features, i.e., random perturbation and another one is awareness possibilities. By adopting all such modifications the new modified algorithm is enhanced the convergence to difficult high multi-model optima. This modified version of CSA is used in different optimization problems for checking its performance and the results received from current approach is highly efficient as compared to other different techniques.

## 4.2  Taxonomy of Various Metaheuristic Approaches

A figure above represents the taxonomy [20] of metaheuristic algorithms. Various Metaheuristic approaches are explained below (Fig. 3.1).

1.   Evolution-based algorithm

Evolutionary algorithm is based on general population-based Metaheuristic algorithm which is a subsection of mutative computation. The technique used in EA algorithm is based on the biological evolution like selection, reproduction, mutation, and recombination. Some EA-based algorithm is mentioned as under:-

- Genetic algorithms: Genetic algorithm contains a natural section process which refers to a larger class of evolutionary algorithms. In order to make high-quality solutions and problems which depends upon biologically inspired operators [40] like crossover, transformation, and selection, genetic algorithm are the type of algorithm which are mostly used. On the basis of the Darwin's theory of evolution, genetic algorithm presents by John Holland in 1960 and further completed by his student in 1989.
- Evolution Stragery: Evolution Strategies based on the nature inspired search and optimization methods and these evolutionary algorithms are used in recombination, selection, and mutation which are useful for better results for making a solution to the population of individuals containing candidate [33].
- Biogeography-based optimizer: The Biogeography optimization algorithm is the type of Metaheuristic algorithm which develop a function through recurrence and speculative improving candidate solutions with the help of quality evaluation

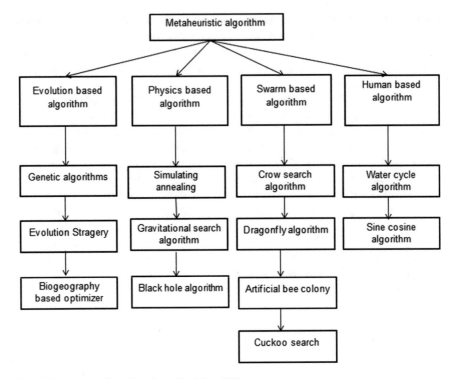

**Fig. 1** Taxonomy of metaheuristic algorithms [20]

and fitness function. This algorithm optimizer includes different alterations and without having any presumption regarding the problem and used in a large range of problems. BBO does not want descent of the function and is not differentiating the function as compared to classic optimization procedures like quasi-newton methods requires gradient of the function. The main purpose of using BBO is to optimized multilayered real-valued functions [39]. BBO is used to generate new candidate solutions and also managing old candidate solutions. This technique uses a simple formula to integrate both the solutions.

2.   Physics-based algorithm

This algorithm is used to represent how to solve optimization problems through these inspirations. These inspirations created through physics [47] and its assertion in the solutions and its progress with time. There are different algorithms which are based upon physics are given as under:-

- Simulating annealing: Simulated annealing is a physics-based metaheuristic technique which is used to obtain better results through a global optimization problem. This technique uses some nearest points to position x and with this possibility, the algorithm selects with the condition that it will stay with x or shift to x and with the progress of the algorithm the process of shifting one point to new which unite to zero. In this technique annealing procedure is used by global time varying element T [1].

- Gravitational search algorithm: The base of this algorithm is gravitation. There is a great impact of the arrangement of variables of swarm-based algorithm into the global optimization capability [3]. Gravitational search algorithm convergence point is decided by gravitational stability.

- Black hole algorithm: In the metaheuristic techniques black hole is a newly generated bio-based approach. Like other bio inspired algorithms this is also a population-based Metaheuristic algorithm. This algorithm [46] is used to solve problems in different areas such as data, image processing, engineering, computer vision, and data mining. This approach deals with complete research on the black hole technique and its applications.

3.   Swarm-based algorithm

Swarm-based algorithm consists of an artificial group of simple agents so basically it contains a system which includes decentralized and behavior of self-organized system. Several swarm-based algorithms described as below:-

- Crow search algorithm: Crow Search Algorithm works as an evaluator. CSA technique used vigilant behavior of crows. In this approach, there are limitations and decision factor and also relates to betterment of six constrained [34] structure plan. Crows are protective of their food and also notice the incompetent behavior of each other so that save their food from others.

- Dragonfly algorithm: Dragonfly algorithm is derived from the imitate behavior of dragonfly. Dragonflies usually move from one place to another in the search of their foods and migration. Swarming based on both static and dynamic. In case of dynamic, dragonfly unites with a group and join together in one path. The DA technique core center is based on the swarming behaviors of dragonfly. The behavior of swarming is related to Metaheuristic optimization phases that are exploitation and exploration [49].

- Artificial bee colony: In this approach, bees are divided into different groups, namely, employed bees, scouts, and onlookers. Food sources are according to the number of employed bees in a colony. The scout bees [36] are those bees whose food source has been deserted. Onlookers select the food sources, according to the dance of employed bees.

- Cuckoo search: This algorithm was proposed by Suash Deb and Xin she Yang. Basically, this algorithm is based upon the cuckoo species put their eggs in the nests very carefully from other birds. Any host bird who finds the eggs, which are not their own, they discarded the nest and eggs or make a new nest to some other place. This technique is also used in different optimization problems [43].

4.   Human-based algorithm

Human-based genetic algorithm (HBGA) is the type genetic algorithm which considers the human suggestions in the process of development. In other words, human-based algorithm utilizes the functions of genetic algorithm through humans. Some HB algorithms are mentioned as under:

- Water cycle algorithm: This process [16] is also called H20 which described the incessant movement of water. Precipitation, evaporation, and surface run off all are the phases which consider in the cycle. Based upon the water cycle study, this algorithm also gives good results in global optimization solution which includes exploitation and exploration phases same as in swarm optimization algorithm and provides new addition in Metaheuristic approaches.
- Sine, cosine algorithm: In this method [9], with the help of numerical method which is based upon cosine functions, multiple candidate solutions generate which varies to find the best solutions.

5.   Comparison of different optimization algorithms in different research field

Constraints handling Optimization problem are the main concern these days, which is gaining its sight so rapidly. Various natural inspired algorithms are being used which copies their behavior from the nature and are being implemented in the algorithms for getting the best results. In this table survey has been done over various optimization algorithms on the basis of their research field (Table 1).

**Table 1** Comparison of different optimization algorithms in different research field

| Author and reference | Name of the optimization technique | Research domain | Purpose |
| --- | --- | --- | --- |
| Beloufa et al. [10] | Artificial Bee Colony Algorithm | Diabetes disease | • Design of fuzzy classifier for diabetes disease<br>• The performances are evaluated through classification rate, sensitivity, and specificity |
| Sreeram et al. [24] | Bat Algorithm | Hybrid bat algorithm | • Bat algorithm developed different techniques<br>• In addition, to give better results in different problems, several modified versions have been developed |
| Delalic et al. [22] | Elephant Herding Algorithm | Traveling salesman problem | • This algorithm is rapidly gaining its importance these days<br>• EHO algorithm is applied to some instance of TSPLIB |
| Kumar et al. [31] | Artificial Bee Colony Algorithm | Prediction of online consumer repurchase intention | • The main purpose is to find out online repurchase intentions of purchaser and for this; artificial bee colony technique is used |
| Majhi et al. [32] | Crow Search Algorithm | Optimization problems | • An STS-CSA algorithm is used to merge space transformation search technique and determine a result of current search space<br>• STS-CSA has been assessed through standard IEEE CEC 2017 benchmark operations |

(continued)

**Table 1** (continued)

| Author and reference | Name of the optimization technique | Research domain | Purpose |
|---|---|---|---|
| Sambariya et al. [38] | Bat Algorithm | Power System Stabilizer for Multi machine Power System | • The working of this algorithm is based on fuzzy logic-based power system stabilizer which is used to improved small signal stability |
| Hegazy et al. [20] | Salp Swarm Algorithm | Feature selection | • In this technique efforts are made for betterment of basic SSA structure so that increase of better results and convergence rate<br>• Another new variable, i.e., weight, introduced for best problem solution |
| Jayabarathi et al. [27] | Bat Algorithm | Different Engineering Applications | • Bat Algorithm is used to solve various optimization problems |
| Saremi et al. [39] | Grasshopper Optimization Algorithm | Theory and application | • Grasshopper optimization algorithm (GOA) deals in difficult problems in structural optimization<br>• It is based on the actions of grasshopper belongs to swarms to figure out the optimization problems |
| Shankar et al. [41] | Elephant Herding Algorithm | Efficient image encryption scheme | • Adaptive Elephant Herding Optimization (AEHO) algorithm is adapting the most beneficial solution for encryption techniques<br>• In signcryption method encryption and digital signature process combines as a whole |

(continued)

**Table 1** (continued)

| Author and reference | Name of the optimization technique | Research domain | Purpose |
|---|---|---|---|
| Sayed [43] | Crow Search Algorithm | Survey over crow search algorithm | • The basic purpose of this chapter is to research deeply about the applications of CSA<br>• Study of exposition of the CSA investigates in different structures |
| Karaboga et al. [29] | Artificial Bee Colony Algorithm | Novel clustering approach | • In the field of data clustering optimization problem ABC algorithm have proved to be better in providing accuracy of the result |
| Dash et al. [12] | Firefly Algorithm | Parkinson's disease diagnosis | • The main purpose of the maps is to choose primary values of the population<br>• Enlarge populations and get a better exploration procedure to attain overall optima ignoring the local optima |
| Hassanien et al. [14] | Elephant Herding Algorithm | Intelligent human emotion recognition | • Elephant herding optimization (EHO) deals with arousal and dominance<br>• Discrete Wavelet transforms which earlier used for feature extraction and to eradicate artifacts EEG data were pre-processed<br>• EHO and SVR are linked with each other to evaluate prediction performance |

(continued)

**Table 1** (continued)

| Author and reference | Name of the optimization technique | Research domain | Purpose |
|---|---|---|---|
| Arora et al. [7] | Crow Search Algorithm | Unconstrained Function Optimization and Feature Selection | • On the basis of the result, it is found that GWOCSA performance in handling feature selection problem is highly recommended to solve real-world complex problems as compared to other techniques |
| Kilic et al. [30] | Lion Optimization Algorithm | Neuro fuzzy inference system | • The lion optimization algorithm gives better results in standard deviation, training time metrics, good, and bad<br>• IALO algorithm performs higher than ALO techniques<br>• Duration of training has been decreasing up to 80% |
| Hussien et al. [23] | Whale Optimization Algorithm | Feature selection | • A new binary version of the whale optimization algorithm is introduced for choosing an optimal subset for decease in dimensions and finds a solution for classification problem<br>• Updated technique is based upon sigmoid transfer function (S-shape) |
| Feng et al. [16] | Grasshopper Optimization Algorithm | Bin packing problem | • EGOA algorithm used in different Engineering problems |
| Jaafari et al. [26] | Firefly Algorithm | Prediction modeling of wildfire probability | • Used for clear prediction of wildfire possibilities |

(continued)

**Table 1** (continued)

| Author and reference | Name of the optimization technique | Research domain | Purpose |
|---|---|---|---|
| Bas et al. [9] | Social Spider Algorithm | Continuous optimization task | • The new binary version of the social spider algorithm is applied for the solution of binary problems<br>• Presently, there is a lack of focal point in binary version<br>• Most important part of binary version is transfer function which has further separated in two parts, i.e., V-shaped and S-shaped |
| Heidari et al. [8] | Grasshopper Optimization Algorithm | Hybrid multilayer perceptron neural network | • GOA is used for multilayer assumption<br>• The technique used in GOA is very promising to solve optimization problems with its updated efficient machinery<br>• The GOA multilayer perception model used in different datasets such as orthopaedic patients, coronary heart disease and breast cancer |
| Mohammed et al. [36] | Whale Optimization Algorithm | 0–1 knapsack problem | • This is the enhanced optimization algorithm which applied to resolving multiple and single dimensional problems |
| Selvi et al. [40] | Lion Optimization Algorithm | Message broadcasting system in VANET | • Lion optimization algorithm (LOA) based upon the characteristics of lions |
| Shirke et al. [42] | Crow Search Algorithm | Evaluation of Optimization in Discrete Applications | • To resolve different engineering design problems<br>• CSA is efficient for solving optimization and provide better results |

(continued)

**Table 1** (continued)

| Author and reference | Name of the optimization technique | Research domain | Purpose |
|---|---|---|---|
| Thalamala et al. [46] | Social Spider Algorithm | Data classifier | • Used for data classification where the entire spider has the prototypes for each database instance<br>• Another algorithm named single prototype social spider optimization is generated for classification of data which helps to decrease the solution space dimensions |
| Johari et al. [28] | Firefly Algorithm | Optimization problem | • Several optimization problems are being solved by Firefly algorithm<br>• Fireflies algorithm is used in different optimization problem like multi-objective, chaotic, discrete, etc. |
| Subanya et al. [45] | Artificial Bee Colony Algorithm | Feature selection for Cardiovascular Disease Classification | • The main function of this updated ABC algorithm is to work for creating and amend automatically association operations and regulation directly from data |
| Wang et al. [47] | Whale Optimization Algorithm | Multi-resource allocation | • Comparison between WOA and IWOA shows that IWOA performance is superior to WOA |
| Chander et al. [11] | Lion Optimization Algorithm | Data clustering | • All the solutions which are produced by the proposed algorithm depend upon the adaptive directive fractional lion algorithm |

(continued)

**Table 1** (continued)

| Author and reference | Name of the optimization technique | Research domain | Purpose |
|---|---|---|---|
| Hichem et al. [21] | Grasshopper Optimization Algorithm | Feature selection problem | • Grasshopper optimization technique is applied for feature subset problem<br>• Feature selection is a process which applies to optimization, classification perfection by searching a small subset of features from the actual group of features<br>• To check the performance in feature selection problems, both Swarm algorithm and Grasshopper methods are compared |
| Sayed et al. [25] | Crow Search Algorithm | Power system problem | • This model considers exciting features and when it applies to make multi-model its exploration tactics founds high difficulties<br>• Based on the behaviors of the crows |
| Mezura-Montes et al. [34] | Whale Optimization Algorithm | Feature selection | • This method is designed for getting the greatest feature subset that is helpful in the increase of the perfection of the classification |

## 5 Conclusion

This chapter studies vital problems regarding metaheuristics and new submissions for possible research opportunities and provides open challenge for population-based and nature-based optimization algorithms. During the scrutinization of these vital issues, the first step is to make a brief comparison between the latest updated Metaheuristic algorithms. In this study it has been noticed that mostly in the new generation algorithm, problem of huge numbers of parameters is the main drawback for metaheuristic algorithms. For getting the best results within a prescribed time period, various parameters used by Metaheuristic algorithms should be tuned specifically for the optimization work. The main purpose of some Metaheuristic algorithms like SOS, GWA, TLBO, and SSO is to utilize less number of parameters. At the same time, another serious problem for Metaheuristic algorithms is the shortage of local search techniques that can attain local optima. One of the most important points for Metaheuristic algorithms' performance is to make stability between both phases, i.e., exploitation and exploration. To provide a technique for solving huge and complex troubles within a limited time through metaheuristic algorithm which is an essential advantage as it is very difficult for exact algorithms to achieve better results in limited time. In addition to this, they executed easily and while solving optimization problem there is no necessity for background knowledge and ground truth. It has come to the notice that after the successful achievement of metaheuristic algorithms, search for making new metaheuristic algorithm will enlarge in the future. Some quality level must be fixed in this area so that metaheuristic performed more objectively and recognized the deficiencies. Besides this, another thing which is importantto chaotic versions of metaheuristic algorithms provide remarkable performance.

# References

1. Abraham, A., Jatoth, R.K., Rajasekhar, A.: Hybrid differential artificial bee colony algorithm. J. Comput. Theor. Nanosci. **9**(2), 249–257 (2012)
2. Al-Obeidat, F., Belacel, N., Spencer, B.: Combining machine learning and metaheuristics algorithms for classification method PROAFTN. In: Enhanced Living Environments, pp. 53–79. Springer, Cham (2019)
3. Ali, E.S., Abd Elazim, S.M., Abdelaziz, A.Y.: Ant lion optimization algorithm for renewable distributed generations. Energy **116**, 445–458 (2016)
4. Ali, M., Prasad, R.: Significant wave height forecasting via an extreme learning machine model integrated with improved complete ensemble empirical mode decomposition. Renew. Sustain. Energy Rev. **104**, 281–295 (2019)
5. Alresheedi, S.S., Lu, S., Abd Elaziz, M., Ewees, A.A.: Improved multiobjective salp swarm optimization for virtual machine placement in cloud computing. Hum.-Centric Comput. Inf. Sci. **9**(1), 15 (2019)
6. Ardabili, S., Mosavi, A., Varkonyi-Koczy, A.R.: Advances in machine learning modeling reviewing hybrid and ensemble methods. In: International Conference on Global Research and Education, pp. 215–227. Springer, Cham (2019, September)
7. Arora, S., Singh, H., Sharma, M., Sharma, S., Anand, P.: A new hybrid algorithm based on grey wolf optimization and crow search algorithm for unconstrained function optimization and feature selection. IEEE Access **7**, 26343–26361 (2019)
8. Heidari, A.A., Faris, H., Aljarah, I., Mirjalili, S.: An efficient hybrid multilayer perceptron neural network with grasshopper optimization. Soft. Comput. **23**(17), 7941–7958 (2019)
9. Bas, E., Ulker, E.: A binary social spider algorithm for continuous optimization task. Soft Comput. 1–27 (2020)
10. Beloufa, F., Chikh, M.A.: Design of fuzzy classifier for diabetes disease using Modified Artificial Bee Colony algorithm. Comput. Methods Programs Biomed. **112**(1), 92–103 (2013)
11. Chander, S., Vijaya, P., Dhyani, P.: Multi kernel and dynamic fractional lion optimization algorithm for data clustering. Alexandria Eng. J. **57**(1), 267–276 (2018)
12. Dash, S., Abraham, A., Luhach, A. K., Mizera-Pietraszko, J., Rodrigues, J.J.: Hybrid chaotic firefly decision making model for Parkinson's disease diagnosis. Int. J. Distrib. Sensor Networks **16**(1) (2020)

13. Dokeroglu, T., Sevinc, E., Kucukyilmaz, T., Cosar, A.: A survey on new generation metaheuristic algorithms. Comput. Ind. Eng. **137**, 106040 (2019)
14. Hassanien, A.E., Kilany, M., Houssein, E.H., AlQaheri, H.: Intelligent human emotion recognition based on elephant herding optimization tuned support vector regression. Biomed. Signal Process. Control **45**, 182–191 (2018)
15. Faris, H., Mirjalili, S., Aljarah, I., Mafarja, M., Heidari, A.A.: Salp swarm algorithm: theory, literature review, and application in extreme learning machines. In: Nature-Inspired Optimizers, pp. 185–199. Springer, Cham (2020)
16. Feng, H., Ni, H., Zhao, R., Zhu, X.: An enhanced grasshopper optimization algorithm to the bin packing problem. J. Control Sci. Eng. (2020)
17. Fister, I., Rauter, S., Yang, X.S., Ljubič, K., Fister Jr., I.: Planning the sports training sessions with the bat algorithm. Neurocomputing **149**, 993–1002 (2015)
18. Fister Jr, I., Fister, D., Fister, I.: Differential evolution strategies with random forest regression in the bat algorithm. In: Proceedings of the 15th Annual Conference Companion on Genetic and Evolutionary Computation, pp. 1703–1706 (2013, July)
19. He, X.S., Fan, Q.W., Karamanoglu, M., Yang, X. S.: Comparison of constraint-handling techniques for metaheuristic optimization. In International Conference on Computational Science, pp. 357–366. Springer, Cham (2019, June)
20. Hegazy, A.E., Makhlouf, M.A., El-Tawel, G.S.: Improved salp swarm algorithm for feature selection. J. King Saud Uni.-Comput. Inf. Sci. **32**(3), 335–344 (2020)
21. Hichem, H., Elkamel, M., Rafik, M., Mesaaoud, M.T., Ouahiba, C.: A new binary grasshopper optimization algorithm for feature selection problem. J. King Saud Uni.-Comput. Inf. Sci. (2019)
22. Delalic, S., Chahin, M., Alihodzic, A.: Optimal City Selection and Concert Tour Planning Based on Heuristic Optimization Methods and the Use of Social Media Analytics. In 2019 XXVII International Conference on Information, Communication and Automation Technologies (ICAT), pp. 1–6. IEEE (2019, October)
23. Hussien, A.G., Hassanien, A.E., Houssein, E.H., Bhattacharyya, S., Amin, M.: S-shaped binary whale optimization algorithm for feature selection. In: Recent Trends in Signal and Image Processing, pp. 79–87. Springer, Singapore
24. Sreeram, I., Vuppala, V.P.K.: HTTP flood attack detection in application layer using machine learning metrics and bio inspired bat algorithm. Appl. Comput. Inf. **15**(1), 59–66 (2019)
25. Sayed, G.I., Hassanien, A.E., Azar, A.T.: Feature selection via a novel chaotic crow search algorithm. Neural Comput. Appl. **31**(1), 171–188 (2019)
26. Jaafari, A., Termeh, S.V.R., Bui, D.T.: Genetic and firefly metaheuristic algorithms for an optimized neuro-fuzzy prediction modeling of wildfire probability. J. Environ. Manage. **243**, 358–369 (2019)
27. Jayabarathi, T., Raghunathan, T., Gandomi, A.H.: The bat algorithm, variants and some practical engineering applications: a review. In: Nature-Inspired Algorithms and Applied Optimization, pp. 313–330. Springer, Cham (2018)
28. Johari, N.F., Zain, A.M., Mustaffa, N.H., Udin, A.: Machining parameters optimization using hybrid firefly algorithm and particle swarm optimization. In: Journal of Physics: Conference Series, vol. 892, p. 012005 (2017)
29. Karaboga, D., Ozturk, C.: A novel clustering approach: Artificial Bee Colony (ABC) algorithm. Appl. Soft Comput. **11**(1), 652–657 (2011)
30. Kilic, H., Yuzgec, U., Karakuzu, C.: Improved antlion optimizer algorithm and its performance on neuro fuzzy inference system. Neural Network World **29**(4), 235–254 (2019)
31. Kumar, A., Kabra, G., Mussada, E.K., Dash, M.K., Rana, P.S.: Combined artificial bee colony algorithm and machine learning techniques for prediction of online consumer repurchase intention. Neural Comput. Appl. **31**(2), 877–890 (2019)
32. Majhi, S.K., Sahoo, M., Pradhan, R.: A space transformational crow search algorithm for optimization problems. Evol. Intell., 1–20 (2019)
33. Mezura-Montes, E., Palomeque-Ortiz, A.G.: Self-adaptive and deterministic parameter control in differential evolution for constrained optimization. In: Constraint-Handling in Evolutionary Optimization, pp. 95–120. Springer, Berlin, Heidelberg (2009)

34. Mezura-Montes, E., Coello, C.A.C.: Constraint-handling in nature-inspired numerical optimization: past, present and future. Swarm Evol. Comput. **1**(4), 173–194 (2011)
35. Michalewicz, Z., Schoenauer, M.: Evolutionary algorithms for constrained parameter optimization problems. Evol. Comput. **4**(1), 1–32 (1996)
36. Mohammed, H.M., Umar, S.U., Rashid, T.A.: A systematic and meta-analysis survey of whale optimization algorithm. Comput. Intell. Neurosci. (2019)
37. Qu, C., Fu, Y.: Crow search algorithm based on neighborhood search of non-inferior solution set. IEEE Access **7**, 52871–52895 (2019)
38. Sambariya, D.K., Prasad, R.: Application of bat algorithm to optimize scaling factors of fuzzy logic-based power system stabilizer for multimachine power system. Int. J. Nonlinear Sci. Numer. Simul. **17**(1), 41–53 (2016)
39. Saremi, S., Mirjalili, S., Lewis, A.: Grasshopper optimisation algorithm: theory and application. Adv. Eng. Softw. **105**, 30–47 (2017)
40. Selvi, M., Ramakrishnan, B.: Lion optimization algorithm (LOA)-based reliable emergency message broadcasting system in VANET. Soft. Comput. **24**(14), 10415–10432 (2020)
41. Shankar, K., Elhoseny, M., Perumal, E., Ilayaraja, M., Kumar, K.S.: An efficient image encryption scheme based on signcryption technique with adaptive elephant herding optimization. In: Cybersecurity and Secure Information Systems, pp. 31–42. Springer, Cham (2019)
42. Shirke, S., Udayakumar, R.: Evaluation of crow search algorithm (CSA) for optimization in discrete applications. In: 2019 3rd International Conference on Trends in Electronics and Informatics (ICOEI), pp. 584–589. IEEE (2019, April)
43. Sayed, G.I., Hassanien, A.E., Azar, A.T.: Feature selection via a novel chaotic crow search algorithm. Neural Comput. Appl. **31**(1), 171–188 (2019)
44. Strumberger, I., Minovic, M., Tuba, M., Bacanin, N.: Performance of elephant herding optimization and tree growth algorithm adapted for node localization in wireless sensor networks. Sensors **19**(11), 2515 (2019)
45. Subanya, B., Rajalaxmi, R.R.: Feature selection using Artificial Bee Colony for cardiovascular disease classification. In: 2014 International Conference on Electronics and Communication Systems (ICECS), pp. 1–6. IEEE (2014, February)
46. Thalamala, R.C., Reddy, A.V.S., Janet, B.: A novel bio-inspired algorithm based on social spiders for improving performance and efficiency of data clustering. J. Intell. Syst. **29**(1), 311–326 (2018)
47. Wang, Z., Deng, H., Zhu, X., Hu, L.: Application of improve whale optimization algorithm in muti-resource allocation. Int. J. Innovative Comput. **15**(3) (2019)
48. Yang, X.S., He, X.: Firefly algorithm: recent advances and applications. Int. J. Swarm Intell. **1**(1), 36–50 (2013)
49. Yazdani, M., Jolai, F.: Lion optimization algorithm (LOA): a nature-inspired metaheuristic algorithm. J. Comput. Des. Eng. **3**(1), 24–36 (2016)
50. Yusta, S.C.: Different metaheuristic strategies to solve the feature selection problem. Pattern Recogn. Lett. **30**(5), 525–534 (2009)
51. Kumar, Y., Sood, K., Kaul, S., Vasuja, R.: Big data analytics and its benefits in healthcare. In Big Data Analytics in Healthcare, pp. 3–21. Springer, Cham (2020)

# Experimental Comparison of Constraint Handling Schemes in Particle Swarm Optimization

**Mehdi Rostamian, Ali R. Kashani, Charles V. Camp, and Amir H. Gandomi**

**Abstract** Nature-inspired optimization algorithms have been designed for unconstrained problems. However, real-world optimization problems usually deal with a lot of limitations, either boundary of design variables, or equality/inequality constraints. Therefore, an extensive number of efforts have been made to make these limitations understandable for the optimization algorithms. Here, a more important fact is that those constraint handling approaches affect the algorithms' performances considerably. In this study, some of the well-known strategies are incorporated into particle swarm optimization algorithms (PSO). The performance of the PSO algorithm is examined through several benchmarks, constrained problems, and the results discussed comprehensively.

## 1 Introduction

Real-world optimization problems are often very complicated, with many decision variables and practical limitations on the range of feasible solutions [10, 14, 15, 17, 26, 27]. These complexities result in optimization problems that are non-convex, discontinuous, have high dimensionality, and pose many challenges in developing algorithms that converge to the optimal global solution. Metaheuristic algorithms are designed for unconstrained optimization problems, so it is crucial to develop methods to account for constraints [6, 11–13, 16, 25, 32]. Most metaheuristic optimization algorithms are based on two crucial phases: 1-diversification, and 2-intensification.

Diversification focuses on exploring the entire search space, often in random and chaotic ways making the algorithm capable of overcoming difficulties related to the discontinuity of the solution space. On the other hand, intensification tends to focus the search on regions identified by the best solutions. Therefore, one of the

M. Rostamian · A. R. Kashani (✉) · C. V. Camp
Department of Civil Engineering, University of Memphis, Memphis, TN 38152, USA
e-mail: Kashani.alireza@ymail.com

A. H. Gandomi
Faculty of Engineering & Information Technology, University of Technology Sydney, Ultimo, NSW 2007, Australia

most critical features of any constraint handling scheme is limiting the influence of infeasible solutions while preserving the stability of the algorithm. Many constrained optimization problems often have the optimal solution located near the boundaries of search space, so any constraint handling approach needs to enable an algorithm to explore the boundaries effectively.

Much research has been devoted to developing efficient methods for handling the constraints in both single- and multi-objective algorithms. Homaifar et al. [22] and Hoffmeister and Sprave [21] proposed methods based on a static penalty function in which a constant penalty term is added to the objective function value. In this approach, for any violated constraint, a penalty term is added to the objective function. The resulting optimization algorithm then attempts to decrease the penalty value while also finding the global optima. Morales and Quezada [33] proposed another version of the static penalty function method called the death penalty where a predetermined large value is set as the objective function for infeasible solutions. Joines and Houck [24], Kazarlis and Petridis [28], and Petridis et al. [35] developed a dynamic penalty function where the penalty term is increased with the iteration of the algorithm. Another iteration-dependent penalty function was proposed by Carlson and Shonkwiler [4] based on simulated annealing called annealing penalty function. In this way, in the course of iterations, the temperature decreases resulting in a higher penalty. Coit and Smith [7], and Gen and Cheng [18] proposed adaptive penalty function methods for handling the constraints by permitting algorithms to explore beyond the feasible search space to some level by adjusting the penalty term during the search. Other researchers have introduced a variety of hybrid methods such as lagrangian multipliers by Adeli and Cheng [1], a hybrid interior-lagrangian penalty-based approach by Myung and Kim [34], and the application of fuzzy logic by Le [30]. Deb [9] proposed a method based on the separation of constraints and objectives. This method is based on a pair-wise comparison of solutions in every iteration. In this context, a tournament selection operator was proposed to compare every candidate solution with the following strategy: 1—any feasible solution overcome the infeasible solution, 2—between two feasible solutions, the fitter solution is the winner, 3—between two infeasible solutions, the one with lower constraint violation is preferred.

The impact of constraint handling methods on the efficiency of algorithms was also the subject of several most recent studies. Li et al. [31] explored the effect of different constraint handling strategies on the performance of evolutionary multi-objective algorithms. In this way, three constraint handling approaches as constrained-domination principle, self-adaptive penalty, and adaptive tradeoff model combined with nondominated sorting genetic algorithm II for solving 18 test problems. Jamal et al. [23] explored three constraint handling methods (i.e., ε-Level comparison, superiority of feasible solutions, and penalty function) for matrix adaptation evolutionary strategy to solve CEC-2010 benchmark constrained problems. Biswas et al. [3] tackled the problem of optimal power flow solutions using differential evolution algorithms. In this study three different constraint handling approaches were utilized as follows: 1—superiority of feasible solutions (SF), 2—self-adaptive penalty (SP), and 3—an ensemble of these two constraint handling techniques (SF and SP). Ameca-Alducin et al. [2] explored the efficiency of four constraint handling schemes (i.e., stochastic ranking, $\varrho$-constrained, penalty, and feasibility rules) for differential evolution algorithm to handle dynamic constrained optimization problems. Zou et al. [37] developed a new constraint handling method for solving the combined heat and power economic dispatch using an improved genetic algorithm. Datta et al. [8] proposed a novel constraint handling approach based on individual penalty approach in which all the constraints were scaled adaptively without a need for specific information from the problem. The proposed approach was examined through solving 23 benchmark test problems and two engineering problems using a hybrid evolutionary algorithm.

In this paper, we presented a review of six well-known penalty function approaches for constraint handling. Each of these schemes is incorporated into a particle swarm optimization (PSO) algorithm to evaluate their effectiveness and efficiency for a set of benchmark constrained optimization problems.

## 2 Particle Swarm Optimization

Kennedy and Eberhart [29] developed the particle swarm optimization (PSO) algorithm based on the behaviors of bird flocks in searching for food. In this context, the PSO algorithm searches the solution space with a population of particles. Each particle in the swarm represents a potential solution to the optimization problem. The PSO algorithm changes the position of the particles within the search space with the aim of finding more appropriate solutions. PSO determines the position of particles within the search space by two primary qualities; position and velocity. In PSO, each particles' position changes iteratively based on its current position and velocity given as:

$$X_i^{t+1} = X_i^t + V_i^{t+1} \tag{1}$$

where $X_i^{t+1}$ is the updated position of the $i$-th particle, $X_i^t$ is the current position, and $V_i^{t+1}$ is the velocity. A targeted search is conducted by a particle movement using a velocity term. Each particles' velocity connected to two important achievements in each iteration: the particles' position relative to the global best-found solution $P_g$ and to its own best solution $P_i$. Clerc and Kennedy [5] proposed updating the velocity term as

$$V_i^{t+1} = \chi \left[ V_i^t + C_1 r_1 \left( P_i - X_i^t \right) + C_2 r_2 \left( P_g - X_i^t \right) \right] \tag{2}$$

where $V_i^{t+1}$ and $V_i^t$ are the new velocity and old velocity of the $i$-th particle, respectively, $C_1$ and $C_2$ control the relative attraction to $P_i$ and $P_g$, respectively, $r_1$ and $r_2$ are random numbers within [0,1], and $\chi < 1$ is the constriction factor that makes convergence rating slower and provides a better exploration of solution space (diversification).

Choosing appropriate values for parameters in the PSO algorithm is vital to obtaining the best performance. In Eq. (2), the values of $\chi$, $C_1$, and $C_2$ impact how each particle will be attracted to its best position and the global best position. Clerc and Kennedy [5] proposed the following values: $C_1 = C_2 = 2.05$ and $\chi = 0.72984$.

## 3 Constraint Handling Approaches

In general, an optimization problem for the objective function $f(x)$ can be described as

$$Minimize\ f(x) \tag{3}$$

subject to

$$h_i(x) = 0, i = 1, 2, 3, \ldots, m \tag{4}$$

$$g_j(x) \leq 0, j = 1, 2, 3, \ldots, p \tag{5}$$

where $x$ is a vector of design variables, $h$ and $g$ are equality and inequality constraints, respectively, $m$ and $p$ are the number of equality and inequality constraints, respectively.

In this study, six different penalty function-based approaches are utilized to incorporate the effects of constraints into the PSO algorithm. Penalty function-based techniques are utilized to transform a constrained problem into an unconstrained one. In this way, the optimization algorithm generates a potential solution without considering its feasibility. At the next step, the constraints are checked, and a penalty value, based on the degree of violations of each constraint, is added to the objective function value. The penalty functions considered in this study are:

1.  A simple static penalty function approach proposed by Homaifar et al. [22] is used to compute the penalized objective function $F(x)$ as

$$F(x) = f(x) + \sum_{i=1}^{p} R_{i,j} g_i(x)^2 \tag{6}$$

where $R_{i,j}$ is the penalty coefficient, $p$ is the number of constraints, and $j = 1, 2, \ldots, l$, where $l$ is the number of levels of a violation defined by the user. In this study, we used the same constant penalty coefficients for all the constraints.

2.  The static penalty function method proposed by Hoffmeister and Sprave [21] is defined as

$$F(x) = f(x) + \sqrt{\sum_{i=1}^{p} \delta(-g_i(x)) g_i(x)^2} \tag{7}$$

where

$$\delta(x) = \begin{cases} 1 \ if \ x > 0 \\ 0 \ if \ x \leq 0 \end{cases} \tag{8}$$

3.  The death penalty function proposed by Morales and Quezada [33] for updating the objective value is given as

$$F(x) = \begin{cases} f(x) \quad if \ x \ is \ feasible \\ K - \sum_{i=1}^{s} \left(\frac{K}{p}\right) otherwise \end{cases} \tag{9}$$

where $K$ is a large constant, $s$ is the number of satisfied constraints, and $p$ is the total number of constraints.

4.  The adaptive penalty function approach proposed by Coit and Smith [7] is utilized to alter the penalty term based on the feedback taken from the evolution process. The penalized objective function is

$$F(x) = f(x) + \left(B_{feasible} - B_{all}\right) \sum_{i=1}^{p} \left(\frac{g_i(x)}{NFT(t)}\right)^k \tag{10}$$

where $B_{feasible}$ is the best known feasible objective function at generation $t$, $B_{all}$ is the best known (unpenalized) objective function at generation $t$, $k$ is a constant that adjusts the "severity" of the penalty, and $NFT$ is the so-called Near Feasibility Threshold, which is defined as the threshold distance from the feasible region as

$$NFT = \frac{NFT_0}{1 + \lambda \times t} \tag{11}$$

where $NFT_0$ is an upper bound for $NFT$, and $\lambda$ is a constant which guarantees a search of the whole region between $NFT_0$ and zero.

5. The dynamic penalty function developed by Joines and Houck [24] is defined as

$$F(x) = f(x) + (C \times t)^{\alpha} \sum_{i=1}^{p} |g_i(x)|^{\beta} \tag{12}$$

where $C$, $\alpha$, and $\beta$ are constants defined by the user.

6. The annealing-based penalty function method developed by Joines and Houck [24] is defined as

$$F(x) = f(x) + \exp\left( (C \times t)^{\alpha} \sum_{i=1}^{p} |g_i(x)|^{\beta} \right) \tag{13}$$

These methods labeled from 1 to 6 are referred to in all tables and figures as Homaifar, Hoffmeister, Death, Adaptive, Dynamic, and Annealing, respectively.

## 4 Numerical Simulation

In this section, we incorporate the above-mentioned constraint handling approaches into a PSO algorithm to solve several benchmark problems. In all the cases, the particle population size is 50, and the number of iterations is 1,000. Metaheuristic optimization algorithms search the solution space stochastically to find the global minimum. Therefore, we evaluated the performance of each constraint handling approach based on the best, mean, and standard deviation (STD) of solutions over a series of 50 independent runs. The best-found solutions are highlighted in bold in their relevant tables.

In all cases, parameter values for each constraint handling method are held constant. For the Homaifar approach, the constant penalty coefficient is 1013 for all the constraints. In the Death method, the $K$ value is 109. In the Adaptive approach $NFT_0$, $\lambda$, and $K$ are equal to 10, 2, and 2, respectively. In the Dynamic scheme, $\alpha$, $\beta$, and $C$ are equal to 2, 1, and 0.5, respectively. In the Annealing method, $\alpha$, $\beta$, and $C$ are equal to 1, 1, and 0.05, respectively.

### 4.1 Test Problems Series 1

In the following section, we solved five single-objective constrained benchmark optimization problems as follows (Wikipedia website [36]: 1—Rosenbrock function constrained with a cubic and a line, 2—Rosenbrock function constrained to a disk,

3—Mishra's Bird function, 4—Townsend function (modified), and 5—Simionescu function. Table 1 lists the objective function and constraints for each problem.

Table 2 lists the best-found solutions to the benchmark functions for each constraint handing method. Table 3 gives values for the mean ± STD of the 50 independent runs for each case. Table 4 lists the values of the constraints for the best solution for each case. The results in Table 4 indicate that none of the algorithms can satisfy the F1 constraints. Also, only two methods were successful in meeting all the F3 constraints. In all the remaining functions, all the constraints are satisfied.

For function F1, the values of the constraint violations recorded by Hoffmeister, Death, Dynamic, and Annealing are negligible. Since the Annealing constraint handing method produced both the best solution and the lowest constraint violations, it is considered the best method for function F1. However, the Hoffmeister method has the lowest mean value over multiple runs. For function F2, the Hoffmeister method produced the best solution and had the lowest mean value. Function F3 posed more of a challenge than the other functions. In this case, the Homaifar and Death methods were able to solve the problem, and recorded similar best and mean values; however, the Death method recorded a slightly lower STD. Results from F4 simulations showed that the Death penalty function method found the lowest best value, while the Dynamic method had the lowest mean value. For the F5 function, all the methods except Dynamic found equal best values; however, the Adaptive method had the lowest STD value.

## 4.2 Test Problems Series 2

In this section, more complicated optimization problems with numerous constraints and design variables are considered to evaluate the performance of constraint handling approaches better. To this end, we selected some benchmark optimization problems presented by Dr. Abdel-Rahman Hedar on his official website [19, 20]. Table 5 lists the objective functions and constraints for the selected optimization problems.

Numerical simulations are conducted on these functions using PSO with previously mentioned constraint handling schemes. Tables 6 and 7 tabulate the results for the best solution, and the mean ± STD over a series of independent runs, respectively. Table 8 provides constraint values from the best solution found using each constraint handling method. The results listed in Table 8 show that for functions G1 to G4, the Homaifar, Death, Adaptive, and Dynamic methods meet all the constraints successfully. A comparison of the best results for G1 shows that the Death method found the lowest objective function value and had a lower mean value than the other successful methods. For the G2 function, the Adaptive method found the highest objective function value and had the best mean and STD values. Results for the G4 function, show the lowest objective function value was recorded by the Adaptive method, while the associated mean and STD values are comparable with other successful methods. For the G6 function, the only approach that satisfied both constraints is the Adaptive

**Table 1** Benchmark optimization problems 1$^{st}$ case

| ID | Function name | Definition | Constraints | Boundary conditions |
|---|---|---|---|---|
| F1 | Rosenbrock function constrained with a cubic and a line | $\min f(x, y) =$ $(1 - x)^2 + 100(y - x^2)^2$ | $g_1 = (x - 1)^3 - y + 1 \leq 0$ $g_2 = x + y - 2 \leq 0$ | $-1.5 \leq x \leq 1.5$ $-0.5 \leq y \leq 2.5$ |
| F2 | Rosenbrock function constrained to a disk | $\min f(x, y) =$ $(1 - x)^2 + 100(y - x^2)^2$ | $g = x^2 + y^2 - 2 \leq 0$ | $-1.5 \leq x \leq 1.5$ $-1.5 \leq y \leq 1.5$ |
| F3 | Mishra's Bird function | $\min f(x, y) =$ $\sin(y)\exp\big[(1 - \cos x)^2\big] +$ $\cos(x)\exp\big[(1 - \sin y)^2\big] +$ $(x - y)^2$ | $g = (x + 5)^2 + (y + 5)^2 - 25 \leq 0$ | $-10 \leq x \leq 0$ $-6.5 \leq y \leq 0$ |
| F4 | Townsend function (modified) | $\min f(x, y) =$ $-[cos((x - 0.1)y)]^2 -$ $x \sin(3x + y)$ | $x^2 + y^2 < [2 \cos t - 0.5 \cos 2t - 0.25 \cos 3t - 0.125 \cos 4t]^2 +$ $[2 \sin t]^2 \leq 0$ where t = arctan2(xy) | $-2.25 \leq x \leq 2.5$ $-2.5 \leq y \leq 1.75$ |
| F5 | Simionescu function | $\min f(x, y) = 0.1xy$ | $x^2 + y^2 \leq \left[ 1 + 0.2 \cos\left(n \arctan \frac{x}{y}\right)\right]^2$ | $-1.25 \leq x, y \leq 1.25$ |

**Table 2** Best results for 1$^{st}$ case

| Constraint handling scheme | F1 | F2 | F3 |
|---|---|---|---|
| Homaifar | 0.00214 | 2.17430e−29 | −48.40602 |
| Hoffmeister | 2.57235e−22 | 1.97215e−31 | −97.05423 |
| Death | 3.24207e−25 | 2.61868e−25 | −48.40602 |
| Adaptive | 0.00032 | 5.85779e−28 | −104.14808 |
| Dynamic | 1.00000 | 1 | −105.09690 |
| Annealing | 2.49045e−26 | 6.24514e−23 | −106.76454 |
| Constraint handling scheme | F4 | F5 | |
| Homaifar | −3.37179 | −0.15625 | |
| Hoffmeister | −3.36867 | −0.15625 | |
| Death | −3.36720 | −0.15625 | |
| Adaptive | −3.37183 | −0.15625 | |
| Dynamic | −2.36984 | 0.84375 | |
| Annealing | −3.36916 | −0.15625 | |

method. The G7 function seemed to be a challenging problem; in that, none of the methods could satisfy all the constraints. In the G8 function study, all the methods except for the Hoffmeister and Annealing methods satisfied the constraints effectively. Among the successful approaches, the Dynamic method had higher best and mean values, while the other methods reached similar best values. A comparison of the mean values of the remaining successful methods demonstrated that Homaifar and Death very close, while Homaifar had a lower STD value. The results listed in Table 8 for the G9 function show that all the methods, except Annealing, were able to satisfy all the constraints. In contrast, the Hoffmeister method found the lowest best objective function value and had the lowest mean and STD values.

## 5  Summary

In this study, the performance of six popular penalty function-based constraint handling methods is explored. A PSO algorithm was selected as the testbed for this study because of its robustness and ability to handle complex optimization problems. Each of the six penalty function methods was incorporated into a PSO algorithm. Twelve benchmark optimization problems were solved to examine the effectiveness of each of the six constraint handling approaches. For each of the constraint handling method and objective function (total of 72 cases), the best solution was reported, and the mean and standard deviation were computed a series of 50 independent runs. For each method, the values of constraint were reported for the best solution. In

**Table 3** Mean and STD for 1$^{st}$ case

| Constraint handling scheme | F1 | F2 | F3 |
|---|---|---|---|
| Homaifar | 0.066480295 ± 7.0153E−02 | 5.5561E−15 ± 3.5830E−14 | −48.4060 ± 1.9735E−14 |
| Hoffmeister | 1.10E−08 ± 2.9197E−08 | 8.8445E−19 ± 3.2715E−18 | −97.0542 ± 5.5671E−14 |
| Death | 0.000122386 ± 3.5167E−04 | 4.7381E−07 ± 2.2950E−06 | −48.4060 ± 9.1355E−15 |
| Adaptive | 0.113585377 ± 1.0286E−01 | 3.7770E−13 ± 2.3308E−12 | −96.6595 ± 9.2323 |
| Dynamic | 1.009659277 ± 3.9075E−02 | 1.0000 ± 2.7925E−14 | −99.3211 ± 8.7576 |
| Annealing | 3.96E−05 ± 1.2847E−04 | 1.3882E−05 ± 4.5058E−05 | −104.8126 ± 9.6777 |

| Constraint handling scheme | F4 | F5 | |
|---|---|---|---|
| Homaifar | −3.2238 ± 0.0864 | −0.1563 ± 9.0765E−17 | |
| Hoffmeister | −3.2168 ± 0.0784 | −0.1563 ± 8.8928E−17 | |
| Death | −3.2517 ± 0.0861 | −0.1563 ± 8.8573E−17 | |
| Adaptive | −3.2788 ± 0.0861 | −0.1563 ± 8.8307E−17 | |
| Dynamic | −2.2564 ± 0.0857 | 0.8438 ± 0 | |
| Annealing | −3.2509 ± 0.0864 | −0.1563 ± 8.8928E−17 | |

**Table 4** Constraint values for the best solution for 1<sup>st</sup> case

| Function | Constraints | Homaifar | Hoffmeister | Death | Adaptive | Dynamic | Annealing |
|---|---|---|---|---|---|---|---|
| F1 | g1 | −0.000634361 | −2.085E−13 | −1.70708E−12 | −9.70381E-5 | −1.89488E−10 | −4.49862E−13 |
|  | g2 | 0.001752213 | 6.03739E−13 | 1.13776E−12 | 0.000629708 | 6.73502E−10 | 2.98428E−13 |
| F2 | g | −2.7978E−14 | −2.6645E−15 | −3.0718E−12 | −1.2390E−13 | −5.7046E−08 | −2.7978E−14 |
| F3 | g | −5.4459 | 9.5974 | −5.4459 | 9.5257 | 9.7568 | 9.8223 |
| F4 | g | −4.2072 | −4.2358 | −4.1844 | −4.0910 | −4.4387 | −4.3576 |
| F5 | g | −1.685 | −1.685 | −1.685 | −1.685 | −1.685 | −1.685 |

**Table 5** Benchmark optimization problems 2nd case

| ID | Definition | Constraints | Boundary conditions |
|---|---|---|---|
| G1 | $\min f(x) = 5 \sum_{i=1}^{4} x_i - 5 \sum_{i=1}^{4} x_i^2 - \sum_{i=5}^{13} x_i$ | $g_1 = 2x_1 + 2x_2 + x_{10} + x_{11} - 10 \leq 0$ <br> $g_2 = 2x_1 + 2x_3 + x_{10} + x_{12} - 10 \leq 0$ <br> $g_3 = 2x_2 + 2x_3 + x_{11} + x_{12} - 10 \leq 0$ <br> $g_4 = -8x_1 + x_{10} \leq 0$ <br> $g_5 = -8x_2 + x_{11} \leq 0$ <br> $g_6 = -8x_3 + x_{12} \leq 0$ <br> $g_7 = -2x_4 - x_5 + x_{10} \leq 0$ <br> $g_8 = -2x_6 - x_7 + x_{11} \leq 0$ <br> $g_9 = -2x_8 - x_9 + x_{12} \leq 0$ | $0 \leq x_i \leq u_i$ <br> $i = 1, 2, \ldots, n$ <br> $u = (1, 1, 1, 1, 1, 1, 1, 1, 1, 100, 100, 100, 1)$ |
| G2 | $\max f(x) = \left\| \dfrac{\sum_{i=1}^{n} cos^4(x_i) - 2 \prod_{i=1}^{n} cos^2(x_i)}{\sqrt{\sum_{i=1}^{n} i x_i^2}} \right\|$ | $g_1 = -\prod_{i=1}^{n} x_i + 0.75 \leq 0$ <br><br> $g_2 = \sum_{i=1}^{n} x_i - 7.5n \leq 0$ | $0 \leq x_i \leq 10$ <br> $i = 1, 2, \ldots, 20$ |

(continued)

**Table 5** (continued)

| ID | Definition | Constraints | Boundary conditions |
|---|---|---|---|
| G4 | $\min f(x) = 5.3578547x_3^2 + 0.8356891x_1x_5 + 37.293239x_1 - 40792.141$ | $g_1 = v(x) - 92 \le 0$ <br> $g_3 = v(x) - 110 \le 0$ <br> $g_4 = -v(x) + 90 \le 0$ <br> $g_5 = \omega(x) - 25 \le 0$ <br> $g_6 = -\omega(x) + 20 \le 0$ <br> $v(x) = 85.334407 + 0.0056858x_2x_5 + 0.0006262x_1x_4 - 0.0022053x_3x_5$ <br> $v(x) = 80.51249 + 0.0071317x_2x_5 + 0.0029955x_1x_2 + 0.0021813x_3^2$ <br> $\omega(x) = 9.300961 + 0.0047026x_3x_5 + 0.0012547x_1x_3 + 0.0019085x_3x_4$ | $l_i \le x_i \le u_i$ <br> $i = 1, 2, \ldots, 5$ <br> $l = (78, 33, 27, 27, 27), u = (102, 45, 45, 45, 45)$ |
| G6 | $\min f(x) = (x_1 - 10)^3 + (x_2 - 20)^3$ | $g_1 = (x_1 - 5)^2 + (x_2 - 5)^2 + 100 \le 0$ <br> $g_2 = (x_1 - 5)^2 + (x_2 - 5)^2 - 82.81 \le 0$ | $l \le x_i \le 100$ <br> $i = 1, 2$ <br> $l = (13, 0)$ |

(continued)

**Table 5** (continued)

| ID | Definition | Constraints | Boundary conditions |
|---|---|---|---|
| G7 | $\min f(x) = x_1^2 + x_2^2 + x_1x_2 - 14x_1 - 16x_2 + (x_3 - 10)^2 + 4(x_4 - 5)^2 + (x_5 - 3)^2 + 2(x_6 - 1)^2 + 5x_7^2 + 7(x_8 - 11)^2 + 2(x_9 - 10)^2 + (x_{10} - 7)^2 + 45$ | $g_1 = 4x_1 + 5x_2 - 3x_7 + 9x_8 - 105 \leq 0$ <br> $g_2 = 10x_1 - 8x_2 - 17x_7 + 2x_8 \leq 0$ <br> $g_3 = -8x_1 + 2x_2 + 5x_9 + 2x_{10} - 12 \leq 0$ <br> $g_4 = 3(x_1 - 2)^2 + 4(x_2 - 3)^2 + 2x_3^2 - 7x_4 - 120 \leq 0$ <br> $g_5 = 5x_1^2 + 8x_2 + (x_3 - 6)^2 - 2x_4 - 40 \leq 0$ <br> $g_6 = 0.5(x_1 - 8)^2 + 2(x_2 - 4)^2 + 3x_5^2 - x_6 - 30 \leq 0$ <br> $g_7 = x_1^2 + 2(x_2 - 2)^2 - 2x_1x_2 + 14x_5 - 6x_6 \leq 0$ <br> $g_8 = -3x_1 + 6x_2 + 12(x_9 - 8)^2 - 7x_{10} \leq 0$ | $-10 \leq x_i \leq 10$ <br> $i = 1, 2, \ldots, 10$ |
| G8 | $\max f(x) = \frac{\sin^3(2\pi x_1)\sin(2\pi x_2)}{x_1^3(x_1+x_2)}$ | $g_1 = x_1^2 - x_2 + 1 \leq 0$ <br> $g_2 = 1 - x_1 + (x_2 - 4)^2 \leq 0$ | $0 \leq x_i \leq 10$ <br> $i = 1, 2$ |
| G9 | $\min f(x) = (x_1 - 10)^2 + 5(x_2 - 12)^2 + 3(x_4 - 11)^2 + x_3^4 + 10x_5^6 + 7x_6^2 + x_7^4 - 4x_6x_7 - 10x_6 - 8x_7$ | $g_1 = v_1 + 3v_2^2 + v_3 + 4v_4^2 + 5v_5 - 127 \leq 0$ <br> $g_2 = 7x_1 + 3x_2 + 10x_3^2 + x_4 - x_5 - 282 \leq 0$ <br> $g_3 = 23x_1 + v_2 + 6x_6^2 - 8x_7 - 196 \leq 0$ <br> $g_4 = 2v_1 + v_2 - 3x_1x_2 + 2x_3^2 + 5x_6 - 11x_7 \leq 0$ <br> $v_1 = 2x_1^2 \leq 0$ <br> $v_2 = x_2^2 \leq 0$ | $-10 \leq x_i \leq 10$ <br> $i = 1, 2, \ldots, 7$ |

**Table 6** Best results for 2nd case

| Constraint handling scheme | G1 | G2 | G4 | G6 |
|---|---|---|---|---|
| Homaifar | −18.7552 | −0.4398 | −33,233.6511 | −6,313.0662 |
| Hoffmeister | −29.5263 | −Inf | −35,202.9264 | −8,840.8022 |
| Death | −75.6721 | −0.4282 | −33,257.6403 | −6,854.7599 |
| Adaptive | −30.0434 | −0.4527 | −33,813.9574 | −1,643.5994 |
| Dynamic | −16.3929 | −0.1928 | −32,955.7715 | 3.3224e+06 |
| Annealing | −383.9712 | −Inf | −34,320.3859 | −7,988.8455 |
| Constraint handling scheme | G7 | G8 | G9 | |
| Homaifar | 129.5706 | −0.0958 | 732.8192 | |
| Hoffmeister | 44.7292 | −1,541.5176 | 693.7655 | |
| Death | 144.4177 | −0.0958 | 721.4620 | |
| Adaptive | −3,381.3957 | −0.0958 | 788.6190 | |
| Dynamic | 1.4013e+07 | −0.0860 | 888.7828 | |
| Annealing | 1.4870e+07 | −1,558.5455 | 539.2907 | |

general, the results indicated the Homaifar and Adaptive methods provide satisfactory performance, while the Hoffmeister and Annealing methods were unsuccessful in satisfying the constraints in all the cases.

**Table 7** Mean and STD for 2nd case

| Constraint handling scheme | G1 | G2 | G4 | G6 |
|---|---|---|---|---|
| Homaifar | $-1.53e+01 \pm 2.26e+00$ | $-2.97e-1 \pm 6.49e-02$ | $-3.29e+04 \pm 1.31e+02$ | $-2.90e+03 \pm 1.87e+03$ |
| Hoffmeister | $-1.97e+1 \pm 2.64e+00$ | $-Inf$ | $-3.51e+04 \pm 5.66e+01$ | $-8.69e+03 \pm 7.25e+01$ |
| Death | $-1.79e+01 \pm 9.36e+00$ | $-2.98e-01 \pm 6.1e-02$ | $-3.28e+04 \pm 1.34e+02$ | $-6.03e+03 \pm 5.17e+02$ |
| Adaptive | $-8.14e+00 \pm 9.36e+00$ | $-4.31e-01 \pm 6.1e-02$ | $-3.26e+04 \pm 1.34e+02$ | $9.66e+05 \pm 4.82e+05$ |
| Dynamic | $-1.49e+01 \pm 6.16e-01$ | $-1.51e-01 \pm 1.32e-02$ | $-3.24e04 \pm 6.14e-02$ | $6.52e+06 \pm 1.56e+06$ |
| Annealing | $3.12e+06 \pm 1.26e+07$ | $-Inf$ | $-3.25e+04 \pm 7.12e+02$ | $2.85e+07 \pm 6.21e+07$ |
| Constraint handling scheme | G7 | G8 | G9 | |
| Homaifar | $3.31e + 02 \pm 1.53e + 2$ | $-1.00e-1 \pm 6.07e-17$ | $8.38e+02 \pm 4.51e+01$ | |
| Hoffmeister | $1.03e + 02 \pm 3.28e + 01$ | $-1.54e + 03 \pm 5.26e-13$ | $7.66e+02 \pm 3.79e+01$ | |
| Death | $3.89e + 02 \pm 2.34e + 02$ | $-1.00e-01 \pm 6.24e-17$ | $8.14e + 02 \pm 5.13e + 01$ | |
| Adaptive | $-3.26e+03 \pm 3.44e+02$ | $-1.00e-01 \pm 6.24e-17$ | $9.77e+02 \pm 5.13e+01$ | |
| Dynamic | $5.20e+07 \pm 3.53e+07$ | $5.00 \pm 3.54e + 04$ | $1.01e+03 \pm 6.43e+01$ | |
| Annealing | $6.33e+07 \pm 2.91e+07$ | $-4.05e+02 \pm 6.91e+02$ | $1.76e+03 \pm 5.87e+02$ | |

**Table 8** Constraint values for the best solution for 2$^{nd}$ case

| ID | Constraints | Homaifar | Hoffmeister | Death | Adaptive | Dynamic | Annealing |
|---|---|---|---|---|---|---|---|
| G1 | g1 | −6.8015 | 4.1138 | −6.6545 | −6.1299 | −0.6733 | 178.9962 |
| | g2 | −6.2117 | 5.9941 | −5.8829 | −4.4511 | −0.5364 | 178.1315 |
| | g3 | −5.7446 | 17.8150 | −5.2382 | −4.3265 | −0.3819 | 184.7554 |
| | g4 | −0.5743 | −0.5907 | −0.4075 | −2.5022 | −5.3720 | 91.1861 |
| | g5 | −2.4217 | 11.0694 | **−0.8515** | −2.6834 | −5.2472 | 97.8100 |
| | g6 | −3.0717 | 11.7691 | −0.7883 | −2.1008 | −5.1138 | 96.9394 |
| | g7 | −1.0595 | −0.3005 | −0.2981 | −1.7997 | −0.4049 | 90.5813 |
| | g8 | −1.5135 | 11.1074 | −0.7523 | −1.7912 | −0.2567 | 97.5370 |
| | g9 | −1.4735 | 12.3086 | −0.8344 | −0.4050 | −0.1208 | 95.2121 |
| G2 | g1 | −0.3483 | 0.7500 | −0.0000 | −0.1744 | −11.3301E08 | 0.7500 |
| | g2 | −105.5281 | −150.0000 | −110.2935 | −110.6437 | −66.9536 | −150.0000 |
| G4 | g1 | −91.2519 | −88.8924 | −91.6257 | −91.7387 | −91.2396 | −89.5925 |
| | g2 | −0.7481 | −3.1076 | −0.3743 | −0.2613 | −0.7604 | −2.4075 |
| | g3 | −6.7130 | −0.6636 | −7.1186 | −6.8308 | −6.1124 | −2.4804 |
| | g4 | −13.2870 | −19.3364 | −12.8814 | −13.1692 | −13.8876 | −17.5196 |
| | g5 | −0.7500 | 4.0644 | −0.9254 | −0.1964 | −0.2006 | 3.9898 |
| | g6 | −4.2500 | −9.0644 | −4.0746 | −4.8036 | −4.7994 | −8.9898 |
| G6 | g1 | −0.3505 | 53.2423 | −0.0577 | −0.7250 | 0.2726 | 23.0374 |
| | g2 | 25.6657 | −25.0682 | 26.3706 | −0.2570 | 9.3178 | −19.2648 |
| G7 | g1 | −50.6637 | −28.2747 | −45.0462 | −124.5271 | −33.2748 | −69.9097 |
| | g2 | −90.9175 | −121.4250 | −110.7612 | −35.9301 | −119.0130 | 19.4864 |
| | g3 | 11.3410 | 11.9677 | 10.8961 | −4.9997 | 3.5965 | 12.7076 |
| | g4 | −122.8382 | −118.8054 | −113.1044 | −23.8555 | −131.0416 | −71.6281 |
| | g5 | −12.2697 | −6.8061 | −23.6437 | 0.9809 | 3.7317 | −13.6003 |
| | g6 | 62.6469 | 23.7481 | 61.9868 | −1.1412 | 38.4242 | 12.4138 |
| | g7 | 32.7484 | 8.5404 | 31.1607 | −83.0901 | 10.2969 | −36.6248 |
| | g8 | −43.8874 | −17.1694 | −41.5682 | −52.4224 | −33.2774 | 14.8685 |
| G8 | g1 | −1.7375 | 0.9996 | −1.7375 | −1.7390 | −1.8584 | 1.0000 |
| | g2 | −0.1678 | 16.9969 | −0.1678 | −0.1671 | −0.1101 | 17.0000 |
| G9 | g1 | −115.6328 | −104.0385 | −107.5731 | −2.3858 | −68.4760 | 6212.5453 |
| | g2 | −286.1398 | −262.3689 | −276.4030 | −253.1911 | −269.3198 | −200.5948 |
| | g3 | −213.0346 | −187.6622 | −215.3880 | −170.7742 | −185.4042 | 195.7844 |
| | g4 | −35.5948 | −45.7608 | −30.0437 | −8.4887 | −14.5741 | 69.3815 |

# References

1. Adeli, H., Cheng, N.T.: Augmented Lagrangian genetic algorithm for structural optimization. J. Aerosp. Eng. **7**(1), 104–118 (1994)
2. Ameca-Alducin, M.Y., Hasani-Shoreh, M., Blaikie, W., Neumann, F., Mezura-Montes, E.: A comparison of constraint handling techniques for dynamic constrained optimization problems.

In: 2018 IEEE Congress on Evolutionary Computation (CEC), pp. 1–8. IEEE (2018)
3. Biswas, P.P., Suganthan, P.N., Mallipeddi, R., Amaratunga, G.A.: Optimal power flow solutions using differential evolution algorithm integrated with effective constraint handling techniques. Eng. Appl. Artif. Intell. **68**, 81–100 (2018)
4. Carlson, S.E., Shonkwiler, R.: Annealing a genetic algorithm over constraints. In: SMC'98 Conference Proceedings. 1998 IEEE International Conference on Systems, Man, and Cybernetics (Cat. No. 98CH36218) (Vol. 4, pp. 3931–3936). IEEE (1998)
5. Clerc, M., Kennedy, J.: The particle swarm-explosion, stability, and convergence in a multidimensional complex space. IEEE Trans. Evol. Comput. **6**(1), 58–73 (2002)
6. Coello, C.A.C.: Theoretical and numerical constraint-handling techniques used with evolutionary algorithms: a survey of the state of the art. Comput. Methods Appl. Mech. Eng. **191**(11–12), 1245–1287 (2002)
7. Coit, D.W., Smith, A.E.: Penalty guided genetic search for reliability design optimization. Comput. Ind. Eng. **30**(4), 895–904 (1996)
8. Datta, R., Deb, K., Kim, J.H.: CHIP: Constraint Handling with Individual Penalty approach using a hybrid evolutionary algorithm. Neural Comput. Appl. **31**(9), 5255–5271 (2019)
9. Deb, K.: An efficient constraint handling method for genetic algorithms. Comput. Methods Appl. Mech. Eng. **186**(2–4), 311–338 (2000)
10. Gandomi, A.H., Kashani, A.R.: Automating pseudo-static analysis of concrete cantilever retaining wall using evolutionary algorithms. Measurement **115**, 104–124 (2018)
11. Gandomi, A.H., Kashani, A.R.: Probabilistic evolutionary bound constraint handling for particle swarm optimization. Oper. Res. Int. J. **18**(3), 801–823 (2018)
12. Gandomi, A.H., Kashani, A.R.: Evolutionary bound constraint handling for particle swarm optimization. In: 2016 4th International Symposium on Computational and Business Intelligence (ISCBI), pp. 148–152. IEEE (2016)
13. Gandomi, A.H., Kashani, A.R., Mousavi, M.: Boundary constraint handling affection on slope stability analysis. In: Engineering and Applied Sciences Optimization, pp. 341–358. Springer, Cham (2015)
14. Gandomi, A.H., Kashani, A.R., Mousavi, M., Jalalvandi, M.: Slope stability analysis using evolutionary optimization techniques. Int. J. Numer. Anal. Meth. Geomech. **41**(2), 251–264 (2017)
15. Gandomi, A.H., Kashani, A.R., Zeighami, F.: Retaining wall optimization using interior search algorithm with different bound constraint handling. Int. J. Numer. Anal. Methods Geomech. (2017)
16. Gandomi, A.H., Yang, X.S.: Evolutionary boundary constraint handling scheme. Neural Comput. Appl. **21**(6), 1449–1462 (2012)
17. Gandomi, A.H., Yang, X.S., Talatahari, S., Alavi, A.H.: Metaheuristic algorithms in modeling and optimization. In: Gandomi et al. (eds.) Metaheuristic Applications in Structures and Infrastructures, pp. 1–24. Elsevier, Waltham, MA (2013)
18. Gen, M., Cheng, R.: Interval programming using genetic algorithms. In: Proceedings of the Sixth International Symposium on Robotics and Manufacturing, Montpellier, France (1996)
19. Hedar, A.-R.: Dr. Abdel-Rahman Hedar's official website (2020). http://www-optima.amp.i.kyoto-u.ac.jp/member/student/hedar/Hedar.html
20. Hedar, A.-R.: Test problems for con-strained global optimization (2020). http://www-opti-ma.amp.i.kyotou.ac.jp/member/student/hedar/Hedar_files/TestGO_files/Page422.htm
21. Hoffmeister, F., Sprave, J. Problem-independent handling of constraints by use of metric penalty functions. In: Proceedings of Evolutionary Programming, pp. 289–294 (1996)
22. Homaifar, A., Qi, C.X., Lai, S.H.: Constrained optimization via genetic algorithms. Simulation **62**(4), 242–253 (1994)
23. Jamal, M.., Ming, F., Zhengang, J.: Solving constrained optimization problems by using covariance matrix adaptation evolutionary strategy with constraint handling methods. In: Proceedings of the 2nd International Conference on Innovation in Artificial Intelligence, pp. 6–15 (201)
24. Joines, J.A., Houck, C.R.: On the use of non-stationary penalty functions to solve nonlinear constrained optimization problems with GA's. In: IEEE, 1994, vol. 572, pp. 579–584 (1994)

25. Jordehi, A.R.: A review on constraint handling strategies in particle swarm optimisation. Neural Comput. Appl. **26**(6), 1265–1275 (2015)
26. Kashani, A.R., Saneirad, A., Gandomi, A.H.: Optimum design of reinforced earth walls using evolutionary optimization algorithms. Neural Comput. Appl., 1–24 (2019)
27. Kashani, A.R., Gandomi, M., Camp, C.V., Gandomi, A.H.: Optimum design of shallow foundation using evolutionary algorithms. Soft. Comput. **24**(9), 6809–6833 (2020)
28. Kazarlis, S., Petridis, V.: Varying fitness functions in genetic algorithms: Studying the rate of increase of the dynamic penalty terms. In: International conference on parallel problem solving from nature, pp. 211–220. Springer, Berlin, Heidelberg (1998)
29. Kennedy, J., Eberhart, R.C.: Swarm Intelligence. Morgan Kaufmann, San Francisco, CA (2001)
30. Le, T.V.: A fuzzy evolutionary approach to constrained optimization problems. In: Proceedings of the Second IEEE Conference on Evolutionary Computation, pp. 274–278. IEEE Perth (1995)
31. Li, J.P., Wang, Y., Yang, S., Cai, Z.: A comparative study of constraint-handling techniques in evolutionary constrained multiobjective optimization. In: 2016 IEEE Congress on Evolutionary Computation (CEC), pp. 4175–4182. IEEE (2016)
32. Michalewicz, Z.: A survey of constraint handling techniques in evolutionary computation methods. Evol. Prog. **4**, 135–155 (1995)
33. Morales, A.K., Quezada, C.V.: A universal eclectic genetic algorithm for constrained optimization. In: Proceedings of the 6th European Congress on Intelligent Techniques and Soft Computing, vol. 1, pp. 518–522 (1998)
34. Myung, H., Kim, J.H.: Hybrid interior-lagrangian penalty based evolutionary optimization. In: International Conference on Evolutionary Programming, pp. 85–94. Springer, Berlin, Heidelberg (1998)
35. Petridis, V., Kazarlis, S., Bakirtzis, A.: Varying fitness functions in genetic algorithm constrained optimization: the cutting stock and unit commitment problems. IEEE Trans. Syst., Man, Cybern., Part B (Cybern.) **28**(5), 629–640 (1998)
36. Wikipedia.: Test functions for optimization (16 April 2020). https://en.wikipedia.org/wiki/Test_functions_for_optimization
37. Zou, D., Li, S., Kong, X., Ouyang, H., Li, Z.: Solving the combined heat and power economic dispatch problems by an improved genetic algorithm and a new constraint handling strategy. Appl. Energy **237**, 646–670 (2019)

# Online Landscape Analysis for Guiding Constraint Handling in Particle Swarm Optimisation

Katherine M. Malan[ID]

**Abstract** Many real-world optimisation problems are constrained in multiple ways. As with other metaheuristics, particle swarm optimisation (PSO) algorithms do not naturally handle constrained search spaces. When PSO is used to solve a constrained problem, then the algorithm has to be modified to incorporate an appropriate constraint handling technique. Previous studies with evolutionary algorithms have shown that the choice of the most appropriate constraint handling technique depends on the features of the problem being solved. This study investigates whether this is also the case with PSO. Results are presented to show that there is performance complementarity between different constraint handling techniques when used with a traditional global best PSO algorithm. A landscape-aware approach is then implemented that uses rules derived from offline machine learning on a training set of problem instances. The rules are used to automatically switch between constraint handling techniques during PSO search. The switching is based on landscape information collected from the particles during search and requires no additional sampling or function evaluations. Results show that the proposed approach of switching techniques performs better than using any one of the individual constraint handling techniques. It is also shown that landscape-aware switching outperforms random switching, illustrating the value of using landscape features to guide the choice of constraint handling technique for PSO.

## Nomenclature

| | |
|---|---|
| $\epsilon$DFR | Takahama and Sakai's [26] modification of Deb's feasibility ranking (constraint handling technique) |
| $\mathbf{x}$ | Multidimensional solution to the problem |
| $\phi(\mathbf{x})$ | Function combining constraint violations into a single value |
| $D$ | Dimension of the problem instance |

K. M. Malan (✉)
Department of Decision Sciences, University of South Africa, Preller Street,
Muckleneuk, Pretoria, South Africa
e-mail: malankm@unisa.ac.za

© The Author(s), under exclusive license to Springer Nature Singapore Pte Ltd. 2021
A. J. Kulkarni et al. (eds.), *Constraint Handling in Metaheuristics and Applications*,
https://doi.org/10.1007/978-981-33-6710-4_5

| $f(\mathbf{x})$ | Objective function |
|---|---|
| $FE$ | Number of function evaluations |
| $g_i(\mathbf{x})$ | Inequality constraint function $i$ |
| $h_j(\mathbf{x})$ | Equality constraint function $j$ |
| $OLA\_limit$ | Size of search history archive (parameter of LA approach) |
| $SW\_freq$ | Switching frequency (parameter of LA approach) |
| 25_IZ | Proportion of solutions in the top 50% percentile for both fitness and violation (fitness landscape metric) |
| 4_IZ | Proportion of solutions in the top 20% percentile for both fitness and violation (fitness landscape metric) |
| CHT | Constraint handling technique |
| DFR | Deb's feasibility ranking (constraint handling technique) |
| DP | Death penalty (constraint handling technique) |
| FsR | Feasibility ratio (fitness landscape metric) |
| FVC | Fitness violation correlation (fitness landscape metric) |
| LA | Landscape aware (constraint handling technique) |
| NCH | No constraint handling (constraint handling technique) |
| PSO | Particle swarm optimisation |
| RFB× | Ratio of feasible boundary crossings (fitness landscape metric) |
| WP | Weighted penalty (constraint handling technique) |

## 1  Introduction

Most real-world optimisation problems have constraints, but metaheuristics in their original form were not designed to handle constraints. To address this, a number of constraint handling techniques have been proposed to be used with metaheuristics [3, 18] and these can be broadly categorised into penalty, repair, feasibility ranking and multi-objective approaches.

As with other metaheuristics, when PSO algorithms are used to solve complex constrained problems, one of these constraint handling approaches has to be chosen to assist the algorithm in finding good feasible solutions. We know from theory [29, 30] and experience that there can be no best algorithm for solving all optimisation problems. The challenge is rather in selecting the most appropriate algorithm for a given set of problems with similar properties. In a similar vein, it has been shown that no single constraint handling technique is the best in all cases when used with evolutionary algorithms [16] and that some properties of search landscapes seem to favour particular constraint handling approaches.

The field of fitness landscape analysis [14] has recently gained momentum in the evolutionary computation community with regular tutorials, workshops and special sessions dedicated to this topic at all the major evolutionary computation conferences. In the continuous optimisation domain, advances have been made in using landscape analysis to select algorithms [1, 15, 19], but all of these studies have been restricted to unconstrained (or only bound constrained) problems.

Recently, a landscape-aware approach to constraint handling was proposed for the differential evolution algorithm [13]. The purpose of this study is to investigate whether a similar approach will be effective in the context of particle swarm optimisation. Firstly, experiments are conducted to compare the performance of commonly used constraint handling techniques when used with a traditional global best PSO. Results are presented to show that there is performance complementarity between different constraint handling techniques. Secondly, a landscape-aware switching approach is applied in the context of PSO. Rather than introducing a new constraint handling technique, the approach utilises commonly used techniques, but selects techniques based on the landscape features experienced by the swarm during the search.

## 2 Particle Swarm Optimisation and Constraint Handling

Particle swarm optimisation (PSO) [5, 10] is a stochastic population-based optimisation technique. Starting with a random swarm of solutions, called particles, the positions of particles in the search space are adjusted at each iteration of the algorithm. The adjustment has random elements, but is largely determined by the distance to the best solution found in the neighbourhood of the particle and the distance from the best solution found by the particle itself during the search process. This section describes the traditional global best PSO model, which is used as the base algorithm in the experimentation. It then defines constrained continuous optimisation problems in general and constraint handling techniques that have been proposed for PSO in the literature.

### 2.1 Traditional Global Best PSO

The traditional global best PSO model (gbest PSO for short) [5, 10] determines the multidimensional position of a particle, $x_i$, at time step $t + 1$ by adding a multidimensional step size (called the *velocity* of the particle), $v_i$, at time step $t + 1$, to the position of the particle at time $t$, using the equation:

$$\mathbf{x}_i(t + 1) = \mathbf{x}_i(t) + \mathbf{v}_i(t + 1). \tag{1}$$

The velocity at time step $t + 1$ is given as:

$$\mathbf{v}_i(t + 1) = w \cdot \mathbf{v}_i(t) + c_1 \cdot \mathbf{r}_1(t) \odot (\mathbf{y}_i(t) - \mathbf{x}_i(t)) + c_2 \cdot \mathbf{r}_2(t) \odot (\hat{\mathbf{y}}(t) - \mathbf{x}_i(t)), \tag{2}$$

where $w$ is the inertia weight (introduced by Shi and Eberhart [23]), $\mathbf{v}_i(t)$ is the velocity of particle $i$ at time stamp $t$, $c_1$ and $c_2$ are the cognitive and social acceleration constants, respectively, $\mathbf{r}_1(t), \mathbf{r}_2(t) \sim U(0, 1)^D$ where $D$ is the dimension of the problem, $\odot$ denotes element-by-element vector multiplication, $\mathbf{y}_i(t)$ refers to particle $i$'s personal best position and $\hat{\mathbf{y}}(t)$ refers to the global best position at time step $t$, being the best solution from the set of personal best positions of all particles.

Equation 2 shows that the position of a particle is influenced by three terms: the particle's previous velocity, the relative position of the personal best particle and the relative position of the swarm's global best particle. The global best is the fittest of all particles' personal bests. The behaviour of the gbest PSO is influenced by the relative weights of these three terms, set using the constants $w$, $c_1$ and $c_2$. The choice of values of these constants has to be made together to ensure convergence of the swarm [28]. Although the optimal choice of parameters is problem dependent, a common choice that works reasonably well for many problems is 0.7298 for $w$ and 1.496 for both acceleration constants [6].

## 2.2  Constrained Continuous Optimisation

A constrained continuous minimisation problem can be expressed in algebraic form as follows:

$$\text{Minimise } f(\mathbf{x}), \quad \mathbf{x} = (x_1, x_2 \dots, x_n) \in \mathbb{R}^n, \tag{3}$$

$$\text{subject to } \begin{matrix} g_i(\mathbf{x}) \le 0, \ i = 1, \dots, p, \\ h_j(\mathbf{x}) = 0, \ j = 1, \dots, q, \ \text{and} \end{matrix} \tag{4}$$

$$\min_k \le x_k \le \max_k, \quad \text{for} \ k = 1, \dots, n, \tag{5}$$

where $f(\mathbf{x})$ is the objective function to be minimised, $\mathbf{x}$ is an $n$-dimensional solution to the problem with boundary constraints (5), and $g_i(\mathbf{x})$ and $h_j(\mathbf{x})$ are the inequality and equality constraints, respectively. Equality constraints are typically re-expressed as inequality constraints for some small error margin $\epsilon$, such as $10^{-4}$ as follows:

$$|h_j(\mathbf{x})| - \epsilon \le 0, \quad j = 1, \dots, q. \tag{6}$$

The feasible set consists of the solutions that satisfy all the inequality constraints $g_i(\mathbf{x})$ and the equality constraints $h_j(\mathbf{x})$ to within $\epsilon$.

To quantify the extent of constraint violation of a solution, the constraints can be combined into a single value as follows [17]:

$$\phi(\mathbf{x}) = \frac{\sum_{i=1}^{p} G_i(\mathbf{x}) + \sum_{j=1}^{q} H_j(\mathbf{x})}{p + q} \tag{7}$$

where

$$G_i(\mathbf{x}) = \begin{cases} g_i(\mathbf{x}) & \text{if} \ g_i(\mathbf{x}) > 0 \\ 0 & \text{if} \ g_i(\mathbf{x}) \le 0 \end{cases} \tag{8}$$

and

$$H_j(\mathbf{x}) = \begin{cases} |h_j(\mathbf{x})| & \text{if} \quad |h_j(\mathbf{x})| - \epsilon > 0 \\ 0 & \text{if} \quad |h_j(\mathbf{x})| - \epsilon \leq 0. \end{cases} \qquad (9)$$

## 2.3 Constraint Handling with PSO

Some of the earliest approaches to constraint handling with PSO included a strategy of preserving feasibility [8], a penalty-based approach [20] and feasibility ranking [21]. Coath and Halgamuge [2] compared the first of these two techniques with PSO on benchmark problems and concluded that the choice of constraint handling method is problem dependent.

Since then, a number of other constraint handling approaches have been proposed with PSOs. In 2015, Jordehi [9] published a survey on constraint handling techniques used with PSO. He claimed that the most common approach was Deb's feasibility ranking approach [4], followed by static penalty and death penalty approaches and that other approaches, such as multi-objective approaches, were rarely used with PSO.

Takahama and Sakai [26] proposed a modification to Deb's feasibility ranking with an $\epsilon$ tolerance that reduces over time. The basic idea is that a solution that violates the constraints within the current tolerance is regarded as feasible. This has the effect of the search being mostly guided by fitness at the beginning of the search (when $\epsilon$ is large), but the constraints increasingly being taken into account as the search progresses (and $\epsilon$ reaches 0). The approach was applied with differential evolution (called $\epsilon$DEg) [27] and won the CEC 2010 Competition on Constrained Real-Parameter Optimization [17].

## 3 Performance Complementarity of Constraint Handling

When there are a number of alternative algorithmic solutions to a problem, algorithm selection refers to the process of selecting the most appropriate algorithm for a given problem. An essential feature of the success of algorithm selection is the existence of performance complementarity between algorithms [11]—when different algorithms have different strengths, these strengths can be exploited so that the best algorithm can be selected for each problem. This section investigates whether there is performance complementarity between constraint handling techniques (CHTs) when used with PSO.

## 3.1 Constraint Handling Approaches

To test the relative performance of different CHTs, the following five approaches were implemented with gbest PSO (as described in Sect. 2.1) with 50 particles, $c_1 = c_2 = 1.496$, and $w = 0.7298$:

1. No constraint handling (NCH): This approach considers only the fitness value when comparing solutions.
2. Death penalty (DP): The death penalty approach rejects infeasible solutions. In PSO terms, this means that the global best and personal best particles are only replaced by feasible solutions with better fitness. Usually, with a death penalty approach, the swarm would be initialised with feasible solutions [9], but this is not always practically possible. Because many of the problems in this study have equality constraints, the swarm was initialised randomly, and could therefore contain infeasible solutions. Infeasible global or personal best particles were replaced with the first feasible solutions found by the swarm.
3. Weighted penalty (WP): A weighted penalty approach combines the fitness and constraints into a single objective function, effectively converting the problem into an unconstrained problem. When the problem is a minimisation problem, both of these objectives are minimised. The approach used in this study was to apply a static weighting of 50% to both the fitness and the constraint violation, where the constraints are combined as in Eq. 7.
4. Deb's feasibility ranking (DFR): This approach compares solutions using the following rules:

   – A feasible solution is preferred to an infeasible one.
   – Two feasible solutions are compared based on fitness.
   – Two infeasible solutions are compared by their level of constraint violation.

   In the context of PSO, the above rules are applied to decide whether the global best and personal best particles should be replaced.
5. Takahama and Sakai's [26] modification of Deb's feasibility ranking (denoted $\epsilon$DFR): This approach uses Deb's rules, but the decision of whether a solution is feasible or not depends on a tolerance level of $\epsilon$. The approach used for adapting $\epsilon$ was as follows [25]: $\epsilon$ is set to zero at a cutoff number of function evaluations, $FE_c$. Before this point, $\epsilon$ was defined as

$$\epsilon = \phi(\mathbf{x}_\theta) \times \left(1 - \frac{FE_i}{FE_c}\right)^{cp}, \tag{10}$$

where $x_\theta$ is the $\theta$-th solution in the swarm ordered by violations $\phi(\mathbf{x})$ from lowest to highest, $FE_i$ is the current number of function evaluations, and $cp$ is a parameter to control the speed of reducing relaxation of constraints. Following [25], $FE_c$ was set to 80% of the computational budget, $\theta$ was set to $0.8 \times$ swarm size and $cp$ was set to 5.

## 3.2 Problem Instances and Performance Ranking

The IEEE CEC 2010 Special Session on Constrained Real-Parameter Optimization [17] defined a set of 18 problems for comparing algorithm performance. The problems have different objective functions and numbers of inequality and equality constraints and are scalable to any dimension. For most problems, the constraints are rotated to prevent feasible patches that are parallel to the axes.

The CEC 2010 problem suite was used in this study as a basis for comparing the performance of different constraint handling techniques applied to PSO. As in the competition, the problems were solved in 10 and 30 dimensions, resulting in 36 problem instances.

Each version of PSO was run 30 times on each problem instance with a computational budget of $20000 \times D$ function evaluations. The success rate of an algorithm on a problem instance was defined as the proportion of feasible runs out of 30, where a feasible run was defined as a run that returned a feasible solution within the given budget. The performance between algorithms on the same problem instance was compared using the CEC 2010 competition rules [24]:

- If two algorithms had different success rates, the algorithm with the higher success rate was the winner.
- If two algorithms had the same success rate $> 0$, the algorithm with the better mean objective value of the feasible runs was the winner.
- If two algorithms had a success rate $= 0$, the algorithm with the lowest mean violation was the winner.

Table 1 shows results of two example problem instances to illustrate the performance ranking. On problem instance C09 in 10D, WP achieved the highest success rate of 0.6 (18 of the 30 runs resulted in feasible solutions) and is given the rank of 1. $\epsilon$DFR is ranked 2, with a success rate of 0.267. NCH, DP and DFR all had a success rate of 0 and so are compared based on the mean violation, achieving ranks of 3, 4 and 5, respectively. On the second example problem, C07 in 30D, all five CHTs achieved a success rate of 1 (all 30 runs finding feasible solutions). The performance is therefore compared based on the mean fitness, resulting in a rank of 1 for $\epsilon$DFR and 5 for WP.

## 3.3 Results

Figure 1 shows the quartiles of ranks of the five CHTs on all 36 problem instances, with the median values as black lines. All approaches had a maximum rank of 5 and a minimum rank of 1. This indicates that each approach achieved the best performance (rank 1) on at least one instance, but also the worst performance (rank 5) on at least one instance. Over all the problem instances, WP was the best performing approach, followed by $\epsilon$DFR, DFR, DP and NCH as the worst performing approach. Table 2

**Table 1** Ranking of PSO constraint handling techniques on two problem instances

|  | Success rate | Mean fitness (feasible runs) | Mean violation | Algorithm rank |
|---|---|---|---|---|
| *CEC 2010 problem C09 in 10 dimensions* | | | | |
| NCH | 0 | n/a | 5.308 | 3 |
| DP | 0 | n/a | 278.491 | 4 |
| WP | 0.600 | 0.000 | 0.000 | 1 |
| DFR | 0 | n/a | 289.129 | 5 |
| $\epsilon$DFR | 0.267 | 0.000 | 0.002 | 2 |
| *CEC 2010 problem C07 in 30 dimensions* | | | | |
| NCH | 1 | 4.401 | 0.000 | 3 |
| DP | 1 | 4.109 | 0.000 | 2 |
| WP | 1 | 7.973 | 0.000 | 5 |
| DFR | 1 | 5.121 | 0.000 | 4 |
| $\epsilon$DFR | 1 | 1.816 | 0.000 | 1 |

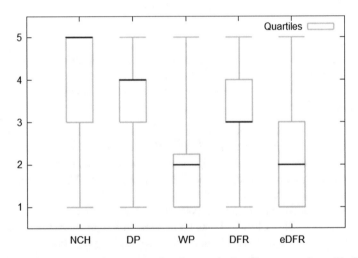

**Fig. 1** Distribution of the performance ranks of constraint handling approaches with gbest PSO over the CEC 2010 problem set in 10 and 30 dimensions. The median is indicated as a black line

shows the mean ranks with the number of times that each approach was the best performing and the worst performing.

These results show that there is performance complementarity between the approaches when used with PSO. Although WP is the best approach overall, it is only the best performing in one-third of the cases. For the other two-thirds of the instances, one of the other approaches performed better than WP. When faced with a new problem to solve, one may be inclined to use a weighted penalty approach, because it performed the best on average. However, there is a chance that it may per-

**Table 2** Performance of five constraint handling approaches on the CEC2020 problem suite in 10 and 30 dimensions (36 problem instances)

| Strategy | Mean rank | Best performing | | Worst performing | |
|---|---|---|---|---|---|
| NCH | 3.72 | 7 instances | (19%) | 20 instances | (56%) |
| DP | 3.69 | 1 instances | (3%) | 8 instances | (22%) |
| WP | **2.00** | **12 instances** | **(33%)** | **1 instance** | **(3%)** |
| DFR | 3.25 | 5 instances | (14%) | 5 instances | (14%) |
| $\epsilon$DFR | 2.25 | 9 instances | (25%) | 2 instances | (6%) |

form badly on the new instance. To reduce this chance, a landscape-aware approach aims to choose the best approach as the features of the problem are revealed to the search algorithm.

## 4 Landscape Analysis of Constrained Search Spaces

Malan et al. [12] introduced violation landscapes as a complementary view to fitness landscapes for analysing constrained search spaces. A violation landscape is defined in the same way as a fitness landscape, except that the level of constraint violation replaces the fitness function. In this study, the constraints were combined into a single function as defined in Eq. 7 to define the violation landscape.

To illustrate the concepts of fitness and violation landscapes, consider the CEC 2010 problem C01, defined with two inequality constraints:

$$\text{Minimize:} \quad f(\mathbf{x}) = -\left| \frac{\sum_{i=1}^{D} \cos^4(z_i) - 2\prod_{i=1}^{D} \cos^2(z_i)}{\sqrt{\sum_{i=1}^{D} i z_i^2}} \right|, \ \mathbf{z} = \mathbf{x} - \mathbf{o}, \ \mathbf{x} \in [0, 10]^D$$

$$\text{Subject to:} \quad g_1(\mathbf{x}) = 0.75 - \prod_{i=1}^{D} z_i \leq 0, \quad g_2(\mathbf{x}) = \sum_{i=1}^{D} z_i - 7.5D \leq 0,$$

$$(11)$$

where $\mathbf{o}$ is a predefined $D$-dimensional constant vector.

Figure 2 plots the fitness function, while Fig. 3 plots the aggregated level of constraint violation. The violation landscape shows that there is a large connected feasible region in the centre (plotted in black). The fitness landscape shows that the optimal solution in terms of fitness is positioned at approximately (1, 1). However, from the violation landscape, it can be seen that this point is in infeasible space.

Visualising the fitness and violation landscapes alongside each other provides insight into the nature of the search challenge, but is obviously not possible for higher dimensional problems. Fitness landscape analysis techniques [14] have been

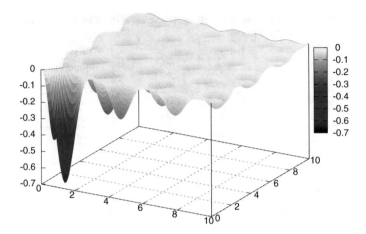

**Fig. 2** Fitness landscape of CEC 2010 C01 benchmark function in two dimensions

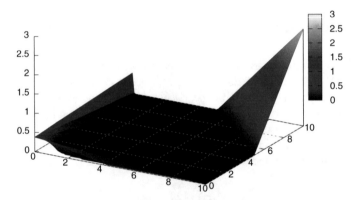

**Fig. 3** Violation landscape of CEC 2010 C01 benchmark function in two dimensions

used extensively to analyse search spaces and a similar approach can be used to analyse violation landscapes.

Given a fitness and violation landscape, metrics can be defined for characterising the search space and providing insight into the problem. The following previously proposed metrics [12] were used in this study:

1. The feasibility ratio (FsR) estimates the proportion of feasible solutions in the search space. Given a sample of $n$ solutions, FsR is defined as FsR $= \frac{n_f}{n}$, where $n_f$ denotes the number of feasible solutions in the sample. For problem C01 in 2D (illustrated in Fig. 3), the FsR is approximately 0.85.
2. The ratio of feasible boundary crossings (RFB$\times$) quantifies the level of disjointedness of the feasible regions. Given a walk through the search space resulting in a sequence of $n$ solutions, $\mathbf{x}_1, \mathbf{x}_2, \ldots, \mathbf{x}_n$, the string of binary values

$\mathbf{b} = b_1, b_2, \ldots, b_n$ is defined such that $b_i = 0$ if $\mathbf{x}_i$ is feasible and $b_i = 1$ if $\mathbf{x}_i$ is infeasible. RFB$\times$ is then defined as

$$\text{RFB}\times = \frac{\sum_{i=1}^{n-1} cross(i)}{n-1} \tag{12}$$

where

$$cross(i) = \begin{cases} 0 \text{ if } b_i = b_{i+1} \\ 1 \text{ otherwise.} \end{cases} \tag{13}$$

For problem C01 in 2D, based on random walks of 400 steps with a maximum step size of 0.5 (5% of the range of the problem) in both dimensions, the RFB$\times$ value is approximately 0.04. This means that only 4 out of every 100 steps cross a boundary between feasible and infeasible space, indicating that the feasible solutions are joined in a common area (as opposed to disjointed).

3. Fitness violation correlation (FVC) quantifies the correlation between the fitness and violation values in the search space. Based on a sample of solutions with resulting fitness-violation pairs, the FVC is defined as the Spearman's rank correlation coefficient between the fitness and violation values. For problem C01 in 2D, the FVC metric is negative, indicating that the fitness and violation values decrease in opposite directions.

4. The ideal zone metrics, 25_IZ and 4_IZ, quantify the proportion of solutions that have both good fitness and low violation. Given a scatterplot of fitness-violation pairs of a sample of solutions for a minimisation problem, the "ideal zone" (IZ) corresponds with the bottom left corner where fitness is good and violations are low. Metric 25_IZ is defined as the proportion of points in a sample that are below the 50% percentile for both fitness and violation, whereas 4_IZ is defined as the proportion of points in a sample that are below the 20% percentile for both fitness and violation. For problem C01 in 2D, both the 4_IZ and 25_IZ metrics are zero, indicating that the points that are the fittest are not the points that are feasible.

The premise of this study is that the landscape metrics defined above can be used to guide the choice of appropriate constraint handling techniques. Consider for example a search algorithm trying to solve the two-dimensional problem illustrated in Figs. 2 and 3. A weighted penalty approach will translate into searching the penalised landscape illustrated in Fig. 4, which will guide the search algorithm to an infeasible solution. In contrast, Deb's feasibility ranking approach will be more effective on this problem as it switches between the fitness and violation landscapes, depending on whether the current solutions are feasible or not. Experiments on this problem confirmed that Deb's feasibility ranking was able to solve this problem, whereas a weighted penalty approached failed to find a feasible solution.

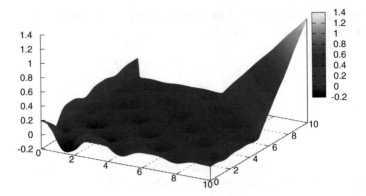

**Fig. 4** Penalised fitness landscape of CEC 2010 C01 benchmark function in two dimensions

## 5 Landscape-Aware Constraint Handling

This section describes the approach used for landscape-aware (LA) constraint handling that was previously proposed in the context of the differential evolution algorithm [13]. The approach involves an offline training phase and an online adaptive phase.

### 5.1 Offline Training Phase

Using a training set of benchmark problem instances, the process is as follows:

1. Landscape characterisation of training data: The training instances are characterised in terms of the landscape metrics described in Sect. 4. These metrics are calculated based on random samples of the search space of each problem instance.
2. Performance analysis of constituent CHTs: Using the same base search algorithm, experiments are run on the training instances to measure the relative performance of the constituent CHTs. A class is associated with each instance indicating the best performing CHT on that instance.
3. Derivation of rules for selecting constraint handling approaches: Using a classification algorithm on the training data, rules are derived for predicting the best performing CHT based on the landscape metrics.

### 5.2 Online Adaptive Phase

A landscape-aware approach is implemented using the rules derived in the offline training phase. Landscape metrics are calculated using the solutions encountered by the algorithm in the recent history of the search. No additional sampling or function evaluations is performed—the sample used for the landscape analysis is simply the stored recent history of the PSO search.

The CHT is changed at set intervals to the technique that is predicted to be the best by the rules, given the landscape features as experienced by the search algorithm. There are two parameters that control the LA constraint handling approach:

- Size of the search history archive ($OLA\_limit$): A search history is stored for each particle in the swarm. The combined search history of all particles forms the sample for performing the online landscape analysis. The parameter, $OLA\_limit$, defines the number of solutions stored for each particle as the search history. The total size of the archive, therefore, equals $OLA\_limit \times$ the swarm size.
- The switching frequency ($SW\_freq$): The switching of CHTs is performed after a set number of iterations of the search algorithm. Every $SW\_freq$ iteration, the landscape characteristics are calculated on the search history (sample of solutions in the archive) and a CHT is chosen using the rules.

A large search history archive ($OLA\_limit$) implies a larger sample for computing the online landscape metrics for choosing the CHT, whereas reducing the $OLA\_limit$ implies using landscape features from more recent search history. A large $SW\_freq$ will result in a more frequent re-evaluation of the landscape metrics and switching between CHTs.

# 6  Experimental Results

This section describes the experiments for testing the LA constraint handling approach with a global best PSO.

## 6.1  Experimental Setup

The CEC 2010 benchmark suite [17] was used as the basis for forming six problem instances at 5, 10, 15, 20, 25 and 30 dimensions ($D$), resulting in 108 problem instances. Two-thirds of these instances were randomly selected as the training set (72 instances) and the remaining one-third of the instances (36) were set aside for testing the online adaptive constraint handling.

### 6.1.1  Landscape Characterisation of Training Data

Each training problem instance was characterised as a feature vector of five metrics: FsR, RFB$\times$, FVC, 25_IZ, and 4_IZ. The metrics were calculated from samples of $200 \times D$ solutions for each instance, generated using multiple hill climbing walks based on fitness only. Each walk started at a random initial position. Random neighbouring solutions were sampled from a Gaussian distribution with the current position as mean and a standard deviation of 5% of the range of the domain of the problem instance. A walk was terminated if no better neighbour could be found after sampling 100 random neighbours.

### 6.1.2 Performance Analysis of Constituent CHTs

The three best performing techniques identified in Sect. 3, namely WP, $\epsilon$DFR, and DFR, were selected as the constituent CHTs for the landscape-aware approach. The performance of these three approaches on the training set instances was contrasted using the same approach as in Sect. 3 to allocate a best performing CHT class to each problem instance.

### 6.1.3 Derivation of Rules

The training set consisting of the landscape feature vector with best performing CHT class was then used to derive decision trees for predicting under which landscape scenarios each constraint handling technique would perform the best. The C4.5 algorithm [22] (implemented in WEKA [7] as J48) was used to induce the models. The following rules were extracted from the trees:

1. WP is predicted to be the best when (FsR > 0 AND FVC > −0.0266 AND RFBx > 0.081) OR (FsR = 0 AND FVC > 0.3704).
2. FR is predicted to be the best when FsR = 0 AND 4_IZ ≤ 0.005 AND 25_IZ > 0.163.
3. $\epsilon$DFR is predicted to be the best when (FsR > 0 AND RFBx ≤ 0.0005) OR (FsR = 0 AND 4_IZ > 0.005 AND FVC ≤ 0.37).

### 6.1.4 Online LA Constraint Handling:

The LA approach was implemented with $OLA\_limit = 2 \times D$, so for a 10-dimensional problem, the maximum sample size for the online landscape analysis was 20 solutions for each particle (1000 for a swarm size of 50). The search history was modelled as a queue of solutions for each particle, with the oldest information being discarded as the limit was reached and new data was added. The switching frequency (parameter $SW\_freq$) was set to $30 \times D$, implying that a CHT was chosen every 300 iterations for a 10-dimensional problem.

## 6.2   Results

The effectiveness of the LA approach was tested by comparing the performance against the constituent CHTs on the testing set (remaining one-third of problem instances not used in the offline training phase to derive the rules). A random switching approach was included in the experiment to measure the benefit of randomly choosing between the three constituent CHTs without considering the landscape information. All approaches used the same base PSO algorithm as defined in Sect. 3. In summary, the five approaches included in the experiment were

1. WP: Weighted penalty strategy as described in Sect. 3.
2. DFR: Deb's feasibility ranking strategy as described in Sect. 3.
3. $\epsilon$DFR: Feasibility ranking with an $\epsilon$ as described in Sect. 3.
4. RD: Random switching between the above three strategies at the same frequency as LA.
5. LA: Landscape-aware switching based on landscape information collected by the swarm using rules from the offline training phase.

Thirty independent runs of each version of PSO were executed and the performances were ranked using the CEC 2010 competition rules as outlined in Sect. 3. The results are given in Table 3.

As before, WP and $\epsilon$DFR performed similarly on average, achieving mean ranks of 2.75 and 2.86, respectively, while DFR performed the worst overall with a rank of 4.11. DFR was the best performing algorithm on only three instances while WP and $\epsilon$DFR were the best on 8 instances each. The LA approach outperformed all of the constituent strategies, achieving a mean rank of 2.06. LA was also the best performing strategy on more instances than any of the other approaches (one third) and was the only strategy that was the worst performing on none of the instances.

The LA approach was able to out-perform the constituent strategies by choosing the strategy that was predicted to be the best for each problem instance being solved. The RS strategy performed better than DFR, but worse than the other two constituent strategies, indicating that there was no benefit to switching randomly between constituent strategies in this case.

**Table 3** Performance of five constraint handling approaches on the test set ($SW\_freq = 30 \times D$, $OLA\_limit = 2 \times D$)

| Strategy | Mean rank | Best performing | | Worst performing | |
|---|---|---|---|---|---|
| WP | 2.75 | 8 instances | (22%) | 4 instances | (11%) |
| DFR | 4.11 | 3 instances | (8%) | 23 instances | (64%) |
| $\epsilon$DFR | 2.86 | 8 instances | (22%) | 3 instances | (8%) |
| RS | 3.06 | 5 instances | (14%) | 6 instances | (17%) |
| LA | **2.06** | **12 instances** | **(33%)** | **0 instances** | **(0%)** |

**Table 4**  Data from a sample run of the LA approach on problem C02 in 5D

| Iteration | Online landscape metrics | | | | | Selected strategy |
|---|---|---|---|---|---|---|
| | FsR | RFBx | FVC | 25_IZ | 4_IZ | |
| 150 | 0 | 0 | −0.081 | 0.288 | 0.014 | εDFR |
| 300 | 0 | 0 | −0.46 | 0.186 | 0 | FR |
| 1050 | 0.118 | 0.102 | 0.26 | 0.376 | 0 | WP |

## 6.3  Illustrative Run of Landscape-Aware Approach

To illustrate how the LA approach works, Table 4 shows some of the data extracted from a sample run on problem C02 in 5D. The table gives the values of the landscape metrics calculated from the sample of solutions in the swarm history at iterations 150, 300 and 1050. At the end of iteration 150, the landscape metrics show that the particles in the swarm had not yet encountered any feasible solutions (FsR = 0). Also, the FVC was a negative value, indicating that the fitness and violation values in the search history were negatively correlated. Considering the selection rules on page 14, the strategy predicted to perform the best with this profile of values was εDFR, which was the strategy selected by the LA algorithm.

By iteration 300, the 4_IZ value had dropped from 0.014 to 0. With this change, the FR strategy was predicted to be the best. The LA algorithm, therefore, selected FR as the strategy. At iteration 1050, the landscape metrics indicated that feasible solutions had been encountered by the swarm (with FsR = 0.118). Also, the FVC had changed to a positive value. With this landscape profile, WP was predicted to be the best performing strategy.

In this way, the LA approach switches between strategies depending on the features of the landscape as experienced by the swarm's recent search history.

## 7  Conclusion

Before metaheuristics can be effectively applied to solving real-world problems in general, they need adaptive ways of handling constraints. It is argued that adaptive techniques are needed because the best approach is problem dependent. This study presented results to show that there is performance complementarity between different constraint handling techniques when used with PSO. A dataset of benchmark problem instances was generated consisting of metrics for characterising the landscape features with class labels identifying the best performing constraint handling technique. In an offline phase, machine learning was used to derive rules for selecting constraint handling techniques based on landscape features. An online landscape-aware approach was implemented by using these rules as a basis for switching between techniques using the swarm history as the sample for computing the features.

Results show that there is value in utilising different constraint handling techniques with PSO and in selecting techniques based on the features of the search landscapes. Future work will include deriving general high-level rules for constraint handling selection that can apply to different search strategies and problem representations.

# References

1. Bischl, B., Mersmann, O., Trautmann, H., Preuß, M.: Algorithm selection based on exploratory landscape analysis and cost-sensitive learning. In: Proceedings of the Genetic and Evolutionary Computation Conference, pp. 313–320 (2012)
2. Coath, G., Halgamuge, S.K.: A comparison of constraint-handling methods for the application of particle swarm optimization to constrained nonlinear optimization problems. In: The 2003 Congress on Evolutionary Computation, CEC'03, vol. 4, pp. 2419–2425 (2003)
3. Coello Coello, C.A.: A survey of constraint handling techniques used with evolutionary algorithms. Technical report, Laboratorio Nacional de Informática Avanzada (1999)
4. Deb, K.: An efficient constraint handling method for genetic algorithms. Comput. Methods Appl. Mech. Eng. **186**(2–4), 311–338 (2000)
5. Eberhart, R., Kennedy, J.: A new optimizer using particle swarm theory. In: Proceedings of the Sixth International Symposium on Micromachine and Human Science, pp. 39–43 (1995)
6. Eberhart, R., Shi, Y.: Comparing inertia weights and constriction factors in particle swarm optimization. In: Proceedings of the IEEE Congress on Evolutionary Computation, vol. 1, pp. 84–88, July 2000
7. Hall, M., Frank, E., Holmes, G., Pfahringer, B., Reutemann, P., Witten, I.H.: The WEKA data mining software: an update. SIGKDD Explor. Newsl. **11**(1), 10–18 (2009)
8. Hu, X., Eberhart, R.: Solving constrained nonlinear optimization problems with particle swarm optimization. In: Proceedings of the Sixth World Multiconference on Systemics, Cybernetics and Informatics (2002)
9. Jordehi, A.R.: A review on constraint handling strategies in particle swarm optimisation. Neural Comput. Appl. **26**(6), 1265–1275 (2015). Jan
10. Kennedy, J., Eberhart, R.: Particle swarm optimization. In: Proceedings of the IEEE International Joint Conference on Neural Networks, pp. 1942–1948 (1995)
11. Kerschke, P., Hoos, H.H., Neumann, F., Trautmann, H.: Automated algorithm selection: survey and perspectives. Evol. Comput. **27**(1), 3–45 (2019). Mar
12. Malan, K.M., Oberholzer, J.F., Engelbrecht, A.P.: Characterising constrained continuous optimisation problems. In: 2015 IEEE Congress on Evolutionary Computation (CEC), pp. 1351–1358, May 2015
13. Malan, K.M.: Landscape-aware constraint handling applied to differential evolution. In: Theory and Practice of Natural Computing, pp. 176–187. Springer International Publishing (2018)
14. Malan, K.M., Engelbrecht, A.P.: A survey of techniques for characterising fitness landscapes and some possible ways forward. Inf. Sci. **241**, 148–163 (2013). Aug
15. Malan, K.M., Engelbrecht, A.P.: Particle swarm optimisation failure prediction based on fitness landscape characteristics. In: Proceedings of IEEE Swarm Intelligence Symposium, pp. 1–9 (2014)
16. Malan, K.M., Moser, I.: Constraint handling guided by landscape analysis in combinatorial and continuous search spaces. Evolutionary Computation p. Just Accepted (2018)
17. Mallipeddi, R., Suganthan, P.N.: Problem definitions and evaluation criteria for the CEC 2010 competition on constrained real-parameter optimization. Technical report, Nanyang Technological University, Singapore (2010)

18. Michalewicz, Z.: A survey of constraint handling techniques in evolutionary computation methods. Evol. Program. **4**, 135–155 (1995)
19. Muñoz, M.A., Kirley, M., Halgamuge, S.K.: The algorithm selection problem on the continuous optimization domain. In: Computational Intelligence in Intelligent Data Analysis. Studies in Computational Intelligence, vol. 445, pp. 75–89. Springer, Berlin, Heidelberg (2013)
20. Parsopoulos, K.E., Vrahatis, M.N.: Particle swarm optimization method for constrained optimization problems. In: Proceedings of the Euro-International Symposium on Computational Intelligence (2002)
21. Pulido, G.T., Coello Coello, C.A.: A constraint-handling mechanism for particle swarm optimization. In: Proceedings of the 2004 Congress on Evolutionary Computation, vol. 2, pp. 1396–1403 (2004)
22. Quinlan, J.R.: C4.5: Programs for Machine Learning. Morgan Kaufmann Publishers Inc., San Francisco, CA, USA (1993)
23. Shi, Y., Eberhart, R.: A modified particle swarm optimizer. In: Proceedings of the 1998 IEEE World Congress on Computational Intelligence, pp. 69–73 (1998)
24. Suganthan, P.: Comparison of results on the 2010 CEC benchmark function set. Technical report, Nanyang Technological University, Singapore. http://www3.ntu.edu.sg/home/epnsugan/index_files/CEC10-Const/CEC2010_COMPARISON2.pdf (2010)
25. Takahama, T., Sakai, S.: Solving constrained optimization problems by the $\epsilon$-constrained particle swarm optimizer with adaptive velocity limit control. In: 2006 IEEE Conference on Cybernetics and Intelligent Systems, pp. 1–7 (2006)
26. Takahama, T., Sakai, S.: Constrained optimization by $\epsilon$ constrained particle swarm optimizer with $\epsilon$-level control. In: Abraham, A., Dote, Y., Furuhashi, T., Köppen, M., Ohuchi, A., Ohsawa, Y. (eds.) Soft Computing as Transdisciplinary Science and Technology: Proceedings of the fourth IEEE International Workshop WSTST'05, pp. 1019–1029. Springer, Berlin, Heidelberg (2005)
27. Takahama, T., Sakai, S.: Constrained optimization by the epsilon constrained differential evolution with an archive and gradient-based mutation. In: IEEE Congress on Evolutionary Computation, pp. 1–9 (2010)
28. Van Den Bergh, F.: An analysis of particle swarm optimizers. Ph.D. thesis, Department of Computer Science, University of Pretoria, Pretoria, South Africa (2002)
29. Wolpert, D.H., Macready, W.G.: No free lunch theorems for search. Technical Report SFI-TR-95-02-010, Santa Fe Institute (1995)
30. Wolpert, D.H., Macready, W.G.: No free lunch theorems for optimization. IEEE Trans. Evol. Comput. **1**(1), 67–82 (1997). April

# On the use of Gradient-Based Repair Method for Solving Constrained Multiobjective Optimization Problems—A Comparative Study

Victor H. Cantú, Antonin Ponsich, and Catherine Azzaro-Pantel

**Abstract** In this chapter, we study the effect of repairing infeasible solutions using the gradient information for solving constrained multiobjective problems (CMOPs) with multiobjective evolutionary algorithms (MOEAs). For this purpose, the gradient-based repair method is embedded in six classical constraint-handling techniques: constraint dominance principle, adaptive threshold penalty function (ATP), C-MOEA/D, stochastic ranking, $\varepsilon$-constrained and improved $\varepsilon$-constrained. The test functions used include classical problems with inequality constraints (CFs and LIRCMOPs functions) as well as six recent problems with equality constraints. The obtained results show that the gradient information coupled with a classical technique is not computationally prohibitive and can make the given classical technique much more robust. Moreover, in highly constrained problems, like those involving equality constraints, the use of the gradient for repairing solutions may not only be useful but also necessary in order to obtain a good approximation of the true Pareto front.

## Symbols and Acronyms

| | |
|---|---|
| AASF | Augmented achievement scalarizing function |
| ATP | Adaptive threshold penalty function method |
| CDP | Constraint dominance principle method |
| C/MOEAD | Constrained MOEA/D method |
| CF | Constrained function suite |
| CMOP | Constrained multiobjective optimization problem |
| DE | Differential evolution algorithm |

V. H. Cantú · C. Azzaro-Pantel (✉)
Laboratoire de Génie Chimique, Université Toulouse, CNRS, INPT, UPS,
Toulouse, France
e-mail: catherine.azzaropantel@toulouse-inp.fr

A. Ponsich
Departamento de Sistemas, Universidad Autónoma Metropolitana, Azcapotzalco,
Mexico City, Mexico

| EQC | Equality constrained function suite |
| $\mathcal{F}$ | Feasible region |
| $\mathbf{f}(\mathbf{x})$ | Objective vector |
| $\mathbf{g}(\mathbf{x})$ | Inequality constraint vector |
| HV | Hypervolume indicator |
| $\mathbf{h}(\mathbf{x})$ | Equality constraint vector |
| $I$ | Feasibility ratio |
| IGD | Inverted generational distance indicator |
| LIRCMOP | Large-infeasible regions function suite |
| MOEA | Multiobjective evolutionnary algorithm |
| MOEA/D | Multiobjective evolutionnary algorithm based on decomposition |
| $\mathcal{PF}$ | Pareto-optimal front |
| $\mathcal{PS}$ | Pareto-optimal set |
| $\mathbb{R}^n$ | $n$-dimensional Euclidean space |
| $\mathbb{R}^n_+$ | Non-negative orthant of $\mathbb{R}^n$ |
| $\mathcal{S}$ | Search space |
| SR | Stochastic ranking method |
| $\mathbf{w}$ | Weight vector |
| $\mathbf{x}$ | Decision (variable) vector |
| $\mathbf{z}^*$ | Ideal objective vector |
| $\mathbf{z}^{nad}$ | Nadir objective vector |

# 1   Introduction

The formulation of many real-world engineering problems involve the simultaneous optimization of several conflicting criteria, frequently with a set of constraints that need to be fulfilled. These problems can be stated mathematically as constrained multiobjective optimization problems (CMOPs), and their solution consists of providing a finite number of well-distributed efficient solutions so that the decision-makers can make the right choice according to their experience.

Regarding the methods for solving these problems, multiobjective evolutionary algorithms (MOEAs) have drawn the attention of researchers due to their working mode that enables them to provide an approximation of the Pareto set in one single run. These population-based algorithms may work mainly under three different paradigms: (1) in dominance-based algorithms, the Pareto dominance relationship is used to establish a ranking among candidate solutions (e.g., [1, 2]); (2) decomposition-based algorithms divide the original multiobjective problem into a collection of scalar optimization subproblems, which are solved in a collaborative way (e.g., [3]), and (3) quality-indicator-based algorithms search to optimize one or several performance metrics (e.g. [4]). Nonetheless, like the majority of metaheuristic techniques, MOEAs lack constraint-handling mechanisms in their canonical design. Consequently, a range of constraint-handling techniques have been proposed in the literature, aiming to guide the search to promising constrained regions. The working principle of these techniques is variable, for instance, some repair infeasible solutions to some extent, others enable infeasible individuals to survive with a certain probability or by means of a relaxation on the constraints, while others prefer feasible solutions over infeasible ones all along the evolutionary process. Besides, constraint-handling techniques that were originally proposed for tackling single-objective problems have been adapted to consider multiple objectives, this is the case for techniques like penalty function [5] or $\varepsilon$-constrained method [6]. Even though a number of works have been devoted to the study of constraint-handling techniques within MOEAs, this issue still constitutes an active research area as the applicability of these techniques for solving highly constrained real-world optimization problems is limited by their efficiency for constraint handling.

Of special interest to our study stands the gradient-based repair method [7]. Proposed by Chootinan and Chen in 2005 in a single-objective optimization framework, this method uses the gradient information of constraints to repair infeasible solutions. It was employed to solve the single-objective constrained problems proposed in CEC 2006 and CEC 2010 competitions, obtaining encouraging results [8–10]. However, to the best of our knowledge, this method has not been explored for the solution of CMOPs.

Therefore, in this work, the gradient-based repair method is considered to solve CMOPs. A comparative study is carried out to demonstrate the benefits of using this repair strategy, when combined with six well-known constraint-handling methods, namely constraint dominance principle (CDP), Adaptive Threshold Penalty function (ATP), C-MOEA/D, stochastic ranking, $\varepsilon$-constrained and improved $\varepsilon$-constrained. Two recent benchmarks used in many related works are considered, namely CF [11] and LIRCMOP functions [12], as well as six problems containing equality constraints.

The remainder of this chapter is organized as follows. Section 2 presents some definitions relevant to this work as well as a general background on constrained multiobjective optimization. Section 3 develops and briefly discusses the six constraint-handling techniques studied in this work, in addition to the gradient-based repair method. The experimental methodology and the computational results are described in Sects. 4 and 5, respectively. Finally, conclusions and perspectives for future work are drawn in Sect. 6.

# 2 Related Background

## 2.1 Basic Concepts

We study a constrained multiobjective optimization problem (CMOP) of the form:

$$\text{minimize} \quad \mathbf{f}(\mathbf{x}), \tag{1}$$
$$\text{subject to} \quad g_i(\mathbf{x}) \leq 0, \quad i = \{1, \ldots, p\}$$
$$h_j(\mathbf{x}) = 0, \quad j = \{1, \ldots, q\}$$
$$l_i \leq x_i \leq u_i, \quad i = \{1, \ldots, n\}.$$

where $\mathbf{x} = [x_1, x_2, \ldots, x_n]^T$ is a $n$-dimensional vector of decision variables (either discrete or continuous), $\mathbf{f}(\mathbf{x}) = [f_1(\mathbf{x}), f_2(\mathbf{x}), \ldots, f_k(\mathbf{x})]^T$ is a $k$-dimensional vector of (conflicting) objective functions to be minimized[1], $p$ is the number of inequality constraints and $q$ is the number of equality constraints. The functions $g_i$ and $h_j$ may be linear or non-linear, continuous or not, real-valued functions. Each variable $x_i$ has upper and lower bounds, $u_i$ and $l_i$, respectively, which define the search space $S \subseteq \mathbb{R}^n$. Inequality and equality constraints define the feasible region $\mathcal{F} \subseteq S$.

In multiobjective optimization, in order to compare two feasible solutions $\mathbf{x}, \mathbf{y} \in \mathcal{F}$, the *Pareto dominance relation* must be defined: A solution $\mathbf{x}$ is said to *dominate* a solution $\mathbf{y}$, denoted by $\mathbf{x} \prec \mathbf{y}$, if and only if $\mathbf{x}$ is at least as good as $\mathbf{y}$ in all objectives ($\forall i \in \{1, \ldots, k\}, f_i(\mathbf{x}) \leq f_i(\mathbf{y})$) and better in at least one objective ($\exists j \in \{1, \ldots, k\}, f_i(\mathbf{x}) < f_i(\mathbf{y})$).

The set of all incomparable (non-dominated) solutions is known as the *Pareto Optimal Set* ($\mathcal{PS}$), and is formally defined as

$$\mathcal{PS} := \{\mathbf{x}^* \in \mathcal{F} : \nexists \mathbf{x} \in \mathcal{F}, \mathbf{x} \prec \mathbf{x}^*\}. \tag{2}$$

The image of these non-dominated vectors is collectively known as the *Pareto Optimal Front* ($\mathcal{PF}$) or *True Pareto Front* (true PF), expressed as

$$\mathcal{PF} := \{\mathbf{f}(\mathbf{x}^*) : \mathbf{x}^* \in \mathcal{PS}\}. \tag{3}$$

Throughout this chapter, we use the term true PF interchangeably with $\mathcal{PF}$. Further, since the $\mathcal{PS}$ may contain an infinite number of solutions, the multiobjective problem is limited to determine a finite number of Pareto-optimal solutions that represents a good approximation of the $\mathcal{PF}$ in terms of both convergence and diversity.

The *ideal objective vector* $\mathbf{z}^* \in \mathbb{R}^k$ is obtained by minimizing each objective function individually subject to the constraints, in this manner, each component of the ideal vector can be represented as $z_i^* = \min\{f_i(\mathbf{x}) : \mathbf{x} \in \mathcal{F}\}$. This ideal (utopian) point constitutes a reference point, and is especially useful in some scalarizing functions used within decomposition-based MOEAs.

The *nadir objective vector* $\mathbf{z}^{\text{nad}} \in \mathbb{R}^k$ constitutes an upper bound of the Pareto-optimal front. It is defined in the objective space as $z_i^{\text{nad}} = \max\{f_i(\mathbf{x}) : \mathbf{x} \in \mathcal{PS}\}$. It is used, along with the ideal point, for performing an objective normalization as shown in (7).

In the context of constrained optimization, the *overall constraint violation* is useful to evaluate the degree of infeasibility of a solution $\mathbf{x}$. It is defined as

---

[1]For the case of maximization, a negative sign is just added to the objective function.

$$\phi(\mathbf{x}) := \sum_{i=1}^{p} \max\{0, g_i(\mathbf{x})\}^{\alpha} + \sum_{j=1}^{q} |h_j(\mathbf{x})|^{\alpha}. \tag{4}$$

where $\alpha$ takes a positive value, in this work $\alpha = 1$. Note that equality constraints can be transformed into inequality constraints as: $\forall j \in \{1, \ldots, q\}, |h_j(\mathbf{x})| - \epsilon \leq 0$, where $\epsilon$ is a small real-valued tolerance, in this work $\epsilon = 1e - 4$.

The *feasibility ratio* $(I)$ indicates the ratio of feasible solutions belonging to an approximation archive $A$ to the $\mathcal{PF}$, it can be represented as

$$I(A) := \frac{|A_f|}{|A|} \tag{5}$$

where $A_f$ is a subset of $A$ containing only feasible solutions ($\{A_f \in A : A_f \in \mathcal{F}\}$), and $|\cdot|$ stands for the cardinality of the given set.

## 2.2 Decomposition-Based Approach

In this chapter, a decomposition-based algorithm is considered as the search engine, namely MOEA/D [13]. In this algorithm, the original CMOP (1) is represented as a number of scalar optimization problems using of a scalarizing function $u$, each subproblem using a different target direction or *weight vector* $\mathbf{w} \in \mathbb{R}_+^k$ such that $\sum_{j=1}^{k} w_j = 1$. This transformation can be represented as

$$\text{minimize} \quad u(\mathbf{f}'(\mathbf{x}); \mathbf{w}) \tag{6}$$

with the same constraints as in (1). In this work, we perform an objective normalization such that

$$\mathbf{f}'(\mathbf{x}) := \frac{\mathbf{f}(\mathbf{x}) - \mathbf{z}^*}{\mathbf{z}^{\text{nad}} - \mathbf{z}^*}. \tag{7}$$

where $\mathbf{z}^*, \mathbf{z}^{\text{nad}} \in \mathbb{R}^k$ are the ideal and nadir points found so far. The Augmented Achievement Scalarizing Function (AASF) [14, 15] employed here is defined as

$$u^{aasf}(\mathbf{f}'; \mathbf{w}) := \max_{i} \left\{ \frac{f_i'}{w_i} \right\} + \alpha \sum_{i=1}^{k} \left\{ \frac{f_i'}{w_i} \right\} \tag{8}$$

where $\alpha$ should take a small value. In this work $\alpha = 1e - 3$. It should be noted that using a decomposition-based algorithm allows considering some constraint-handling techniques that were originally proposed on a single-objective framework, so that the utility function $u$ may be regarded as the objective function to be minimized. Besides, in MOEA/D the original optimization problem is divided into a number of scalar optimization subproblems equal to the number of individuals in the population, that is, each individual is assigned to solve a specific subproblem according to its weight vector associated. Also, the optimization procedure is performed in a collaborative way, since the population is divided into neighbourhoods which can share a given solution if a given neighbour advances in its search with it. In this work, we use Differential Evolution algorithm in its classical version $DE/rand/1/bin$ as the search operator.

## 2.3 Constraint-Handling Strategies within MOEAs

In the past decade, the development of constraint-handling strategies for the solution of CMOPs has drawn considerable interest from the evolutionary multiobjective optimization (EMO) community. It must be emphasized that, commonly, most MOEAs are adapted for treating CMOPs through the constraint dominance principle (CDP, explained in detail in the next section) [2], however, the disadvantages of this strategy, well known for single-objective optimization, is also evident in multiobjective optimization, where diversity among solutions is important in order to achieve a good approximation to the true Pareto front.

Consequently, with the purpose of tackling the drawbacks of CDP for NSGA-II, the Infeasibility-Driven Evolutionary Algorithm (IDEA) [16] proposes the use of a parameter $\alpha$ standing for the ratio of infeasible solutions to survive in the population. A constraint violation function is used as an additional objective and then, non-dominated ranking is applied to infeasible and feasible individuals separately. In this way, the new population (containing $N$ individuals) holds, at most, $\alpha \cdot N$ best-ranked infeasible solutions. Besides, in a dominance-based framework also, a parameterless adaptive penalty function has been proposed in [17], in which a modified objective function is computed considering a "distance" measure and an adaptive penalty function. These two components of the new objective function are tailored aiming to guide the search towards feasible regions first, and then, infeasible individuals' information is used to explore promising infeasible regions. In addition, in [18], the authors propose adaptations of the CDP and Stochastic Ranking (SR) methods for working in MOEA/D-DE. The proposed SR multiobjective version selects a solution according to the utility function value with probability $p_f$, otherwise the CDP is considered. Results show that MOEA/D-CDP outperforms MOEA/D-SR for the studied problems. Another decomposition-based algorithm adapted for solving CMOPs is proposed in [19], by means of a function that combines the overall constraint violation and the number of active constraints. The mean value of this function, weighted by the ratio of feasible individuals in the population, allows computing a threshold on the allowed constraint violation. Solutions below this threshold are considered as feasible and compared in terms of their scalarizing function. In [6] the $\varepsilon$-constraint method [20] within MOEA/D is explored in which the $\varepsilon$ level is now defined using a normalized constraint violation and an additional $\varepsilon$-comparison rule is proposed accounting for slightly infeasible promising solutions. Further, in [12], another modification to the original $\varepsilon$ level function is presented, to be employed within MOEA/D. This $\varepsilon$ function enables the $\varepsilon$ level to increase/decrease depending on the ratio of feasible solutions in the population so that the search is strengthened in both feasible and infeasible regions throughout the optimization. More recently, a two-stage procedure (called, Push and Pull) was proposed [21]. In the first stage (push stage), the algorithm explores the unconstrained search space, which allows the population to get across infeasible disconnected regions, and then, in the pull stage, original constraints are considered along with the improved $\varepsilon$ constraint to gradually pull the population towards feasible regions. Stagnation of the identified ideal and nadir points determines the switching point between the "push" and "pull" phases. Nevertheless, it must be highlighted that this constraint-handling method has only been tested for LIRCMOP test problems, which contain only inequality constraints and some true Pareto fronts are identical to their counterparts in the unconstrained problems. In [22], diversity preservation in the population is explicitly handled through a modification of MOEA/D-CDP. When two solutions are compared, if at least one of them is infeasible, the similarity measure is computed based on the angle of their corresponding objective vectors with respect to the ideal point. Similar solutions are compared according to their overall constraint violation, otherwise the scalarizing function value is used as the comparison criterion. In [23], the original CMOP is modified considering the overall constraint violation as an additional objective. Two weight vector sets are generated, accounting for infeasible and feasible solutions, respectively, in order that infeasible individuals are well distributed along the Pareto front, and may lead the search to promising regions. In [24], the authors use a modified $\varepsilon$-constraint strategy in a decomposition-based framework. The feasibility ratio and the minimum constraint violation value of the current population are employed in order to compute the $\varepsilon$ level at each generation. Also, the scaling factor $F$ within DE is adjusted dynamically in order to promote local search in late generations. Finally, MOEA/D is modified in [25] so that two solutions (one feasible and one infeasible) are assigned to each weight vector. This strategy is used to consider an individual on each side of the feasibility boundary. During the selection phase, an offspring solution is compared, considering the scalarizing function and the overall constraint violation. For the two-objective problem, dominance is used to select two surviving individuals among the three considering whether all solutions are non-dominated. Thus, a solution with a high constraint value is discarded, while a solution survives if it dominates the two others.

## 3 A Portfolio of Constraint-Handling Strategies

In this section, the six constraint-handling techniques employed in this work are presented and discussed in more detail. It can be appreciated that they constitute easy-to-implement and simple constraint-handling techniques. Also, even if some original methods have been proposed to tackle only problems with inequality constraints, they can be easily adjusted for considering equality constraints as well.

## 3.1 CDP

This constraint-handling technique establishes the superiority of feasible solutions over infeasible ones. Proposed by Deb in [26], the feasibility rules (also called lexicographical order) consist in a binary tournament selection according to the following criteria:

1. Any feasible solution is preferred to any infeasible solution.
2. Among two feasible solutions, that with better objective function value is preferred.
3. Among two infeasible solutions, that with smaller constraint violation is preferred.

Deb's feasibility rules represent an easy-to-implement, parameter-free technique to handle constraints. Their extension to CMOPs, called the constraint dominance principle (CDP), reformulates condition 2 by a dominance-based comparison, and thus is stated as "among two feasible solutions, that which dominates the other is preferred", where a diversity operator is used if no solution dominates the other. In a decomposition-based algorithm, the utility function is used for condition 2.

Further, due to its simplicity and its overall good performance, the CDP method is usually the first constraint-handling technique tested for treating a given problem within MOEAs. However, one of the main drawbacks of this method appears when dealing with problems with a reduced and disconnected feasible region: since any feasible solution is preferred over an infeasible one, once the algorithm has converged to a feasible region, it might be difficult to escape from there to explore the rest of the search space, i.e., once the constraints are fulfilled, the algorithm is likely to get trapped prematurely in some subregion of the search space.

## 3.2 ATP (Adaptive Threshold Penalty)

This adaptive threshold penalty function (called ATP by the authors in [5]) is particularly adapted to be used within MOEA/D. It uses a threshold value, $\tau$, for dynamically controlling the amount of penalty. The threshold value $\tau$ is then defined as

$$\tau = V_{min} + 0.3(V_{max} - V_{min}) \tag{9}$$

where $V_{min} = \min\{\phi(\mathbf{x}^i)\}, \forall i \in P$ and $V_{max} = \max\{\phi(\mathbf{x}^i)\}, \forall i \in P$, and $P$ represents a given neighborhood in MOEA/D. Then, according to ATP method, the new scalarizing function $u$ is defined as

$$u(\mathbf{f}'(\mathbf{x}); \mathbf{w}) = \begin{cases} u(\mathbf{f}'(\mathbf{x}); \mathbf{w}) + s_1 \phi^2(\mathbf{x}), & \text{if } \phi(\mathbf{x}) < \tau, \\ u(\mathbf{f}'(\mathbf{x}); \mathbf{w}) + s_1 \tau^2 + s_2(\phi(\mathbf{x}) - \tau), & \text{otherwise} \end{cases}$$

where $s_1$ and $s_2$ are two scaling parameters.

## 3.3   C-MOEA/D

This constraint-handling method separates the objective function and the violation of constraints, and proposes a *violation threshold* allowing for a relaxation of constraints under which solutions are considered as feasible [19]. Once the relaxation is carried out and therefore feasible and infeasible solutions have been identified, the constraint dominance principle is employed for the selection step. The violation threshold, $\varphi$, proposed by the authors is calculated as follows:

$$\varphi = \frac{\sum\limits_{j=1}^{|C|} \phi(\mathbf{x}_j)}{|C|} \cdot I_G(C). \tag{10}$$

The first part of the right term designates the average overall constraint violation of the current population $C$ and $I_G$ is the feasibility ratio at generation $G$ of the evolutionary algorithm.

## 3.4   SR

Stochastic ranking (SR) [27] has been proposed as an attempt to balance the relative weights of the objective and the constraint violation that occurs in penalty functions. In this method, the population is sorted following a probabilistic procedure: two individuals are compared according to their objective function with a probability $p_f$, otherwise, the overall constraint violation is used for comparisons. Once the population has been sorted by SR, a part of the population assigned with the highest rank is selected for recombination. In this way, the search is directed by the minimization of the objective function and by feasibility concepts at the same time. Since this method was designed in a single-objective framework, its generalization to a multiobjective problem is not straightforward even when aggregation functions are considered as in decomposition-based algorithms. One attempt of using SR for multiobjective optimization is introduced in [18]. In the present study, the same implementation is considered. Nevertheless, it must be emphasized that no actual *ranking* of the population is performed at all, but instead a stochastic comparison between the parent and the offspring: the comparison is performed firstly according to their scalarizing function with a probability $p_f$, otherwise, the constraint dominance principle is used.

## 3.5   ε-Constrained

This method integrates a relaxation of constraints up to a so-called $\varepsilon$ level, under which solutions are regarded as feasible [20]. Once the feasible and infeasible solutions are identified, feasibility rules are employed for selecting the survival individuals. This technique has proven to be especially efficient in highly constrained problems, such as those involving equality constraints, as it promotes the exploration of regions that would be impossible to reach by simple feasibility rules. The authors of [20] proposed a dynamic control of $\varepsilon$ level, according to

$$\varepsilon(0) = \phi(\mathbf{x}_\theta) \tag{11}$$

$$\varepsilon(t) = \begin{cases} \varepsilon(0)(1 - \frac{t}{T_c})^{cp}, & 0 < t < T_c, \\ 0, & t \geq T_c \end{cases}$$

where $\mathbf{x}_\theta$ is the best $\theta$th individual (in terms of constraint violation) in the first generation, $cp$ is a parameter to control the decreasing speed of the $\varepsilon$ level and $T_c$ represents the generation after which the $\varepsilon$ level is set to 0, and then feasibility rules are considered. Note that CDP is used instead of feasibility rules for the solution of CMOPs.

## 3.6 Improved ε-Constrained

In [12], another function for controlling the $\varepsilon$ level is proposed, aiming to tackle some drawbacks found in the previous $\varepsilon$-constrained method. For instance, in problems with large feasible regions, all population might be feasible and thus the value for $\varepsilon(0)$ becomes zero. This would be equivalent to using CDP (with its drawbacks) all along the evolutionary process. The proposed function for $\varepsilon$ level permits increasing its value if the feasible ratio is above a given threshold ($\alpha$ parameter), according to

$$\varepsilon(0) = \phi(\mathbf{x}_\theta) \tag{12}$$

$$\varepsilon(t) = \begin{cases} \varepsilon(t-1)(1-\rho), & \text{if } I_t(C) < \alpha \text{ and } t < T_c, \\ \phi_{max}(1+\rho), & \text{if } I_t(C) \geq \alpha \text{ and } t < T_c, \\ 0, & t \geq T_c \end{cases}$$

where $\mathbf{x}_\theta$ has the same meaning as in Eq. 11, $I_t(C)$ is the ratio of feasible individuals of the current population $C$ at generation $t$, parameter $\rho$ is to control the speed of reducing the $\varepsilon$ level (it ranges between 0 and 1), parameter $\alpha$ controls the searching preference between the feasible and infeasible regions and $\phi_{max}$ is the maximum overall constraint violation found so far.

## 3.7 Gradient-Based Repair

This constraint-handling method uses the gradient information derived from the constraint set to systematically repair infeasible solutions, in other words, the gradient of constraint violation is used to direct infeasible solutions towards the feasible region [7]. The vector of constraint violations $\Delta C(\mathbf{x})$ is defined as:

$$\Delta C(\mathbf{x}) = [\Delta g_1(\mathbf{x}), \dots, \Delta g_p(\mathbf{x}), \\ \Delta h_1(\mathbf{x}), \dots, \Delta h_q(\mathbf{x})]^T \tag{13}$$

where $\Delta g_i(\mathbf{x}) = \max\{0, g_i(\mathbf{x})\}$ and $\Delta h_j(\mathbf{x}) = h_j(\mathbf{x})$. This information, additionally to the gradient of constraints $\nabla C(\mathbf{x})$, is used to determine the step $\Delta \mathbf{x}$ to be added to the solution $\mathbf{x}$, according to

$$\nabla C(\mathbf{x}) \Delta \mathbf{x} = -\Delta C(\mathbf{x}) \tag{14}$$
$$\Delta \mathbf{x} = -\nabla C(\mathbf{x})^{-1} \Delta C(\mathbf{x}) \tag{15}$$

Although the gradient matrix $\nabla C$ is not invertible in general, the Moore–Penrose inverse or pseudoinverse $\nabla C(\mathbf{x})^+$ [28], which gives an approximate or best (least square) solution to a system of linear equations, can be used instead in Eq. (15). Therefore, once the step $\Delta \mathbf{x}$ is computed, the infeasible point $\mathbf{x}$ moves to a less infeasible point $\mathbf{x} + \Delta \mathbf{x}$. This repair operation is performed with a probability $P_g$ and repeated $R_g$ times while the point is infeasible, for each individual.

   In this work, this repair-based method is combined with any of the six previously mentioned constraint-handling techniques in order to highlight the effect it may have on the performance level of MOEAs for solving CMOPs. In our implementation, the computation of the gradient $\nabla C(\mathbf{x})$ is done numerically using forward finite differences for all problems. Also, it is worth noting that only non-zero elements of $\Delta C(\mathbf{x})$ are repaired, thus, the gradient is only computed for constraints that are being violated. Note that this procedure can produce situations where some variables lie outside their allowed variation range. In [7], two inequality constraints are added for each variable, accounting for their bounds. However, due to the associated computational burden in real-world optimization problems where the number of variables may be high, these additional constraints are not considered here. Instead, an additional repair process, performed at each iteration, sets the variable value to the violated bound if necessary.

# 4 Experimental Methodology

## 4.1 Test Problems

We investigated the performance of the six above-mentioned constraint-handling techniques, and particularly, the effect of embedding the gradient-based repair method within each of these techniques. For this purpose, we carried out a comprehensive study over 29 test problems which include inequality constraints: ten constrained function (CF) test problems [11], fourteen CMOPs with large infeasible regions (LIRCMOPs) [12], as well as six problems with equality constraints (named here EQC) [29–32], among which the last one is a real-world CMOP with application in process engineering.

The reason for the choice of CF and LIRCMOP function suites is related to the interesting features they present (numerous local optima, disconnected Pareto fronts, and inequality constraints that are difficult to satisfy), but also due to the fact that several recent works have studied these problems [21, 24, 25, 33], and thus our results can be compared to those obtained with state-of-the-art algorithms.

## 4.2 Performance Indicators

To assess the performance of the different algorithms, we use the inverted generational distance indicator (IGD) [34] as well as the hypervolume indicator (HV) [35]. Again, the choice of using these two performance indicators is made in order to offer the reader comparable results to that of state-of-the-art algorithms.

The IGD indicates how far the discretized Pareto-optimal front is from the approximation set, i.e. it is the average distance from each reference point to its nearest solution in the approximation set. This non-Pareto compliant indicator measures both convergence and diversity. A smaller value of IGD indicates a better performance of the algorithm. To generate the reference set for CF and LIRCMOP test suites, 1 000 points are sampled uniformly from the true PF for two-objective problems; whereas 10 000 points are sampled uniformly from the true PF for three-objective problems. With respect to the EQC test suite, the number of points sampled from the true PF varies for each problem, depending on the data available.

Concerning the HV, it is the only performance indicator known to be Pareto-compliant [36]. A large HV value shows that a given solution set approximates the Pareto-optimal front well in terms of both convergence and diversity in the objective space. The $k$-dimensional reference vector for the HV computation is set to $[1.1, 1.1, \ldots, 1.1]^T$ in the normalized objective space $[0, 1]^k$, for all problems. To obtain the ideal and nadir points for the real-world problem (EQC6), five independent runs using jDE with $\varepsilon$-constrained method coupled with gradient-based repair were performed for each single-objective problem, setting the number of function evaluations to 500 000, in order to obtain the extreme points.

## 4.3 Experimental Settings

The detailed parameter settings for the experiments carried out here are summarized below.

1. MOEA/D parameters. Scalarizing function: augmented achievement (AASF), probability of choosing parents locally: $\delta = 0.9$, neighborhood size: $T = 0.1N$, maximum number of replacements $n_r = 2$. To maintain diversity, the crowding distance is used if the external archive contains more solutions than the population size ($N$).
2. DE parameters. $CR = 1$, $F = 0.5$. With polynomial mutation parameters: $p_m = 1/n$, $\eta_m = 20$.
3. Population size ($N$). For CFs, $N = 200$ for two-objective problems, $N = 300$ for three-objective problems. For LIRCMOPs, $N = 300$. For EQCs, $N = 100$ for two-objective problems, $N = 300$ for three-objective problems.
4. Number of function evaluations ($NFE$). For CFs, $NFE = 1e5$ for two-objective problems, $NFE = 1.5e5$ for three-objective problems. For LIRCMOPs, $NFE = 1.5e5$. For EQCs, $NFE = 1e5$ for two-objective problems, $NFE = 1.5e5$ for EQC4 and $NFE = 0.5e5$ for EQC5. Also, for the gradient-based repair method, each computation of the gradient of constraints counts is counted as 1 function evaluation.
5. Parameter settings of constraint-handling techniques (are set identical to that proposed in the related articles):

   - ATP: $s_1 = 0.01$, $s_2 = 20$.
   - Stochastic ranking: $p_f = 0.05$.
   - $\varepsilon$-constrained: $\theta = 0.2N$, $cp = 5$, $T_c = 0.2T_{max}$.
   - Improved $\varepsilon$-constrained: $\theta = 0.05N$, $\rho = 0.1$, $\alpha = 0.95$, $T_c = 0.8T_{max}$.
   - Gradient-based repair: $P_g = 1$, $R_g = 3$, step size for finite differences: $1e - 6$.

For each problem, 31 independent runs were performed. The algorithms previously presented were implemented in MATLAB R2019a and the computational experiments were performed with a processor Intel Xeon E3-1505M v6 at 3.00 GHz and 32 Go RAM.

## 5 Results and Discussion

The results obtained with the six above-mentioned algorithms (CDP-MOEA/D, ATP-MOEA/D, C-MOEA/D, SR-MOEA/D, $\varepsilon$-MOEA/D and Improved $\varepsilon$-MOEA/D) without and with gradient-based repair are analysed through the mean and standard deviation values of each performance indicator (see Tables 1 to 6). The overall computational time (in seconds) for performing the 31 runs is displayed in each table presenting the IGD values. Binary comparisons are carried out for each constraint-handling technique in order to show the effect of repairing infeasible solutions using the gradient information. Then, statistical tests are performed for each comparison over the 31 runs, according to the Wilcoxon rank-sum test with $p < 0.05$. Significantly better results are represented in boldface. At the bottom of each table, a summary of these statistical tests is displayed, where I, D and S represent the number of instances for which the gradient-based repair process, respectively, achieves statistically inferior (I), equivalent (no significant difference, D) or statistically superior (S) results when compared with the original technique without repair. Finally, the results regarding the feasibility ratio indicator of the last population ($I_F$) (i.e., that at the last generation, not that in the external archive) are presented in Table 6 for EQC test problems. We considered only to present the $I_F$ for these problems since some canonical algorithms have difficulties for obtaining feasible solutions when tackling problems with equality constraints. Note that no statistical test is implemented for this indicator, as a high value of this indicator does not necessarily mean a good approximation to the $\mathcal{PF}$.

## 5.1 CF Test Problems

Tables 1 and 2 highlight that the overall performance of the gradient-based repair is satisfactory for CF test problems, i.e., globally, the coupling of the gradient-based repair method produces at least as good solutions as the canonical method. More precisely, considering the statistical results presented in Tables 1 and 2 as a whole, it can be stated that the use of the gradient-based repair significantly improves the canonical algorithm in 37.5% of the instances, and significantly deteriorates the canonical algorithm only in 5.8% of the cases. Besides, with respect to the computational times, it can be observed from Table 1 that no additional cost is to be paid for using this gradient-based repair, that is, the computation of the pseudoinverse of the gradient that might be a priori time-consuming, is offset by the fact that less generations needed (since the halt condition for the algorithm is the number of function evaluations). It is noteworthy, consequently, that when the repair of infeasible solutions is taken into account, the evolutionary process is somewhat accelerated, in regarding the number of generations needed to converge to the $\mathcal{PF}$.

However, it should be emphasized that most CF functions constitute difficult problems for all the studied techniques, even when the gradient-based repair is included. The cause of these difficulties arises not only from the constraints but also from the objective functions which involve a significant number of locally optimal fronts. To develop this, let us consider CF5 function. Figure 1 illustrates the convergence difficulty observed when trying to approximate the true Pareto front using C-MOEA/D for this problem. Note that all solutions in Figure 1 constitute feasible solutions. Also, Figures 1 and 2 (right) show that the true Pareto front of the constrained problem differs only in its lower part ($\{f_1 \in \mathbb{R} : 0.5 < f_1 \leq 1\}$) from the unconstrained true Pareto front. The similarity between both figures regarding the poor convergence tends to confirm that the lack of convergence observed in Figure 1 may not be attributable to difficulties in fulfilling the constraints, but to difficulties in escaping

**Table 1** IGD values and CPU time (in seconds) obtained on CF test problems

| | | CDP-MOEA/D | | ATP-MOEA/D | | C-MOEA/D | | SR-MOEA/D | | ε-MOEA/D | | Imp ε-MOEA/D | |
|---|---|---|---|---|---|---|---|---|---|---|---|---|---|
| | | — | w/grad | — | w/grad | — | w/grad | — | w/grad | — | w/grad | — | w/grad |
| CF1 | mean | 5.68e−03 | **2.34e−05** | 5.68e−03 | **1.89e−05** | 4.99e−04 | **1.99e−05** | 6.94e−03 | **2.66e−05** | 5.68e−03 | **2.34e−05** | 5.68e−03 | **2.61e−04** |
| | std | 1.39e−03 | 3.12e−05 | 1.89e−03 | 3.39e−05 | 3.68e−04 | 3.35e−05 | 1.52e−03 | 2.77e−05 | 1.39e−03 | 3.12e−05 | 1.39e−03 | 8.31e−05 |
| | CPU time | 4.78e+01 | 4.25e+01 | 5.17e+01 | 3.73e+01 | 4.67e+01 | 4.44e+01 | 4.96e+01 | 3.50e+01 | 4.85e+01 | 4.23e+01 | 4.60e+01 | 3.98e+01 |
| CF2 | mean | 4.10e−03 | 4.53e−03 | 4.24e−03 | 4.14e−03 | 5.71e−03 | **3.88e−03** | 7.17e−03 | 8.93e−03 | 4.10e−03 | 4.53e−03 | 3.76e−03 | 4.37e−03 |
| | std | 2.86e−03 | 3.29e−03 | 2.64e−03 | 2.54e−03 | 3.63e−03 | 2.07e−03 | 1.19e−02 | 1.78e−02 | 2.86e−03 | 3.29e−03 | 2.68e−03 | 2.22e−03 |
| | CPU time | 5.22e+01 | 6.36e+01 | 5.18e+01 | 5.13e+01 | 5.02e+01 | 6.77e+01 | 4.87e+01 | 4.57e+01 | 4.68e+01 | 6.22e+01 | 4.81e+01 | 6.00e+01 |
| CF3 | mean | 2.22e−01 | 2.07e−01 | 2.14e−01 | 2.09e−01 | 2.07e−01 | 2.05e−01 | 2.32e−01 | 2.46e−01 | 2.22e−01 | 2.07e−01 | 2.03e−01 | 2.06e−01 |
| | std | 1.05e−01 | 9.29e−02 | 7.88e−02 | 8.79e−02 | 8.77e−02 | 9.46e−02 | 1.20e−01 | 1.18e−01 | 1.05e−01 | 9.29e−02 | 8.73e−02 | 8.91e−02 |
| | CPU time | 4.45e+01 | 5.97e+01 | 5.11e+01 | 4.45e+01 | 4.84e+01 | 5.96e+01 | 4.88e+01 | 4.31e+01 | 4.84e+01 | 6.46e+01 | 4.54e+01 | 6.44e+01 |
| CF4 | mean | 5.31e−02 | 4.58e−02 | 5.27e−02 | 4.60e−02 | 6.73e−02 | 4.51e−02 | 4.69e−02 | 4.78e−02 | 5.31e−02 | 4.58e−02 | 6.01e−02 | 4.72e−02 |
| | std | 3.22e−02 | 3.18e−02 | 2.83e−02 | 1.60e−02 | 5.39e−02 | 1.96e−02 | 2.78e−02 | 1.99e−02 | 3.22e−02 | 3.18e−02 | 4.27e−02 | 2.79e−02 |
| | CPU time | 4.99e+01 | 5.40e+01 | 4.97e+01 | 4.32e+01 | 4.70e+01 | 5.94e+01 | 4.69e+01 | 3.99e+01 | 4.63e+01 | 5.49e+01 | 4.81e+01 | 5.28e+01 |
| CF5 | Mean | 2.64e−01 | 2.59e−01 | 2.27e−01 | 2.14e−01 | 2.50e−01 | 2.70e−01 | 2.59e−01 | 2.10e−01 | 2.64e−01 | 2.59e−01 | 2.56e−01 | 2.42e−01 |
| | std | 9.77e−02 | 1.23e−01 | 8.54e−02 | 9.86e−02 | 1.12e−01 | 1.17e−01 | 1.11e−01 | 9.16e−02 | 9.77e−02 | 1.23e−01 | 1.02e−01 | 1.23e−01 |
| | CPU time | 4.62e+01 | 6.15e+01 | 5.29e+01 | 4.81e+01 | 4.56e+01 | 5.86e+01 | 4.68e+01 | 3.61e+01 | 4.54e+01 | 5.64e+01 | 4.92e+01 | 5.29e+01 |

(continued)

**Table 1** (continued)

| | | CDP-MOEA/D | | ATP-MOEA/D | | C-MOEA/D | | SR-MOEA/D | | ε-MOEA/D | | Imp ε-MOEA/D | |
|---|---|---|---|---|---|---|---|---|---|---|---|---|---|
| | | — | w/grad | — | w/grad | — | w/grad | — | w/grad | — | w/grad | — | w/grad |
| CF6 | mean | 1.07e-01 | **7.42e-02** | 6.26e-02 | **3.53e-02** | 1.13e-01 | **5.02e-02** | 1.16e-01 | **7.71e-02** | 1.07e-01 | **7.42e-02** | 9.92e-02 | **5.95e-02** |
| | std | 3.39e-02 | 4.20e-02 | 3.99e-02 | 2.97e-02 | 3.68e-02 | 3.21e-02 | 2.64e-02 | 4.60e-02 | 3.39e-02 | 4.20e-02 | 3.78e-02 | 4.85e-02 |
| | CPU time | 4.80e+01 | 6.24e+01 | 5.24e+01 | 4.43e+01 | 5.53e+01 | 6.37e+01 | 5.27e+01 | 4.17e+01 | 5.29e+01 | 6.09e+01 | 5.38e+01 | 5.84e+01 |
| CF7 | mean | 2.49e-01 | 2.38e-01 | 2.24e-01 | 1.99e-01 | 2.51e-01 | **2.00e-01** | 2.80e-01 | **2.08e-01** | 2.49e-01 | 2.38e-01 | 2.68e-01 | **2.28e-01** |
| | std | 1.11e-01 | 1.26e-01 | 9.42e-02 | 8.72e-02 | 9.89e-02 | 9.55e-02 | 1.29e-01 | 8.49e-02 | 1.11e-01 | 1.26e-01 | 1.09e-01 | 1.13e-01 |
| | CPU time | 4.87e+01 | 5.90e+01 | 4.71e+01 | 7.28e+01 | 4.49e+01 | 7.01e+01 | 4.45e+01 | 4.64e+01 | 4.55e+01 | 6.33e+01 | 5.00e+01 | 6.29e+01 |
| CF8 | mean | 5.59e+08 | **3.77e-01** | **1.06e-01** | 1.13e-01 | 5.59e+08 | **1.35e-01** | 4.27e-01 | **3.81e-01** | 6.74e-01 | **3.65e-01** | 4.64e-01 | **2.23e-01** |
| | std | 3.11e+09 | 7.33e-02 | 2.50e-02 | 1.39e-02 | 3.11e+09 | 2.73e-02 | 4.97e-02 | 5.96e-02 | 1.06e+00 | 6.68e-02 | 9.15e-02 | 1.20e-01 |
| | CPU time | 8.38e+01 | 5.56e+01 | 1.17e+02 | 7.28e+01 | 7.66e+01 | 6.23e+01 | 7.40e+01 | 4.98e+01 | 8.44e+01 | 5.42e+01 | 8.54e+01 | 6.32e+01 |
| CF9 | mean | 5.41e-02 | 5.33e-02 | **4.66e-02** | 5.33e-02 | 6.05e-02 | **5.50e-02** | 6.38e-02 | **5.71e-02** | 5.35e-02 | 5.33e-02 | 5.41e-02 | 5.33e-02 |
| | std | 6.26e-03 | 4.85e-03 | 3.78e-03 | 6.13e-03 | 6.20e-03 | 6.38e-03 | 8.89e-03 | 4.58e-03 | 6.29e-03 | 4.85e-03 | 6.26e-03 | 4.85e-03 |
| | CPU time | 1.47e+02 | 9.31e+01 | 1.43e+02 | 9.15e+01 | 1.35e+02 | 8.90e+01 | 1.37e+02 | 8.76e+01 | 1.45e+02 | 9.84e+01 | 1.44e+02 | 1.02e+02 |
| CF10 | mean | 3.71e+09 | **1.33e+09** | 3.04e-01 | 5.00e-01 | 3.71e+09 | **1.33e+09** | 3.85e+09 | 1.99e+09 | 2.12e+09 | **2.94e+00** | 3.18e+09 | 1.19e+09 |
| | std | 1.24e+09 | 1.95e+09 | 1.48e-01 | 8.67e-01 | 1.24e+09 | 1.95e+09 | 1.03e+09 | 2.09e+09 | 2.09e+09 | 1.11e+01 | 1.75e+09 | 1.90e+09 |
| | CPU time | 7.31e+01 | 4.38e+01 | 9.41e+01 | 5.64e+01 | 7.07e+01 | 3.93e+01 | 7.78e+01 | 4.15e+01 | 8.18e+01 | 4.62e+01 | 7.18e+01 | 4.32e+01 |
| Wilc. test (I-D-S) | | 0-6-4 | | 2-6-2 | | 0-3-7 | | 0-5-5 | | 0-6-4 | | 0-6-4 | |

**Table 2** HV values obtained on CF test problems

| | | CDP-MOEA/D | | ATP-MOEA/D | | C-MOEA/D | | SR-MOEA/D | | ε-MOEA/D | | Imp ε-MOEA/D | |
|---|---|---|---|---|---|---|---|---|---|---|---|---|---|
| | | — | w/grad | — | w/grad | — | w/grad | — | w/grad | — | w/grad | — | w/grad |
| CF1 | mean | 6.76e-01 | **6.85e-01** | 6.76e-01 | **6.85e-01** | 6.84e-01 | **6.85e-01** | 6.74e-01 | **6.85e-01** | 6.76e-01 | **6.85e-01** | 6.76e-01 | **6.85e-01** |
| | std | 2.24e-03 | 4.69e-05 | 2.83e-03 | 7.33e-05 | 6.13e-04 | 7.02e-05 | 2.38e-03 | 4.11e-05 | 2.24e-03 | 4.69e-05 | 2.24e-03 | 1.45e-04 |
| CF2 | mean | 8.14e-01 | 8.16e-01 | 8.13e-01 | **8.17e-01** | 8.13e-01 | **8.17e-01** | 8.13e-01 | 8.14e-01 | 8.14e-01 | 8.16e-01 | 8.16e-01 | 8.15e-01 |
| | std | 8.31e-03 | 6.09e-03 | 7.43e-03 | 7.43e-03 | 6.28e-03 | 4.70e-03 | 9.56e-03 | 1.14e-02 | 8.31e-03 | 6.09e-03 | 3.52e-03 | 5.33e-03 |
| CF3 | mean | 2.25e-01 | 2.33e-01 | 2.25e-01 | 2.32e-01 | 2.31e-01 | 2.33e-01 | 2.24e-01 | 2.13e-01 | 2.25e-01 | 2.33e-01 | 2.33e-01 | 2.33e-01 |
| | std | 5.56e-02 | 4.75e-02 | 4.22e-02 | 4.94e-02 | 4.64e-02 | 4.98e-02 | 5.51e-02 | 5.43e-02 | 5.56e-02 | 4.75e-02 | 4.42e-02 | 4.93e-02 |
| CF4 | mean | 6.35e-01 | 6.42e-01 | 6.35e-01 | 6.38e-01 | 6.21e-01 | 6.42e-01 | 6.43e-01 | 6.37e-01 | 6.35e-01 | 6.42e-01 | 6.26e-01 | 6.37e-01 |
| | std | 3.61e-02 | 2.87e-02 | 3.54e-02 | 2.51e-02 | 5.24e-02 | 1.80e-02 | 3.60e-02 | 2.64e-02 | 3.61e-02 | 2.87e-02 | 4.04e-02 | 2.96e-02 |
| CF5 | mean | 4.14e-01 | 4.13e-01 | 4.46e-01 | 4.56e-01 | 4.26e-01 | 4.14e-01 | 4.17e-01 | 4.41e-01 | 4.14e-01 | 4.13e-01 | 4.20e-01 | 4.25e-01 |
| | std | 7.54e-02 | 9.93e-02 | 6.25e-02 | 7.70e-02 | 1.00e-01 | 9.43e-02 | 9.83e-02 | 8.22e-02 | 7.54e-02 | 9.93e-02 | 8.72e-02 | 1.06e-01 |
| CF6 | mean | 7.66e-01 | **7.86e-01** | 7.98e-01 | **8.10e-01** | 7.71e-01 | **8.01e-01** | 7.64e-01 | **7.86e-01** | 7.66e-01 | **7.86e-01** | 7.75e-01 | **7.98e-01** |
| | std | 1.62e-02 | 2.45e-02 | 2.23e-02 | 1.97e-02 | 1.92e-02 | 2.01e-02 | 9.18e-03 | 2.72e-02 | 1.62e-02 | 2.45e-02 | 1.81e-02 | 2.80e-02 |
| CF7 | mean | 5.59e-01 | 5.59e-01 | 5.77e-01 | 5.91e-01 | 5.74e-01 | 5.91e-01 | 5.56e-01 | 5.98e-01 | 5.59e-01 | 5.59e-01 | 5.67e-01 | 5.72e-01 |
| | std | 1.05e-01 | 1.19e-01 | 1.02e-01 | 7.64e-02 | 7.80e-02 | 8.33e-02 | 1.03e-01 | 6.97e-02 | 1.05e-01 | 1.19e-01 | 8.26e-02 | 1.03e-01 |
| CF8 | mean | 3.64e-01 | **4.34e-01** | 5.94e-01 | 5.97e-01 | 4.83e-01 | **5.73e-01** | 3.56e-01 | **4.27e-01** | 3.57e-01 | **4.34e-01** | 3.81e-01 | **5.29e-01** |
| | std | 6.90e-02 | 3.56e-02 | 3.46e-02 | 1.83e-02 | 9.62e-02 | 2.78e-02 | 2.00e-01 | 2.31e-02 | 9.65e-02 | 2.34e-02 | 2.21e-02 | 7.64e-02 |
| CF9 | mean | **6.72e-01** | 6.58e-01 | **6.83e-01** | 6.59e-01 | 6.46e-01 | 6.50e-01 | **6.55e-01** | 6.48e-01 | **6.75e-01** | 6.58e-01 | **6.72e-01** | 6.58e-01 |
| | std | 2.07e-02 | 2.29e-02 | 1.87e-02 | 2.03e-02 | 1.82e-02 | 2.32e-02 | 2.78e-02 | 1.63e-02 | 1.87e-02 | 2.29e-02 | 2.07e-02 | 2.29e-02 |
| CF10 | mean | 2.03e-02 | 3.24e-02 | 2.47e-01 | 2.87e-01 | 8.94e-03 | 0.00e+00 | 1.67e-02 | 1.63e-02 | 9.49e-02 | 8.70e-02 | 3.35e-02 | 3.94e-02 |
| | std | 7.45e-02 | 6.03e-02 | 9.64e-02 | 1.90e-01 | 3.52e-02 | 0.00e+00 | 7.26e-02 | 5.26e-02 | 1.29e-01 | 5.82e-02 | 6.49e-02 | 6.24e-02 |
| Wilc. test (I-D-S) | | 1-6-3 | | 1-6-3 | | 0-6-4 | | 1-6-3 | | 1-6-3 | | 1-6-3 | |

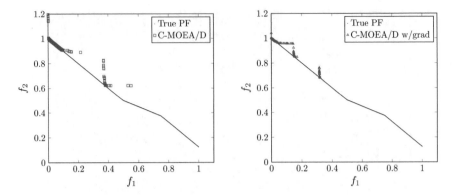

**Fig. 1** Pareto front approximation of the median run considering HV of CF5 function. C-MOEA/D without (left) and with (right) gradient-based repair

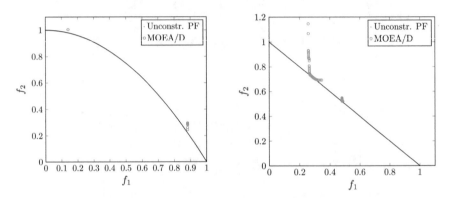

**Fig. 2** Pareto front approximation of the median run considering HV of unconstrained functions CF3 (left) and CF5 (right)

from local optima. This could also explain the high percentage of cases (56.7%) where there is no significant difference while using the repair method. Figure 2 (left) displays the same phenomenon for problem CF3.

## 5.2 LIRCMOP Test Problems

These problems contain large infeasible regions as well as disconnected feasible regions (islands), which means difficulties for the search algorithm, as the search might be stuck in a suboptimal island enclosed by infeasible regions. The results obtained are presented in Tables 3 and 4. For this test suite, it can be appreciated that the use of the gradient-based repair significantly improves the performance of the six constraint-handling techniques.

**Table 3** IGD values and CPU time (in seconds) obtained on LIRCMOP test problems

| | | CDP-MOEA/D | | ATP-MOEA/D | | C-MOEA/D | | SR-MOEA/D | | ε-MOEA/D | | Imp ε-MOEA/D | |
|---|---|---|---|---|---|---|---|---|---|---|---|---|---|
| | | — | w/grad | — | w/grad | — | w/grad | — | w/grad | — | w/grad | — | w/grad |
| LIRCMOP1 | mean | 2.11e-01 | **4.99e-02** | 7.98e+07 | **5.25e-02** | 1.51e-01 | **1.23e-02** | 2.07e-01 | **3.16e-02** | 5.65e-02 | **7.94e-03** | 7.84e-02 | **4.24e-02** |
| | std | 4.86e-02 | 2.15e-02 | 9.05e+07 | 3.47e-02 | 6.23e-02 | 1.12e-02 | 5.36e-02 | 2.53e-02 | 4.52e-02 | 1.76e-03 | 4.69e-02 | 1.79e-02 |
| | CPU time | 6.92e+01 | 4.81e+01 | 7.68e+01 | 4.65e+01 | 7.02e+01 | 5.36e+01 | 7.70e+01 | 5.58e+01 | 7.90e+01 | 5.24e+01 | 7.68e+01 | 5.02e+01 |
| LIRCMOP2 | mean | 1.61e-01 | **8.85e-02** | 1.48e-08 | **3.71e-02** | 1.13e-01 | **3.71e-02** | 1.27e-01 | **5.82e-02** | 5.89e-02 | **1.37e-02** | 7.09e-02 | **2.21e-02** |
| | std | 4.46e-02 | 3.45e-02 | 1.62e-08 | 2.03e-02 | 4.76e-02 | 2.80e-02 | 5.66e-02 | 3.45e-02 | 3.85e-02 | 1.25e-02 | 4.35e-02 | 1.99e-02 |
| | CPU time | 7.65e+01 | 5.29e+01 | 7.49e+01 | 4.62e+01 | 6.89e+01 | 4.53e+01 | 6.65e+01 | 4.93e+01 | 6.88e+01 | 4.48e+01 | 6.85e+01 | 4.80e+01 |
| LIRCMOP3 | mean | 2.08e-01 | **1.03e-02** | 3.38e+09 | **1.90e-01** | 1.95e-01 | **7.08e-03** | 1.83e-01 | **9.70e-03** | 1.34e-01 | **5.41e-03** | 1.66e-01 | **3.47e-03** |
| | std | 3.38e-02 | 9.14e-03 | 2.87e+09 | 1.54e-01 | 4.56e-02 | 8.84e-04 | 5.61e-02 | 1.09e-02 | 7.98e-02 | 2.56e-03 | 6.65e-02 | 5.65e-04 |
| | CPU time | 6.69e+01 | 5.46e+01 | 7.51e+01 | 5.29e+01 | 7.71e+01 | 5.82e+01 | 8.07e+01 | 6.21e+01 | 7.78e+01 | 5.85e+01 | 7.37e+01 | 5.29e+01 |
| LIRCMOP4 | mean | 1.99e-01 | **9.08e-03** | 4.30e+09 | **1.27e-01** | 1.83e-01 | **8.00e-03** | 1.73e-01 | **6.72e-03** | 1.71e-01 | **4.81e-03** | 1.77e-01 | **3.26e-03** |
| | std | 3.16e-02 | 6.78e-03 | 2.83e+09 | 8.10e-02 | 3.74e-02 | 3.30e-03 | 4.49e-02 | 3.83e-03 | 6.13e-02 | 1.79e-03 | 5.05e-02 | 4.73e-04 |
| | CPU time | 7.04e+01 | 5.46e+01 | 7.78e+01 | 5.56e+01 | 6.91e+01 | 5.33e+01 | 6.82e+01 | 5.33e+01 | 7.13e+01 | 5.12e+01 | 6.97e+01 | 6.00e+01 |
| LIRCMOP5 | mean | 1.22e+00 | **3.39e-02** | 7.43e-01 | **2.76e-03** | 1.20e+00 | **5.28e-03** | 1.25e+00 | **2.88e-02** | 1.22e+00 | **3.39e-02** | 1.00e-02 | **2.90e-03** |
| | std | 6.15e-02 | 7.04e-02 | 5.11e-01 | 3.48e-04 | 1.49e-02 | 3.30e-03 | 2.46e-01 | 1.30e-02 | 6.15e-02 | 7.04e-02 | 1.21e-02 | 4.13e-04 |
| | CPU time | 7.97e+01 | 7.59e+01 | 7.90e+01 | 7.71e+01 | 7.28e+01 | 7.31e+01 | 7.94e+01 | 7.63e+01 | 7.77e+01 | 7.94e+01 | 8.04e+01 | 8.31e+01 |
| LIRCMOP6 | mean | 1.35e+00 | **3.04e-02** | 1.35e+00 | **2.88e-03** | 1.35e+00 | **4.30e-03** | 1.35e+00 | **3.75e-02** | 1.35e+00 | **3.04e-02** | 6.93e-03 | **2.90e-03** |
| | std | 1.79e-04 | 6.32e-02 | 2.74e-04 | 2.63e-04 | 1.54e-04 | 1.38e-03 | 2.20e-04 | 4.16e-02 | 1.79e-04 | 6.32e-02 | 1.08e-02 | 3.02e-04 |
| | CPU time | 6.51e+01 | 7.23e+01 | 7.73e+01 | 7.78e+01 | 7.10e+01 | 7.00e+01 | 6.76e+01 | 6.86e+01 | 6.37e+01 | 6.95e+01 | 7.42e+01 | 7.62e+01 |
| LIRCMOP7 | mean | 1.53e+00 | **6.64e-03** | 6.59e-01 | **2.88e-03** | 1.48e+00 | **3.12e-03** | 1.53e+00 | **5.77e-03** | 1.53e+00 | **6.64e-03** | 9.38e-03 | **2.99e-03** |
| | std | 4.76e-01 | 1.26e-02 | 7.73e-01 | 1.53e-03 | 5.37e-01 | 2.86e-03 | 4.75e-01 | 9.84e-03 | 4.76e-01 | 1.26e-02 | 1.06e-02 | 6.12e-04 |
| | CPU time | 7.65e+01 | 8.44e+01 | 8.91e+01 | 8.72e+01 | 8.35e+01 | 8.35e+01 | 7.42e+01 | 7.47e+01 | 6.90e+01 | 7.14e+01 | 6.98e+01 | 6.87e+01 |

(continued)

**Table 3** (continued)

| | | CDP-MOEA/D | | ATP-MOEA/D | | C-MOEA/D | | SR-MOEA/D | | ε-MOEA/D | | Imp ε-MOEA/D | |
|---|---|---|---|---|---|---|---|---|---|---|---|---|---|
| | | — | w/grad | — | w/grad | — | w/grad | — | w/grad | — | w/grad | — | w/grad |
| LIRCMOP8 | mean | 1.56e+00 | **3.19e-03** | 1.26e+00 | **2.47e-03** | 1.64e+00 | **2.51e-03** | 1.63e+00 | **2.48e-03** | 1.56e+00 | **3.19e-03** | 7.80e-03 | **2.67e-03** |
| | std | 3.72e-01 | 2.95e-03 | 6.82e-01 | 1.50e-04 | 2.36e-01 | 2.98e-04 | 2.11e-01 | 2.12e-04 | 3.72e-01 | 2.95e-03 | 3.52e-03 | 1.71e-04 |
| | CPU time | 6.76e+01 | 7.38e+01 | 7.73e+01 | 7.73e+01 | 7.06e+01 | 7.26e+01 | 7.18e+01 | 7.17e+01 | 6.80e+01 | 7.08e+01 | 7.12e+01 | 7.22e+01 |
| LIRCMOP9 | mean | 5.84e-01 | **3.72e-01** | **2.75e-01** | 3.64e-01 | 4.28e-01 | **3.65e-01** | 6.93e-01 | **3.73e-01** | 5.84e-01 | **3.72e-01** | 2.20e-01 | **1.00e-02** |
| | std | 8.51e-02 | 5.83e-03 | 1.17e-01 | 6.77e-02 | 6.08e-02 | 6.41e-02 | 7.93e-02 | 6.63e-03 | 8.51e-02 | 5.83e-03 | 1.14e-01 | 3.32e-03 |
| | CPU time | 7.65e+01 | 7.67e+01 | 8.91e+01 | 8.06e+01 | 7.87e+01 | 7.60e+01 | 8.19e+01 | 6.65e+01 | 6.68e+01 | 6.69e+01 | 6.74e+01 | 6.60e+01 |
| LIRCMOP10 | mean | 5.28e-01 | **2.34e-01** | **4.43e-03** | 6.20e-03 | 1.33e-01 | 2.54e-02 | 4.88e-01 | **2.54e-01** | 5.28e-01 | **2.34e-01** | **3.71e-03** | 5.50e-03 |
| | std | 1.85e-01 | 8.60e-02 | 9.55e-04 | 2.25e-03 | 8.10e-02 | 1.71e-02 | 1.45e-01 | 8.39e-02 | 1.85e-01 | 8.60e-02 | 5.16e-04 | 7.23e-04 |
| | CPU time | 7.30e+01 | 7.15e+01 | 8.90e+01 | 8.03e+01 | 7.43e+01 | 7.31e+01 | 7.34e+01 | 7.39e+01 | 7.02e+01 | 7.18e+01 | 8.66e+01 | 7.79e+01 |
| LIRCMOP11 | mean | 6.12e-01 | **2.54e-01** | 7.06e-03 | 2.27e-02 | 1.43e-01 | **7.97e-02** | 6.28e-01 | **2.43e-01** | 6.12e-01 | **2.54e-01** | **3.33e-03** | 8.18e-03 |
| | std | 1.58e-01 | 1.12e-01 | 2.02e-03 | 2.81e-02 | 4.77e-02 | 4.80e-02 | 1.15e-01 | 8.73e-02 | 1.58e-01 | 1.12e-01 | 1.49e-03 | 2.83e-03 |
| | CPU time | 7.30e+01 | 8.10e+01 | 8.85e+01 | 7.49e+01 | 8.01e+01 | 7.33e+01 | 8.08e+01 | 7.68e+01 | 7.75e+01 | 7.75e+01 | 7.73e+01 | 6.85e+01 |
| LIRCMOP12 | mean | 4.25e-01 | **1.75e-01** | 1.28e-01 | 1.54e-01 | 2.75e-01 | **1.61e-01** | 4.76e-01 | **1.90e-01** | 4.25e-01 | **1.75e-01** | 7.14e-02 | 7.04e-02 |
| | std | 7.50e-02 | 2.48e-02 | 4.90e-02 | 2.02e-02 | 2.81e-02 | 3.35e-02 | 9.25e-02 | 2.84e-02 | 7.50e-02 | 2.48e-02 | 5.84e-02 | 4.68e-02 |
| | CPU time | 8.35e+01 | 8.16e+01 | 9.38e+01 | 7.60e+01 | 7.84e+01 | 7.05e+01 | 7.92e+01 | 7.52e+01 | 8.19e+01 | 8.03e+01 | 7.78e+01 | 6.65e+01 |
| LIRCMOP13 | mean | 1.30e+00 | **6.40e-02** | 8.76e-01 | **6.41e-02** | 1.17e+00 | **6.38e-02** | 1.25e+00 | **6.40e-02** | 1.30e+00 | **6.40e-02** | 6.78e-02 | **6.39e-02** |
| | std | 5.11e-02 | 1.27e-03 | 5.94e-01 | 9.31e-04 | 3.54e-01 | 1.24e-03 | 1.89e-01 | 1.04e-03 | 5.11e-02 | 1.27e-03 | 1.04e-03 | 8.75e-04 |
| | CPU time | 1.13e+02 | 1.90e+02 | 1.47e+02 | 1.93e+02 | 1.43e+02 | 1.92e+02 | 1.43e+02 | 1.94e+02 | 1.11e+02 | 1.89e+02 | 1.70e+02 | 1.98e+02 |
| LIRCMOP14 | mean | 1.25e+00 | **6.45e-02** | 8.43e-01 | **6.42e-02** | 1.13e+00 | **6.39e-02** | 1.21e+00 | **6.42e-02** | 1.25e+00 | **6.45e-02** | 6.51e-02 | **6.41e-02** |
| | std | 4.68e-02 | 1.20e-03 | 5.79e-01 | 1.42e-03 | 3.47e-01 | 1.11e-03 | 1.85e-01 | 1.50e-03 | 4.68e-02 | 1.20e-03 | 8.47e-04 | 1.15e-03 |
| | CPU time | 1.20e+02 | 1.91e+02 | 1.52e+02 | 1.82e+02 | 1.40e+02 | 1.75e+02 | 1.38e+02 | 1.61e+02 | 9.84e+01 | 1.57e+02 | 1.53e+02 | 1.54e+02 |
| Wilc. test (I-D-S) | | 0-0-14 | | 2-2-10 | | 0-0-14 | | 0-0-14 | | 0-0-14 | | 2-1-11 | |

**Table 4** HV values obtained on LIRCMOP test problems

| | | CDP-MOEA/D | | ATP-MOEA/D | | C-MOEA/D | | SR-MOEA/D | | ε-MOEA/D | | Imp ε-MOEA/D | |
|---|---|---|---|---|---|---|---|---|---|---|---|---|---|
| | | — | w/grad | — | w/grad | — | w/grad | — | w/grad | — | w/grad | — | w/grad |
| LIRCMOP1 | mean | 3.16e-01 | **4.65e-01** | 6.14e-02 | **4.72e-01** | 3.71e-01 | **5.20e-01** | 3.19e-01 | **4.89e-01** | 4.66e-01 | **5.30e-01** | 4.43e-01 | **4.81e-01** |
| | std | 4.27e-02 | 2.77e-02 | 8.95e-02 | 3.81e-02 | 6.05e-02 | 2.02e-02 | 4.80e-02 | 3.43e-02 | 5.35e-02 | 4.81e-03 | 5.13e-02 | 2.36e-02 |
| LIRCMOP2 | mean | 6.51e-01 | **7.37e-01** | 1.11e-01 | **8.23e-01** | 7.24e-01 | **8.16e-01** | 6.96e-01 | **7.78e-01** | 7.89e-01 | **8.55e-01** | 7.75e-01 | **8.42e-01** |
| | std | 5.71e-02 | 5.01e-02 | 1.95e-01 | 2.22e-02 | 6.31e-02 | 4.51e-02 | 7.36e-02 | 5.25e-02 | 5.18e-02 | 1.59e-02 | 5.81e-02 | 2.54e-02 |
| LIRCMOP3 | mean | 3.23e-01 | **5.06e-01** | 2.83e-02 | **3.42e-01** | 3.34e-01 | **5.12e-01** | 3.43e-01 | **5.07e-01** | 3.87e-01 | **5.15e-01** | 3.58e-01 | **5.19e-01** |
| | std | 2.97e-02 | 1.33e-02 | 7.50e-02 | 8.62e-02 | 3.90e-02 | 2.06e-03 | 5.13e-02 | 1.54e-02 | 7.26e-02 | 2.65e-03 | 5.84e-02 | 1.19e-03 |
| LIRCMOP4 | mean | 5.87e-01 | **8.11e-01** | 0.00e+00 | **6.59e-01** | 6.06e-01 | **8.13e-01** | 6.22e-01 | **8.15e-01** | 6.24e-01 | **8.17e-01** | 6.16e-01 | **8.20e-01** |
| | std | 3.66e-02 | 6.56e-03 | 0.00e+00 | 8.34e-02 | 4.67e-02 | 3.48e-03 | 5.44e-02 | 3.82e-03 | 6.98e-02 | 2.06e-03 | 5.60e-02 | 1.29e-03 |
| LIRCMOP5 | mean | 0.00e+00 | **8.12e-01** | 2.88e-01 | **8.71e-01** | 0.00e+00 | **8.67e-01** | 0.00e+00 | **8.28e-01** | 0.00e+00 | **8.12e-01** | 8.58e-01 | **8.71e-01** |
| | std | 0.00e+00 | 1.14e-01 | 3.30e-01 | 6.08e-04 | 0.00e+00 | 6.07e-03 | 0.00e+00 | 2.41e-02 | 0.00e+00 | 1.14e-01 | 2.52e-02 | 6.52e-04 |
| LIRCMOP6 | mean | 0.00e+00 | **5.13e-01** | 0.00e+00 | **5.38e-01** | 0.00e+00 | **5.35e-01** | 0.00e+00 | **4.97e-01** | 0.00e+00 | **5.13e-01** | 5.33e-01 | **5.38e-01** |
| | std | 0.00e+00 | 5.60e-02 | 0.00e+00 | 5.23e-04 | 0.00e+00 | 2.41e-03 | 0.00e+00 | 2.97e-02 | 0.00e+00 | 5.60e-02 | 1.08e-02 | 6.19e-04 |
| LIRCMOP7 | mean | 5.49e-02 | **6.22e-01** | 3.55e-01 | **6.26e-01** | 7.18e-02 | **6.26e-01** | 5.30e-02 | **6.23e-01** | 5.49e-02 | **6.22e-01** | 6.18e-01 | **6.26e-01** |
| | std | 1.71e-01 | 1.17e-02 | 2.71e-01 | 2.28e-03 | 1.91e-01 | 2.27e-03 | 1.65e-01 | 1.05e-02 | 1.71e-01 | 1.17e-02 | 9.79e-03 | 1.26e-03 |
| LIRCMOP8 | mean | 3.97e-02 | **6.26e-01** | 1.47e-01 | **6.27e-01** | 1.54e-02 | **6.27e-01** | 1.64e-02 | **6.27e-01** | 3.97e-02 | **6.26e-01** | 6.19e-01 | **6.27e-01** |
| | std | 1.26e-01 | 4.15e-03 | 2.36e-01 | 5.98e-04 | 8.59e-02 | 6.22e-04 | 6.93e-02 | 5.85e-04 | 1.26e-01 | 4.15e-03 | 3.38e-03 | 3.65e-04 |
| LIRCMOP9 | mean | 3.79e-01 | **5.20e-01** | 5.77e-01 | 5.26e-01 | 4.89e-01 | **5.17e-01** | 2.80e-01 | **5.17e-01** | 3.79e-01 | **5.20e-01** | 6.09e-01 | **6.78e-01** |
| | std | 7.55e-02 | 1.35e-02 | 4.32e-02 | 3.11e-02 | 4.66e-02 | 3.27e-02 | 6.03e-02 | 1.29e-02 | 7.55e-02 | 1.35e-02 | 3.65e-02 | 3.50e-03 |
| LIRCMOP10 | mean | 4.55e-01 | **8.46e-01** | 8.55e-01 | **9.49e-01** | 7.74e-01 | **9.39e-01** | 4.97e-01 | **8.35e-01** | 4.55e-01 | **8.46e-01** | 8.56e-01 | **9.49e-01** |
| | std | 2.02e-01 | 4.36e-02 | 8.66e-04 | 1.30e-03 | 4.97e-02 | 5.74e-03 | 1.64e-01 | 4.29e-02 | 2.02e-01 | 4.36e-02 | 4.90e-04 | 4.45e-04 |
| LIRCMOP11 | mean | 3.62e-01 | **6.43e-01** | 8.36e-01 | 8.26e-01 | 7.30e-01 | **7.84e-01** | 3.31e-01 | **6.47e-01** | 3.62e-01 | **6.43e-01** | 8.39e-01 | 8.35e-01 |
| | std | 1.18e-01 | 9.54e-02 | 1.61e-03 | 1.61e-02 | 4.06e-02 | 3.94e-02 | 8.64e-02 | 6.82e-02 | 1.18e-01 | 9.54e-02 | 3.20e-04 | 2.09e-03 |
| LIRCMOP12 | mean | 5.24e-01 | **6.55e-01** | 6.81e-01 | 6.69e-01 | 6.21e-01 | **6.64e-01** | 4.84e-01 | **6.46e-01** | 5.24e-01 | **6.55e-01** | 7.15e-01 | 7.11e-01 |
| | std | 4.85e-02 | 1.25e-02 | 2.86e-02 | 1.24e-02 | 6.78e-03 | 2.00e-02 | 5.74e-02 | 1.17e-02 | 4.85e-02 | 1.25e-02 | 3.11e-02 | 2.58e-02 |
| LIRCMOP13 | mean | 2.65e-03 | **7.49e-01** | 2.49e-01 | **7.49e-01** | 6.64e-02 | **7.49e-01** | 1.70e-02 | **7.50e-01** | 2.65e-03 | **7.49e-01** | 7.37e-01 | **7.50e-01** |
| | std | 1.29e-02 | 2.41e-03 | 3.42e-01 | 2.41e-03 | 1.80e-01 | 2.05e-03 | 7.11e-02 | 1.65e-03 | 1.29e-02 | 2.41e-03 | 2.39e-03 | 2.02e-03 |
| LIRCMOP14 | mean | 3.97e-03 | **7.55e-01** | 2.61e-01 | **7.55e-01** | 7.28e-02 | **7.56e-01** | 2.10e-02 | **7.55e-01** | 3.97e-03 | **7.55e-01** | 7.50e-01 | **7.55e-01** |
| | std | 1.47e-02 | 1.52e-02 | 3.57e-01 | 1.96e-03 | 1.93e-01 | 2.24e-03 | 7.96e-02 | 1.78e-03 | 1.47e-01 | 1.52e-03 | 1.93e-03 | 1.78e-03 |
| Wilc. test (1-D-S) | | 0-0-14 | | 1-2-11 | | 0-0-14 | | 0-0-14 | | 0-0-14 | | 1-1-12 | |

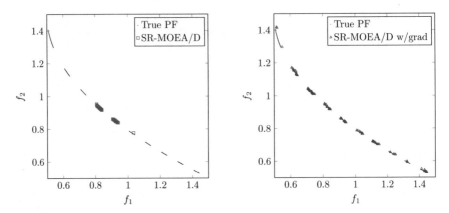

**Fig. 3** Pareto front approximation of the median run considering HV of LIRCMOP4 function. SR-MOEA/D without (left) and with (right) gradient-based repair

Problems LIRCMOP1-4 present large infeasible regions and their $\mathcal{PF}$ are situated on their constraint boundaries. For these problems, each algorithm in their canonical form is able to find at least a part of the true PFs, however, because of the constrained search spaces, they are not capable of identifying the whole $\mathcal{PF}$. That is, the search is stressed in some parts of the Pareto front, in particular those that are found first and which are difficult to move away from. In addition, the incorporation of the gradient-based repair encourages the algorithm to continue searching promising regions even once some Pareto-optimal solutions have been found. Because the true Pareto fronts are surrounded by infeasible regions, new solutions proposed from Pareto-optimal solutions found so far are likely to be slightly infeasible, and thus easily repaired using the constraints' gradient information. Figure 3 illustrates this phenomenon for LIRCMOP4 function. It can be observed from Tables 3 and 4, that the repair process significantly improves every canonical algorithm for all the problems under study.

Problems LIRCMOP5-8 have infeasible regions that may be difficult to cross when approaching the true PFs. These problems are particularly difficult for all the constraint-handling techniques studied here (with the exception of the improved $\varepsilon$-constrained), none of them is able to reach the $\mathcal{PF}$, as they get stuck in subopti-mal fronts which are bounded by constraints. When gradient-based repair is used, every algorithm, however, approaches the true PFs efficiently. Figures 4 and 5 show clearly how the gradient-based repair permits the individuals to pass over the infea-sible region to attain the true PFs. Note that the regions dividing the true PFs and the obtained approximations (left Figures 4 and 5), constitute infeasible regions. Similar to problems LICRMOP1-4, the gradient-based repair significantly improves every canonical algorithm for all these problems. Furthermore, even the improved $\varepsilon$-constrained method, that has been designed to be well-suited for this kind of prob-lems, exhibits a significant enhancement with the use of this repair.

Problems LIRCMOP9-12 contain, in addition to large infeasible regions, con-straints that divide the PFs into a number of disconnected segments. Once more,

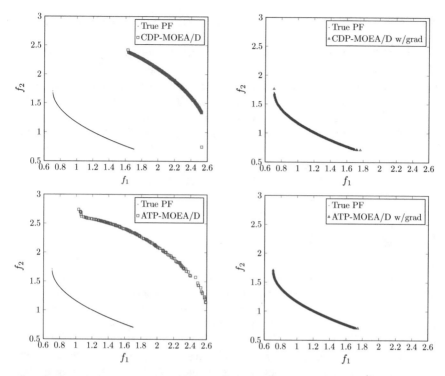

**Fig. 4** Pareto front approximation of the median run considering HV of LIRCMOP5 function. CDP-MOEA/D and ATP-MOEA/D without (left) and with (right) gradient-based repair

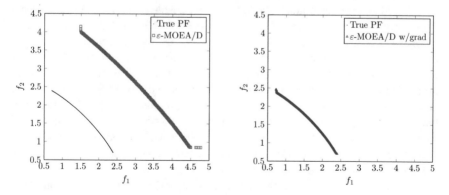

**Fig. 5** Pareto front approximation of the median run considering HV of LIRCMOP7 function. $\varepsilon$-MOEA/D without (left) and with (right) gradient-based repair

the results show that the repair of constraints presents overall good performance for these problems.

LIRCMOP13-14 functions consist of three-objective problems, where the PFs are situated in the boundaries of their constraints. The numerical results show that each algorithm embedded with the gradient-based repair obtained a good approximation of the true PFs, and in fact, virtually the highest possible HV values (for LIRCMOP13, 14: $\approx 7.50e - 01$ and $\approx 7.55e - 01$, respectively) are obtained for every algorithm in every run (according to the diversity criterion used: crowding distance). It must be highlighted that even the methods with a bad performing canonical form (i.e., ATP, SR, CDP), achieve an excellent performance (better than that of canonical improved $\varepsilon$-method) once the repair of infeasible solutions is carried out.

### 5.3  EQC Test Problems

The results obtained for EQC problems are presented in Tables 5 and 6. For Problem EQC6 in Table 5, the "—" means no information is available. The results according to the IGD and HV metrics show that the implementation of the gradient-based repair significantly outperforms the canonical constraint-handling methods for all problems. Besides, the repair of infeasible solutions for problems with equality constraints seems to be necessary in order to obtain acceptable quality solutions, or even more, feasible solutions. For these problems, the mean feasibility ratio of the final population over the 31 runs is also presented in Table 6.

Problem EQC1 [29] has only two equality constraints, however, none of the six constraint-handling techniques studied here is capable of approaching the whole Pareto front. For these canonical techniques, it seems that the exploration is stopped to some extent once a feasible non-dominated solution is found, due to their inability of escaping from a subregion enclosed by equality constraints. In contrast, if the repair of infeasible solutions is carried out, the exploration is pursued all along the evolutionary process and in this way, each one of the six techniques approximates the Pareto front obtaining the maximum HV value ($\approx 9.09e - 01$). It must be emphasized that even the ATP method, which was unable to find a single feasible solution in its canonical form, now converges to the true PF. Figure 6 plots the Pareto front of the median run according to the HV indicator for CDP and $\varepsilon$-constrained methods, without and with the implementation of the gradient-based repair.

Problems EQC2-3 [30] are modifications of the ZDT1 problem containing only one equality (quadratic) constraint, which only considers two variables out of the 30 decision variables. Problem EQC2 is solved by every algorithm when considering the repair of infeasible solutions, obtaining a converged well-distributed approximation of the $\mathcal{PF}$. According to the numerical results of the IGD/HV indicators, the use of the constraints' gradient information is much more important than the choice of the canonical algorithm used. Concerning the problem EQC3, it constitutes a more difficult problem as the true Pareto front is disconnected. Even though the gradient-

**Table 5** IGD values and CPU time (in seconds) obtained on EQC test problems

| | | CDP-MOEA/D | | ATP-MOEA/D | | C-MOEA/D | | SR-MOEA/D | | ε-MOEA/D | | Imp ε-MOEA/D | |
|---|---|---|---|---|---|---|---|---|---|---|---|---|---|
| | | — | w/grad | — | w/grad | — | w/grad | — | w/grad | — | w/grad | — | w/grad |
| EQC1 | mean | 4.80e+00 | **4.29e-02** | 9.56e+09 | **4.09e-02** | 2.63e+00 | **4.16e-02** | 4.19e+00 | **4.30e-02** | 1.66e+00 | **4.29e-02** | 3.08e+00 | **4.87e-02** |
| | std | 2.03e+00 | 2.52e-03 | 3.76e+09 | 3.73e-03 | 9.19e-01 | 2.91e-03 | 1.63e+00 | 3.50e-03 | 1.18e+00 | 2.52e-03 | 1.34e+00 | 7.16e-03 |
| | CPU time | 6.38e+01 | 4.71e+01 | 6.78e+01 | 4.27e+01 | 5.55e+01 | 4.25e+01 | 5.69e+01 | 4.30e+01 | 6.98e+01 | 4.22e+01 | 5.88e+01 | 3.75e+01 |
| EQC2 | mean | 7.24e-01 | **6.34e-03** | 2.27e-02 | **6.38e-03** | 6.25e-01 | **6.34e-03** | 6.44e-01 | **6.48e-03** | 5.81e-01 | **6.34e-03** | 1.16e-01 | **6.33e-03** |
| | std | 6.43e-01 | 3.98e-04 | 7.65e-03 | 3.99e-04 | 2.26e-01 | 4.69e-04 | 4.58e-01 | 4.59e-04 | 1.34e-01 | 3.98e-04 | 2.45e-02 | 4.48e-04 |
| | CPU time | 5.81e+01 | 3.88e+01 | 6.05e+01 | 4.04e+01 | 5.34e+01 | 3.56e+01 | 5.40e+01 | 3.45e+01 | 5.14e+01 | 3.73e+01 | 5.58e+01 | 4.14e+01 |
| EQC3 | mean | 5.87e-01 | 2.84e-01 | 8.00e-02 | **9.29e-02** | 4.58e-01 | **2.99e-01** | **4.46e-01** | 3.38e-01 | 6.61e-01 | **2.84e-01** | 1.36e-01 | **9.58e-02** |
| | std | 9.18e-01 | 1.39e-01 | 1.70e-02 | 8.82e-02 | 3.26e-01 | 1.23e-01 | 5.82e-01 | 7.80e-02 | 1.69e-01 | 1.39e-01 | 3.23e-02 | 1.04e-01 |
| | CPU time | 6.07e+01 | 4.47e+01 | 6.49e+01 | 4.09e+01 | 5.53e+01 | 3.98e+01 | 5.55e+01 | 4.03e+01 | 5.46e+01 | 3.98e+01 | 5.34e+01 | 2.99e+01 |
| EQC4 | mean | 2.68e-01 | **1.17e-01** | 5.22e+05 | **1.22e-01** | 2.36e-01 | **1.16e-01** | 2.85e-01 | **1.17e-01** | 2.25e-01 | **1.17e-01** | 2.25e-01 | **1.18e-01** |
| | std | 4.45e-02 | 3.33e-03 | 2.05e+06 | 4.11e-03 | 3.71e-02 | 3.61e-03 | 3.82e-02 | 3.44e-03 | 3.60e-02 | 3.94e-03 | 3.68e-02 | 4.08e-03 |
| | CPU time | 1.99e+02 | 1.19e+02 | 8.17e+01 | 1.55e+02 | 1.45e+02 | 1.79e+02 | 1.46e+02 | 1.26e+02 | 1.55e+02 | 1.37e+02 | 1.51e+02 | 1.21e+02 |
| EQC5 | mean | 1.06e+00 | **7.86e-03** | 1.87e+08 | **1.02e-02** | 1.03e+00 | **8.15e-03** | 2.03e+06 | **7.93e-03** | 1.11e+00 | **7.94e-03** | 1.05e+00 | **7.88e-03** |
| | std | 7.24e-01 | 2.62e-04 | 1.25e+08 | 1.22e-03 | 6.76e-01 | 2.77e-04 | 4.76e+06 | 2.08e-04 | 6.77e-01 | 2.67e-04 | 7.02e-01 | 2.41e-04 |
| | CPU time | 6.46e+01 | 3.60e+01 | 3.00e+01 | 3.82e+01 | 3.36e+01 | 3.65e+01 | 5.84e+01 | 3.93e+01 | 6.93e+01 | 3.73e+01 | 6.39e+01 | 3.55e+01 |
| EQC6 | mean | — | — | — | — | — | — | — | — | — | — | — | — |
| | std | — | — | — | — | — | — | — | — | — | — | — | — |
| | CPU time | 4.55e+01 | 3.89e+01 | 5.00e+01 | 3.61e+01 | 4.27e+01 | 3.57e+01 | 4.28e+01 | 3.92e+01 | 4.57e+01 | 3.68e+01 | 4.46e+01 | 3.61e+01 |
| Wilc. test (I-D-S) | | 0-1-4 | | 0-0-5 | | 0-0-5 | | 1-0-4 | | 0-0-5 | | 0-0-5 | |

**Table 6** HV values and $I_F$ mean values obtained on EQC test problems

| | | CDP-MOEA/D | | ATP-MOEA/D | | C-MOEA/D | | SR-MOEA/D | | ε-MOEA/D | | Imp ε-MOEA/D | |
|---|---|---|---|---|---|---|---|---|---|---|---|---|---|
| | | — | w/grad | — | w/grad | — | w/grad | — | w/grad | — | w/grad | — | w/grad |
| EQC1 | mean | 3.41e-01 | **9.09e-01** | 0.00e+00 | **9.09e-01** | 5.04e-01 | **9.09e-01** | 3.65e-01 | **9.09e-01** | 7.65e-01 | **9.09e-01** | 6.32e-01 | **9.08e-01** |
| | std | 2.21e-01 | 1.17e-04 | 0.00e+00 | 2.20e-04 | 1.53e-01 | 1.60e-04 | 2.15e-01 | 1.28e-04 | 7.73e-02 | 1.17e-04 | 1.04e-01 | 5.77e-04 |
| | $I_F$ | 1.0000 | 1.0000 | 0.0000 | 0.9242 | 0.0865 | 0.9497 | 0.9687 | 1.0000 | 1.0000 | 1.0000 | 1.0000 | 0.4539 |
| EQC2 | mean | 1.68e-01 | **9.08e-01** | 8.97e-01 | **9.08e-01** | 2.08e-01 | **9.08e-01** | 1.71e-01 | **9.07e-01** | 3.40e-01 | **9.08e-01** | 7.32e-01 | **9.08e-01** |
| | std | 1.06e-01 | 5.46e-04 | 3.58e-03 | 6.96e-04 | 1.50e-01 | 6.25e-04 | 1.19e-01 | 6.62e-04 | 1.41e-01 | 5.46e-04 | 4.89e-02 | 5.62e-04 |
| | $I_F$ | 1.0000 | 1.0000 | 0.0535 | 1.0000 | 0.4297 | 1.0000 | 0.9994 | 1.0000 | 1.0000 | 1.0000 | 1.0000 | 1.0000 |
| EQC3 | mean | 2.25e-01 | **6.71e-01** | 7.51e-01 | **7.57e-01** | 3.54e-01 | **6.67e-01** | 2.37e-01 | **6.49e-01** | 1.81e-01 | **6.71e-01** | 6.93e-01 | **7.43e-01** |
| | std | 1.25e-01 | 5.54e-02 | 7.55e-03 | 3.84e-02 | 2.64e-01 | 5.10e-02 | 1.12e-01 | 3.21e-02 | 1.20e-01 | 5.54e-02 | 8.30e-02 | 4.03e-02 |
| | $I_F$ | 1.0000 | 1.0000 | 0.0119 | 0.8035 | 0.4926 | 0.9552 | 0.9984 | 1.0000 | 1.0000 | 1.0000 | 1.0000 | 0.4410 |
| EQC4 | mean | 7.71e-01 | **8.30e-01** | 3.76e-01 | **8.30e-01** | 7.84e-01 | **8.30e-01** | 7.65e-01 | **8.30e-01** | 7.95e-01 | **8.30e-01** | 7.95e-01 | **8.30e-01** |
| | std | 2.12e-02 | 4.16e-04 | 1.47e-01 | 5.99e-04 | 1.18e-02 | 7.48e-04 | 1.71e-02 | 5.81e-04 | 9.65e-03 | 6.67e-04 | 7.95e-03 | 4.67e-04 |
| | $I_F$ | 1.0000 | 1.0000 | 0.0013 | 0.8053 | 0.2989 | 0.9931 | 0.3395 | 1.0000 | 1.0000 | 1.0000 | 1.0000 | 0.9725 |
| EQC5 | mean | 5.75e-01 | **7.77e-01** | 0.00e+00 | **7.77e-01** | 5.91e-01 | **7.77e-01** | 3.12e-01 | **7.77e-01** | 5.78e-01 | **7.77e-01** | 5.88e-01 | **7.77e-01** |
| | std | 1.13e-01 | 4.86e-05 | 0.00e+00 | 2.79e-04 | 1.03e-01 | 3.43e-05 | 1.99e-01 | 3.51e-05 | 8.43e-02 | 4.41e-05 | 8.89e-02 | 4.79e-05 |
| | $I_F$ | 1.0000 | 1.0000 | 0.0000 | 0.2027 | 0.1068 | 0.3392 | 0.1591 | 0.9920 | 1.0000 | 1.0000 | 1.0000 | 1.0000 |
| EQC6 | mean | 0.00e+00 | **1.04e+00** | 0.00e+00 | **1.04e+00** | 0.00e+00 | **1.04e+00** | 0.00e+00 | **1.04e+00** | 0.00e+00 | **1.04e+00** | 0.00e+00 | **7.62e-01** |
| | std | 0.00e+00 | 6.23e-04 | 0.00e+00 | 1.12e-03 | 0.00e+00 | 4.48e-04 | 0.00e+00 | 2.79e-04 | 0.00e+00 | 1.50e-04 | 0.00e+00 | 2.88e-01 |
| | $I_F$ | 0.0000 | 1.0000 | 0.0000 | 0.5568 | 0.0000 | 0.5623 | 0.0000 | 0.9990 | 0.0000 | 1.0000 | 0.0000 | 0.1613 |
| Wilc. test (I-D-S) | | 0-0-6 | | 0-0-6 | | 0-0-6 | | 0-0-6 | | 0-0-6 | | 0-0-6 | |

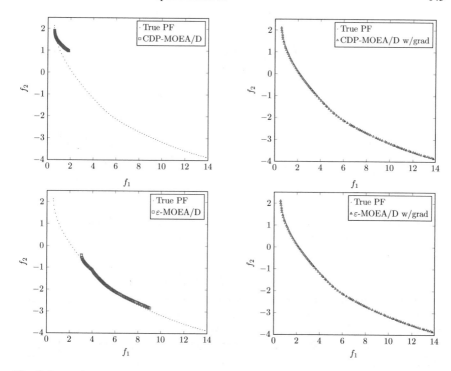

**Fig. 6** Pareto front approximation of the median run considering HV of EQC1 function. CDP-MOEA/D and $\varepsilon$-MOEA/D without (left) and with (right) gradient-based repair

based repair method improves the performance of the algorithms, the disconnected segment of the $\mathcal{PF}$ is difficult to be obtained in all cases.

Problems EQC4-5 [30] involve three-objective functions, with only three decision variables, and one and two equality constraints, respectively. For these problems, the same issue observed for problems EQC1-3 concerning diversity, is presented. As above stated, the canonical constraint-handling techniques lack the capacity of continuing exploring other regions different from those found first corresponding to a specific part of the true PF. They get bounded in equality constrained-search space regions. This phenomenon is displayed in Figure 7 (left) for problem EQC5. The gradient-based repair of infeasible solutions improves the canonical methods so that newly proposed individuals may survive and explore in a smooth way these equality-constrained search spaces, with the possibility of visiting promising regions, as observed in Figure 7 (right) for problem EQC5. The numerical results show that every algorithm, when embedded with the gradient-based repair, achieves the best possible approximation of the true PF, for the population size and diversity criterion used. It must be highlighted that even the methods that exhibited a poor performance in their canonical form, like ATP or SR, have now an excellent performance once the repair process is carried out.

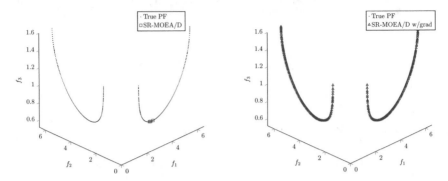

**Fig. 7** Pareto front approximation of the median run considering HV of EQC5 function. SR-MOEA/D without (left) and with (right) gradient-based repair

Problem EQC6 [31, 32] is the so-called Williams–Otto process optimization problem in chemical engineering. It involves ten decision variables and six equality constraints, and thus can be considered as the most difficult problem treated here (with respect to constraint satisfaction). The formulation considered involves the maximization of two objectives: the net present value (NPV) and the profit before taxes (PBT). It is observed from Table 6 that, none of the canonical constraint-handling method is able to find any single feasible solution in any run, not even the CDP method which priority is to search for feasible solutions all along the evolutionary process. The satisfaction of six equality constraints involving ten decision variables seems to be a very difficult task for these algorithms. But once again, the gradient information of constraints permits to find the Pareto-optimal front. A revisited formulation of this problem was used elsewhere for problem solution [32, 37]: at each evaluation of the objective function, all equality constraints are satisfied by solving a system of non-linear equations. This methodology, although computationally intensive, has proven to be efficient for solving the problem. For comparison purposes, we have implemented this solution strategy and it showed CPU times approximately 30 times higher than those reported with the repair performed here. Besides, that reformulation methodology cannot be applied as a general-purpose method (Fig. 8).

## 6 Conclusions

This work has presented a comparative study evidencing the effect of using the constraints' gradient information to repair infeasible solutions when solving CMOPs. This strategy has been embedded with six constraint-handling schemes classically used within MOEA/D: constraint dominance principle, adaptive threshold penalty function (ATP), C-MOEA/D, stochastic ranking, $\varepsilon$-constrained and improved $\varepsilon$-constrained. The use of the constraints' gradient for repairing infeasible solutions

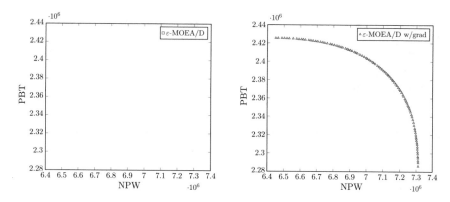

**Fig. 8** Pareto front approximation of the median run considering HV of EQC6 function. $\varepsilon$-MOEA/D without (left; no feasible solution found) and with (right) gradient-based repair

significantly enhances the canonical algorithms under the test problems suites studied (CF, LIRCMOP and EQC). These instances allowed to explore the performance of these algorithms under different landscapes of the search space: functions containing a significant number of local optima (both in the objective space and in the constrained search space), which may be bounded by inequality constrained regions which may produce islands in the search space or disconnected true Pareto fronts enclosed by infeasible regions, but as well equality constrained regions, which yield particular difficulties to reach the whole $\mathcal{PF}$. The obtained results have shown that the gradient information of constraints can make the canonical constraint-handling method much more robust, enabling the population to get across infeasible regions and promoting diversity on the construction of the Pareto front approximation. Besides, the results observed in EQC test problems allow us to conclude that the repair of infeasible individuals is particularly useful for problems with narrow feasible spaces, as those produced by equality constraints, in which canonical methods do not work properly. That is, equality-constrained spaces might be very difficult to reach by simply search operators like DE or SBX, i.e., proposed individuals are likely to be infeasible even if feasible individuals participate to create new offspring, thus the exploration is somewhat stopped in specific parts of the Pareto front (or suboptimal fronts) and the MOEA cannot efficiently evolve because of the lack of new promising solutions, despite the constraint-handling technique used. The repair procedure considered here repairs those promising infeasible solutions so that the multiobjective algorithm can continue constructing a well-distributed approximation of the Pareto front. Moreover, when the computation of the gradient is performed the computational time does not significantly increase compared to that of canonical algorithms, at least for problems containing up to 30 decision variables.

The approach will now be extended to larger size instances where equality and inequality constraints may be present simultaneously, so that limited cases (if any), for the use of the gradient-based repair may be identified. Also, to complement the present study, the gradient's repair within a dominance-based or indicator-based MOEA will also be explored. We think this work may encourage the research on the use of mathematical properties that can be exploited in order to enhance the population and guide it towards promising regions. Finally, we consider that this study could extend the field of applicability of multiobjective evolutionary algorithms to a wider range of real-world optimization problems.

**Acknowledgements** The first author gratefully thanks the Mexican Council of Science and Technology (CONACyT) for scholarship support to pursue PhD studies.

# References

1. Zitzler, E., Laumanns, M., Thiele, L.: TIK-report **103**, (2001)
2. Deb, K., Pratap, A., Agarwal, S., Meyarivan, T.: IEEE Transactions on Evolutionary Computation **6**(2), 182 (2002). https://doi.org/10.1109/4235.996017
3. Zhang, Q., Li, H.: IEEE Transactions on Evolutionary Computation **11**(6), 712 (2007). https://doi.org/10.1109/TEVC.2007.892759
4. Beume, N., Naujoks, B., Emmerich, M.: European Journal of Operational Research **181**(3), 1653 (2007). https://doi.org/10.1016/j.ejor.2006.08.008
5. M.A. Jan, Q. Zhang, in *2010 UK Workshop on computational intelligence (UKCI)* (IEEE, 2010), pp. 1–6
6. S. Zapotecas-Martinez, C.A. Coello Coello, in *2014 IEEE Congress on evolutionary computation (CEC)* (IEEE, 2014), pp. 429–436
7. Chootinan, P., Chen, A.: Computers & operations research **33**(8), 2263 (2006)
8. T. Takahama, S. Sakai, in *2006 IEEE International Conference on Evolutionary Computation* (IEEE, 2006), pp. 1–8
9. T. Takahama, S. Sakai, in *Evolutionary Computation (CEC), 2010 IEEE Congress on* (IEEE, 2010), pp. 1–9
10. S. Sardar, S. Maity, S. Das, P.N. Suganthan, in *2011 IEEE Symposium on Differential Evolution (SDE)* (IEEE, 2011), pp. 1–8
11. Q. Zhang, A. Zhou, S. Zhao, P.N. Suganthan, W. Liu, S. Tiwari, Multiobjective optimization test instances for the CEC 2009 special session and competition. Tech. rep. (2008)
12. Fan, Z., Li, W., Cai, X., Huang, H., Fang, Y., You, Y., Mo, J., Wei, C., Goodman, E.: Soft Computing **23**(23), 12491 (2019)
13. Q. Zhang, W. Liu, H. Li, in *2009 IEEE congress on evolutionary computation* (IEEE, 2009), pp. 203–208
14. K. Miettinen, *Nonlinear multiobjective optimization*, vol. 12 (Springer Science & Business Media, 2012)
15. M. Pescador-Rojas, R. Hernández Gómez, E. Montero, N. Rojas-Morales, M.C. Riff, C.A. Coello Coello, in *International Conference on Evolutionary Multi-Criterion Optimization* (Springer, 2017), pp. 499–513
16. H.K. Singh, A. Isaacs, T.T. Nguyen, T. Ray, X. Yao, in *2009 IEEE congress on evolutionary computation* (IEEE, 2009), pp. 3127–3134
17. Woldesenbet, Y.G., Yen, G.G., Tessema, B.G.: IEEE Transactions on Evolutionary Computation **13**(3), 514 (2009)
18. Jan, M.A., Khanum, R.A.: Applied Soft Computing **13**(1), 128 (2013)
19. M. Asafuddoula, T. Ray, R. Sarker, K. Alam, in *2012 IEEE Congress on Evolutionary Computation* (IEEE, 2012), pp. 1–8
20. Takahama, T.: S. Sakai. In: Abraham, A., Dote, Y., Furuhashi, T., Köppen, M., Ohuchi, A., Ohsawa, Y. (eds.) Soft Computing as Transdisciplinary Science and Technology, pp. 1019–1029. Springer, Berlin Heidelberg, Berlin, Heidelberg (2005)

21. Fan, Z., Li, W., Cai, X., Li, H., Wei, C., Zhang, Q., Deb, K., Goodman, E.: Swarm and evolutionary computation **44**, 665 (2019)
22. Fan, Z., Fang, Y., Li, W., Cai, X., Wei, C., Goodman, E.: Applied Soft Computing **74**, 621 (2019)
23. Peng, C., Liu, H.L., Gu, F.: Applied Soft Computing **60**, 613 (2017)
24. Yang, Y., Liu, J., Tan, S.: Applied Soft Computing **89**, 106104 (2020). https://doi.org/10.1016/j.asoc.2020.106104
25. Ishibuchi, H., Fukase, T., Masuyama, N., Nojima, Y.: Proceedings of the Genetic and Evolutionary Computation Conference 665–672 (2018)
26. Deb, K.: Computer Methods in Applied Mechanics and Engineering **186**(2–4), 311 (2000)
27. Runarsson, T.P., Yao, X.: IEEE Transactions on Evolutionary Computation **4**(3), 284 (2000)
28. S.L. Campbell, C.D. Meyer, *Generalized inverses of linear transformations* (SIAM, 2009)
29. Das, I., Dennis, J.E.: SIAM journal on optimization **8**(3), 631 (1998)
30. Cuate, O., Ponsich, A., Uribe, L., Zapotecas-Martínez, S., Lara, A., Schütze, O.: Mathematics **8**(1), 7 (2020)
31. Pintaric, Z.N., Kravanja, Z.: Industrial & engineering chemistry research **45**(12), 4222 (2006)
32. Rangaiah, G.P.: Multi-objective optimization: techniques and applications in chemical engineering, vol. 1. World Scientific (2009)
33. Z. Fana, Z. Wanga, W. Lia, Y. Yuana, Y. Youa, Z. Yanga, F. Suna, J. Ruana, Swarm and Evolutionary Computation (2020)
34. Bosman, P.A., Thierens, D.: IEEE transactions on evolutionary computation **7**(2), 174 (2003)
35. E. Zitzler, L. Thiele, in *International conference on parallel problem solving from nature* (Springer, 1998), pp. 292–301
36. E. Zitzler, D. Brockhoff, L. Thiele, in *International Conference on Evolutionary Multi-Criterion Optimization* (Springer, 2007), pp. 862–876
37. A. Ouattara, Méthodologie d'éco-conception de procédés par optimisation multiobjectif et aide à la décision multicritère. Ph.D. thesis, INPT (2011)

# MAP-Elites for Constrained Optimization

Stefano Fioravanzo and Giovanni Iacca

**Abstract** Constrained optimization problems are often characterized by multiple constraints that, in the practice, must be satisfied with different tolerance levels. While some constraints are hard and as such must be satisfied with zero-tolerance, others may be soft, such that non-zero violations are acceptable. Here, we evaluate the applicability of MAP-Elites to "illuminate" constrained search spaces by mapping them into feature spaces where each feature corresponds to a different constraint. On the one hand, MAP-Elites implicitly preserves diversity, thus allowing a good exploration of the search space. On the other hand, it provides an effective visualization that facilitates a better understanding of how constraint violations correlate with the objective function. We demonstrate the feasibility of this approach on a large set of benchmark problems, in various dimensionalities, and with different algorithmic configurations. As expected, numerical results show that a basic version of MAP-Elites cannot compete on all problems (especially those with equality constraints) with state-of-the-art algorithms that use gradient information or advanced constraint handling techniques. Nevertheless, it has a higher potential at finding constraint violations versus objectives trade-offs and providing new problem information. As such, it could be used in the future as an effective building-block for designing new constrained optimization algorithms.

## 1 Introduction

Several real-world applications, for instance, in engineering design, control systems and health care, can be described in the form of constrained continuous optimization problems, i.e., problems where a certain objective/cost function must be optimized

S. Fioravanzo
Arrikto Inc., 60 E 3rd Ave, Ste 262, San Mateo, CA 9441, USA
e-mail: stefano@arrikto.com

G. Iacca (✉)
University of Trento, Trento 38122, Italy
e-mail: giovanni.iacca@unitn.it

© The Author(s), under exclusive license to Springer Nature Singapore Pte Ltd. 2021
A. J. Kulkarni et al. (eds.), *Constraint Handling in Metaheuristics and Applications*,
https://doi.org/10.1007/978-981-33-6710-4_7

within a certain search space, subject to some problem-dependent constraints. Without loss of generality, these problems can be formulated as

$$\begin{aligned}
\underset{\mathbf{x} \in \mathbf{D}}{\text{minimize}} \quad & f(\mathbf{x}) \\
\text{subject to:} \quad & g_i(\mathbf{x}) \le 0, \quad i = 1, 2, \ldots, m \\
& h_j(\mathbf{x}) = 0, \quad j = 1, 2, \ldots, p
\end{aligned}$$

where (1) $\mathbf{x} \in \mathbf{D} \subseteq \mathbb{R}^n$ is a candidate solution to the problem, being $n$ the problem dimensionality, and $\mathbf{D}$ the search space, typically defined in terms of bounding box constraints $lb_k \le x_k \le ub_k \; \forall k \in \{1, 2, \ldots, n\}$, where $lb_k$ and $ub_k$ are the lower and upper bound, respectively, for each kth variable; (2) $f(\mathbf{x}) : \mathbb{R}^n \to \mathbb{R}$ is the objective function; (3) $g_i(\mathbf{x})$ and $h_j(\mathbf{x})$ (both defined as: $\mathbb{R}^n \to \mathbb{R}$) are, respectively, inequality and equality constraints. A summary of the aforementioned symbols, as well as the symbols used in the rest of the paper, is reported in Table 1.

**Table 1** List of symbols used in the paper (symbols in boldface indicate vectors)

| Symbol | Meaning |
|---|---|
| $n$ | Problem dimensionality (no. of variables) |
| $N$ | Number of features (i.e., no. of tolerance levels) |
| $lb_k$ | Lower bound of kth variable ($k = 1, 2, \ldots, n$) |
| $ub_k$ | Upper bound of kth variable ($k = 1, 2, \ldots, n$) |
| $\mathbf{D}$ | Search space ($\subseteq \mathbb{R}^n$) |
| $f(\mathbf{x})$ | Objective function calculated in $\mathbf{x}$ |
| $g_i(\mathbf{x})$ | ith inequality constraint calculated in $\mathbf{x}$ ($i = 1, 2, \ldots, m$) |
| $h_j(\mathbf{x})$ | jth equality constraint calculated in $\mathbf{x}$ ($j = 1, 2, \ldots, p$) |
| $\mathbf{x}$ | Candidate solution ($\in \mathbb{R}^n$) |
| $\mathbf{x}'$ | Candidate solution ($\in \mathbb{R}^n$) |
| $\mathbf{b}'$ | Feature descriptor calculated in $\mathbf{x}'$ ($\in \mathbb{R}^N$), i.e., constraint violations within each of the $N$ predefined tolerance levels |
| $\mathcal{P}()$ | Associative map <feature descriptor, performance> |
| $\mathcal{X}()$ | Associative map <feature descriptor, solution> |
| $\mathcal{P}(\mathbf{b}')$ | Best performance associated to the feature descriptor $\mathbf{b}'$ (it can be empty) |
| $\mathcal{X}(\mathbf{b}')$ | Best solution associated to the feature descriptor $\mathbf{b}'$ (it can be empty) |
| $p'$ | Performance calculated in $\mathbf{x}'$, i.e., $p' = f(\mathbf{x}')$ |
| $g$ | Current number of iterations ($1, 2, \ldots, I$) |
| $I$ | Number of function evaluations (NFEs) |
| $G$ | Number of function evaluations (NFEs) for map initialization ($G < I$) |

In the past three decades, a large number of computational techniques have been proposed to solve efficiently this class of problems, among which Evolutionary Algorithms (EAs) [17] have shown great potential due to their general applicability and effectiveness. So far, most of the research in the field has focused on how to improve the *feasible* results obtained by EAs, for instance, developing *ad hoc* evolutionary operators, specific constraint repair mechanisms, or constraint handling techniques (CHTs). However, in various real-world applications it could be desirable, or at least acceptable, to consider also *infeasible* solutions. This could be obtained for instance by defining different *tolerance levels* for each constraint, so to reason on the effect of relaxing a certain constraint (and, if so, how much to do that) in order to obtain an improvement on the objective function, and therefore find different trade-offs in terms of constraint violations versus objective [25]. Despite these application needs, to date little research effort has been put on how to allow EAs to identify, rather than a single optimal solution, a *diverse* set of solutions characterized by different trade-offs of this kind. In this sense, the most notable exceptions that explicitly addressed this problem—although with contrasting results—have focused on multiobjective approaches, where the constraint violations were considered as additional objectives to be minimized [5, 10, 26, 33, 35], or surrogate methods [2].

In this paper, our goal is to evaluate the applicability of the Multi-dimensional Archive of Phenotypic Elites (MAP-Elites) [3, 19], an EA recently introduced in the literature in the context of robotic tasks, for tackling these problems, specifically to provide trade-off solutions in constrained optimization. Differently from conventional EAs, MAP-Elites conducts the search by mapping the highest-performing solutions found during the search (elites) into another multi-dimensional *discretized* space, defined by problem-specific features (the latter space is separate from the original search space, and typically of a lower dimensionality). These features are uncorrelated to the actual objective function, and describe some domain-specific properties of the candidate solutions. By means of this mapping, the algorithm "illuminates" the search space by showing the potential value of each area of the feature space, and the corresponding trade-off between the objective and the features of interest.

The functioning of MAP-Elites is simple and intuitive. First, the multi-dimensional feature space is discretized into a multi-dimensional grid, where each bin (i.e., a cell in the grid, which is, in general, a hyper-rectangle) represents a different "niche". Then, an EA-like search is performed by means of selection and variation (mutation and crossover), but instead of keeping a population of solutions that may or may not be diverse, MAP-Elites explicitly maintains diversity by keeping in each niche one elite, which identifies the best solution characterized by the corresponding feature values. At the end of the optimization procedure, a full *map* of possible solutions is provided (rather than a single optimal solution, as in conventional single-objective EAs), each characterized by different features. This map is shown in the form of a multidimensional heatmap, which allows for an easy visual inspection of how the objective function changes across the feature space.

In order to apply MAP-Elites to constrained optimization, the main idea we propose here is to define the feature space based on a discretization of the constraint violations. It is worth noting that in practical applications the discretization has a concrete, domain-dependent meaning: it can be seen as a set of tolerance levels, which as we mentioned can be different for each constraint. With this approach, we are able to produce a visual representation of the objective values in the feature space (in this case, space of constraint violations), thus uncovering possible correlations between the constraints and the objective. Thanks to this visualization, it is indeed easy to understand "where", with respect to the boundaries of the constraints, the best solutions lie. It also is easy to inspect the best overall solution, and check if the algorithm was able to produce particularly interesting solutions violating some of the constraints. As said, this insight can be helpful in cases where the violation of some constraints (within a certain tolerance level) can be an acceptable trade-off for a better overall performance.

As we will see in detail in the paper, despite its simplicity the proposed approach has various advantages: (1) it can be easily adapted/extended to include custom evolutionary operators; (2) it does not necessarily need explicit CHTs, but it can also include them; (3) it implicitly preserves diversity; (4) it allows the user to easily define custom tolerance levels, different for each constraint; (5) it "illuminates" the search space as it provides additional information on the correlation between constraints and objective, which might be of interest in practical applications; (6) it facilitates the interpretation of results through an intuitive visualization.

The rest of the paper is structured as follows. In Sect. 2, we will briefly summarize the most recent works on MAP-Elites and constrained optimization. Then, Sect. 3 describes the basic MAP-Elites algorithm and how it can be applied to constrained optimization. In Sect. 4, we describe the experimental setup (benchmark and algorithmic settings), followed by the analysis of the numerical results, reported in Sect. 5. Finally, Sect. 6 concludes this work and suggests possible future developments.

## 2   Related Work

The study of MAP-Elites and, more in general, EAs explicitly driven by *novelty* [12, 18, 24] or *diversity* [4, 20], rather than the objective alone, is a relatively new area of research in the Evolutionary Computation community. Among these algorithms, MAP-Elites [3, 19] has attracted quite some attention in the field, due to its simplicity and general applicability. Since its introduction in 2015, MAP-Elites has been mostly used as a means to identify *repertoires* of different agent behaviors, e.g., in evolutionary robotics setups. Various examples of applications to maze navigation, legged robot gait optimization, and anthropomorphic robot trajectory optimization can be found in [1, 3, 4, 19, 23, 30–32]. More recently, MAP-Elites has been applied also to Workforce Scheduling and Routing Problem (WSRP) [29] and Genetic Programming [6]. To the best of our knowledge, no prior work exists on the explicit use of MAP-Elites for solving constrained optimization problems.

Evolutionary constrained optimization is, on the other hand, a much more mature area of research: hundreds of papers have shown in the past three decades various algorithmic solutions and real-world problems where EAs were successfully applied to constrained optimization. Summarizing all the recent advances in this area would be impossible, and is obviously outside the scope of this paper. A thorough survey of the literature is performed, for instance, in [13], to which we refer the interested reader for a comprehensive analysis of the state of the art updated to 2016. Another interesting study, published at the end of 2018 by Hellwig and Beyer [9], covers all the aspects related to benchmarking EAs for constrained optimization, including a thorough analysis of the most important benchmark suites available in the literature. Among these, the CEC 2010 benchmark [15] has attracted in the past few years a large body of works that showed how to solve its functions efficiently, and is often used nowadays for benchmarking new algorithms. Currently, the state-of-the-art results on this benchmark have been obtained by $\varepsilon$DEag, an $\varepsilon$ constrained Differential Evolution algorithm with an archive and gradient-based mutation proposed by Takahama and Sakai [28], followed by ECHT-DE, another variant of Differential Evolution that includes an ensemble of four constraint handling techniques, proposed by Mallipeddi and Suganthan [14]. These two works are also good examples of two of the most successful recent trends in the field, which use of gradient-based information (if available, or at least approximable), and the combination of multiple CHTs into a single evolutionary algorithm.

## 3 Methodology

The basic version of MAP-Elites, as introduced in [3, 19], is shown in Algorithm 1. In the pseudocode, $\mathbf{x}$ and $\mathbf{x}'$ are candidate solutions (i.e., $n$-dimensional vectors defined in the search space $\mathbf{D}$); $\mathbf{b}'$ is a *feature descriptor*, which is a location in a user-defined *discretized* feature space, corresponding to the candidate solution $\mathbf{x}'$, (i.e., an $N$-dimensional vector of user-defined features that characterize $\mathbf{x}'$, typically with $N < n$); $p'$ is the performance of the candidate solution $\mathbf{x}'$ (i.e., the scalar value returned by the objective function $f(\mathbf{x}')$; the function itself is assumed to be a black-box, that is its mathematical formulation, if any, is unknown to the algorithm); $\mathcal{P}$ is a <feature descriptor, performance> map (i.e., an associative table that stores the best performance associated to each feature descriptor encountered by the algorithm); $\mathcal{X}$ is a <feature descriptor, solution> map (i.e., an associative table that stores the best solution associated to each feature descriptor encountered by the algorithm); $\mathcal{P}(\mathbf{b}')$ is the best performance associated to the feature descriptor $\mathbf{b}'$ (it can be empty); $\mathcal{X}(\mathbf{b}')$ is the best solution associated to the feature descriptor $\mathbf{b}'$ (it can be empty).

Following the pseudocode, the algorithm first creates the two maps $\mathcal{P}$ and $\mathcal{X}$, which are initially empty. Then, a loop of $I$ iterations (i.e., function evaluations) is executed. For each of the first $G$ iterations, $G$ solutions are randomly sampled in the search space $\mathbf{D}$, which are used for initializing the two maps $\mathcal{P}$ and $\mathcal{X}$. Then, starting from the iteration $G + 1$, a solution $\mathbf{x}$ is randomly selected from the current map $\mathcal{X}$, and a randomly modified copy of it, $\mathbf{x}'$, is generated. The feature descriptor $\mathbf{b}'$ and performance $p'$ associated to this new, perturbed solution are then evaluated.

---

**Algorithm 1** MAP-Elites algorithm, taken from [19]

---
$\mathcal{P} \leftarrow \emptyset, X \leftarrow \emptyset$
**for** $g = 1 \rightarrow I$ **do**
  **if** $g < G$ **then**
    $\mathbf{x}' \leftarrow$ random_solution()
  **else**
    $\mathbf{x} \leftarrow$ random_selection($X$)
    $\mathbf{x}' \leftarrow$ random_variation($\mathbf{x}$)
  $\mathbf{b}' \leftarrow$ feature_descriptor($\mathbf{x}'$)        ▷ for constrained optimization, $\mathbf{b}'$ is a
                                                      vector of constraint violations
  $p' \leftarrow$ performance($\mathbf{x}'$)
  **if** $\mathcal{P}(\mathbf{b}') = \emptyset \vee \mathcal{P}(\mathbf{b}') > p'$ **then**
    $\mathcal{P}(\mathbf{b}') \leftarrow p'$
    $X(\mathbf{b}') \leftarrow \mathbf{x}'$
**return** $\mathcal{P}$ and $X$

---

At this point, the two maps $\mathcal{P}$ and $X$ are updated: if the performance associated to $\mathbf{b}'$, $\mathcal{P}(\mathbf{b}')$, is empty (which can happen if this is the first time that the algorithm generates a solution with that feature descriptor), or if it contains a value that is worse than the performance $p'$ of the newly generated solution (in Algorithm 1, we assume a minimization problem, therefore we check the condition $\mathcal{P}(\mathbf{b}') > p'$), the new solution $\mathbf{x}'$ and its performance $p'$ are assigned to the elements of the maps corresponding to its feature descriptor $\mathbf{b}'$, namely $\mathcal{P}(\mathbf{b}')$ and $X(\mathbf{b}')$. Once the loop terminates, the algorithm returns the two maps $\mathcal{P}$ and $X$, which can be later analyzed for further inspection and post-processing.

It can be immediately noted how simple the algorithm is. With reference to the pseudocode, in order to apply MAP-Elites to a specific problem the following methods must be defined:

- random_solution(): returns a randomly generated solution;
- random_selection($X$): randomly selects a solution from $X$;
- random_variation($\mathbf{x}$): returns a modified copy of $\mathbf{x}$;
- feature_descriptor($\mathbf{x}$): maps a candidate solution $\mathbf{x}$ to its representation in the feature space, $\mathbf{b}$;
- performance($\mathbf{x}$): evaluates the objective function corresponding to the candidate solution $\mathbf{x}$.

The first three methods are rather standard, i.e., they can be based on general-purpose operators typically used in EAs. However, it is possible to customize them according to the specific need. For instance, the basic version MAP-Elites randomly selects at each iteration one solution, and applies only Gaussian mutation operator; on the other hand, the algorithm can be easily configured to use a different selection mechanism (e.g., an informed operator that introduces some selection pressure/bias) or select multiple solutions at each iteration so to apply a recombination operator (crossover) or some other search mechanism such as a local search. We will see in Sect. 4 the details of three different algorithm configurations that we have used in our experimentation.

As for what concerns feature_descriptor($\mathbf{x}$) and performance($\mathbf{x}$), these are obviously problem-dependent: the first one, being dependent on how the user defines the features of interest and the corresponding feature space; the latter, being dependent on the specific objective function at hand.

The application of MAP-Elites to constrained optimization is then quite straightforward: here, we map each constraint of a constrained optimization problem to a different feature in the feature space explored by MAP-Elites, such that each candidate solution is associated to a feature descriptor that is basically a vector of constraint violations. In this specific case then, the user does not necessarily have to define any additional feature, but the features themselves are already part of the problem definition. Leaving aside the algorithmic details (selection and variation) and parameters (the only two parameters of the algorithm are the total and initial number of iterations, respectively, $I$ and $G$, which can be easily set by the user based on computing resources and/or time constraints), the only input required from the user is the discretization of the features (constraint violations) space.

An intuitive way of discretizing this space is to define, for each constraint, a certain number of *tolerance levels*, i.e., amounts of constraint violation used as discretization steps. These can be easily expressed in absolute terms (based on the values of $g_i(\mathbf{x})$ and $h_j(\mathbf{x})$ in case of violations), or normalized w.r.t. known minimum and maximum violations. A simple example of discretization steps is $\{0, \varepsilon, 2\varepsilon, \dots\}$, where $\varepsilon$ is a user-defined parameter. However, as we will show in Sect. 4, also non-linear discretization is possible. In general, the discretization strategy should be based on domain knowledge and defined in such a way that solutions whose violations are equivalent, from a practical point of view, are grouped in the same bin. This would allow to "illuminate" the relation between objective function and constraint violations in a significant, meaningful way. Finally, we must note that while in general a different set of *tolerance levels* can be defined for each constraint (especially if these are expressed in absolute terms), if all constraints have the same codomain (or, if they are normalized), the same tolerance levels can be used for all of them.

# 4   Experimental Setup

We evaluated the performance of the proposed approach on the benchmark functions defined for the CEC 2010 Competition on Constrained Real-Parameter Optimization [15]. This benchmark presents 18 problems with different landscape characteristics, subject to a varying number (up to four) of equality and/or inequality constraints. To assess the scalability of MAP-Elites, we tested these problems in 10 and 30 dimensions. All the details of the experimental setup were set according to the CEC indications [15], with the main parameters listed in Table 2.

As for the discretization steps, the selected values correspond to having each feature discretized into 5 bins, namely: $\{0\}$, $(0, 0.0001]$, $(0.0001, 0.01]$, $(0.01, 1.0]$ and $(1.0, \inf)$. This last aspect deserves attention, since as we have seen in Sect. 3 this is what allows the application of MAP-Elites to constrained optimization. Here, we

**Table 2** Parameters used in the experimental setup

| Parameter | Value |
|---|---|
| Number of benchmark problems | 18 {C1, C2, ..., C18} |
| Number of dimensions ($n$), for each problem | 10D and 30D |
| Number of runs, for each problem/dimensionality | 25 |
| NFEs, per run | $I = 2.0\text{e}5$ for 10D |
| | $I = 6.0\text{e}5$ for 30D |
| NFEs for map initialization, per run | $G = 2000$ |
| Discretization steps for every feature | {0, 0.01, 0.0001, 1} |

have defined as discretization steps the three tolerance levels defined in [15], i.e., 0.0001, 0.01, and 1.0 (see Sect. 5), in addition to an explicit step corresponding to zero-tolerance (corresponding to solutions with $g_i(\mathbf{x}) \leq 0$ and $h_j(\mathbf{x}) = 0$, respectively for inequality and equality constraints).

The features used in MAP-Elites follow the same order as they appear in the corresponding problem definition, with inequality constraints considered before equality constraints. Since the problems contained in the CEC 2010 benchmark have a variable number of equality/inequality constraints, we define a variable-sized feature space, where for each problem there are as many MAP-Elites features as constraints. For visualization purposes, we represent the map $\mathcal{P}(\mathbf{b}')$ obtained in each run of MAP-Elites in the form of a multi-dimensional heatmap, as explained in [3, 19]. The color represents the objective value corresponding to the solution contained in each bin. The first axis (abscissa) corresponds to the first constraint violation, the second axis (ordinate) corresponds to the second constraint violation, and so on for the third and fourth constraints. Feature dimensions are "nested" such that the feature space is first discretized along the 1st and 2nd axes (so to obtain a 2D grid of bins), while the following features (if any) are represented by an "inner" (1D or 2D) discretization inside each bin in the "outer" grid. Obviously, this visualization procedure can be easily extended to handle more than four constraints, although the visual interpretability of the results tends to decrease with the number of features shown in the heatmap.

It should be noted that according to the CEC 2010 benchmark definition [15], a solution $\mathbf{x}$ is considered feasible iff $g_i(\mathbf{x}) \leq 0 \ \forall i \in \{1, 2, \ldots, m\}$ and $|h_j(\mathbf{x})| - \epsilon \leq 0 \ \forall j \in \{1, 2, \ldots, p\}$, where $\epsilon$ is the equality constraint tolerance, set to 0.0001. Otherwise, the solution is considered infeasible. From what we have just discussed, it follows then that feasible solutions in the sense of the CEC 2010 definition can be found: (1) for what concerns inequality constraints, in the first bin ({0}) along each feature dimension; (2) for what concerns equality constraints, in the first two bins ({0}, (0, 0.0001]). In plain terms, this means that we can easily identify feasible solutions found by MAP-Elites by simply looking at the lower-left corner of the heatmap, while solutions with increasing constraint violations are found scanning the heatmap (and each inner bin in case of more than two constraints) towards the

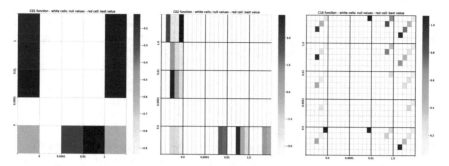

**Fig. 1** Final heatmaps found in a single run of MAP-Elites on C1, Configuration 1 (left), C2, Configuration 1 (center), C16, Configuration 3 (right) in 10D. The three benchmark functions are characterized, respectively, by 2I, 2I-1E, and 2I-2E, where 'I' and 'E' stand for inequality and equality constraints, respectively. In each heatmap, the color of each bin is proportionate to the objective value corresponding to the solution present in it (assuming minimization, the lower the better), while the red bin indicates the solution with the best objective value (regardless its feasibility). It can be observed that the maps allow to "illuminate" the search space of each problem, identifying various trade-off solutions in terms of objective value versus constraint violations, such as solutions with a high performance but with some violated constraints. Note that in case of 3 or 4 constraints, the discretization along the first two (outer) dimensions of the heatmap is indicated by a thicker black line, while the discretization along the "nested" (inner) dimensions are indicated by a thinner black line

upper right side. Some examples of heatmaps obtained by MAP-Elites are shown in Fig. 1.

As for the evolutionary operators (selection and variation, as shown in Algorithm 1), we defined three different algorithmic settings. In all cases, selection is performed according to a uniform distribution over the current map. Variation is instead applied according to the following configurations:

- Configuration 1: mutation ($\sigma = 0.1$), without crossover
- Configuration 2: mutation ($\sigma = 0.5$), without crossover
- Configuration 3: mutation ($\sigma = 0.1$), with crossover.

In all three cases, mutation is implemented by applying to the selected solution (with probability 0.5 for each variable) a Gaussian mutation with $\mu = 0$ and the given value of $\sigma$. Boundary constraints are handled according to a toroidal mechanism: given a decision variable $x$ constrained to the interval $[a, b]$, if the corresponding mutated variable $x'$ exceeds the upper bound $b$ (i.e., $x' = b + \zeta$), $x'$ is transformed into $x' = a + \zeta, \zeta > 0$. Similarly, if $x' = a - \zeta$, $x'$ is transformed into $x' = b - \zeta$, $\zeta > 0$.

In Configuration 3, at each iteration, two solutions are randomly selected from the current map, after which uniform crossover (with probability 0.5 for each variable) is applied by swapping the corresponding variables from the two parents. Then, the first of the two offspring generated by crossover undergoes Gaussian mutation, as in Configurations 1 and 2, and is evaluated in terms of feature descriptor and performance, as shown in Algorithm 1.

The entire experimental setup was implemented in Python 3,[1] and the experimentation was performed on a Ubuntu 18.10 workstation, with a CPU Intel Core i9-7940X @3.10 GHz and 64 GB DDR4.

## 5  Numerical Results

We present here the results obtained on the experimental setup described in Sect. 4. In Tables 3, 4, 5, 6, 7 and 8, we report the results for all the CEC 2010 functions in 10 and 30 dimensions, for the three algorithm settings described above.[2] In the tables, we report the results as suggested in [15], where for each function we show:

1. The objective value corresponding to the best, worst and median solution[3] (over 25 runs) obtained at the end of the computational budget; next to each objective value, we show in parenthesis the no. of violated constraints corresponding to each of these three solutions.
2. The number of violated constraints at the median solution, $c = (c_1, c_2, c_3)$ (where each element $c_i, i = 1, 2, 3$ represents the number of violations higher than three tolerance levels set to 1, 0.01, and 0.0001, respectively), and the corresponding mean violation $\bar{v}$, calculated as

$$\bar{v} = \frac{\sum_{i=1}^{m} G_i(x) + \sum_{j=1}^{p} H_j(x)}{m + p} \tag{1}$$

   where $G_i(x) = g_i(x)$ if $g_i(x) > 0$ (otherwise zero), and $H_j(x) = |h_j(x)|$ if $|h_j(x)| - \epsilon > 0$ (otherwise zero), being $\epsilon$ is the equality constraint tolerance (as seen earlier, 0.0001).
3. The average objective value (over 25 runs) of the final solutions obtained at the end of the budget, and its std. dev.
4. The Feasibility Rate (FR), that is, for each function, the ratio between the number of runs during which at least one feasible solution was found within the budget, and the total number of runs (in our case, 25).

For reference, we report in Tables 9 and 10 the results for all the CEC 2010 functions after $I = 2.0e5$ and $I = 6.0e5$ NFEs, respectively for 10D and 30D, obtained by $\varepsilon$DEag [28], the best algorithm on the CEC 2010 benchmark.

From the tables, we can observe that in all three configurations, MAP-Elites solves with 100% FR C1, C7, C8, C14, C15 in 10D, i.e., all the functions with

---

[1] Code available at: https://github.com/StefanoFioravanzo/MAP-Elites.

[2] The complete set of numerical results and the final heatmaps for each problem and dimensionality are available as Supplementary Material at: https://bit.ly/2BQIR8B.

[3] The final solutions are sorted according to these three criteria: (1) feasible solutions are sorted in front of infeasible solutions; (2) feasible solutions are sorted according to their objective value; (3) infeasible solutions are sorted according to their mean value of constraint violation, calculated as in Eq. (1).

**Table 3** Numerical results (rank based on CEC rules) obtained from 25 independent of MAP-Elites (Configuration 1) on 10D problems

| Function | Best | Worst | Median | $c$ | $\bar{v}$ | Mean | Std | FR |
|---|---|---|---|---|---|---|---|---|
| C01 | −5.900e−01(0) | −3.500e−01(0) | −4.400e−01(0) | (0, 0, 0) | 0 | −4.600e−01 | 7.000e−02 | 1.0 |
| C02 | 8.900e−01(0) | 2.860e+00(2) | 1.290e+00(0) | (0, 0, 0) | 0 | 1.780e+00 | 1.870e+00 | 0.52 |
| C03 | 1.254e+14(1) | 9.654e+14(1) | 5.118e+14(1) | (0, 0, 1) | 2.477e+06 | 5.208e+14 | 2.730e+14 | 0.0 |
| C04 | −1.512e+01(4) | 3.930e+01(4) | 1.646e+01(3) | (0, 0, 3) | 7.993e+03 | 1.487e+01 | 1.249e+01 | 0.0 |
| C05 | −1.795e+02(2) | 3.301e+02(2) | 8.052e+01(2) | (1, 0, 1) | −8.064e+01 | 6.719e+01 | 1.716e+02 | 0.0 |
| C06 | −1.222e+02(2) | 5.043e+02(1) | 1.705e+02(1) | (0, 0, 1) | −1.300e+02 | 1.885e+02 | 1.669e+02 | 0.0 |
| C07 | 1.761e+09(0) | 3.347e+11(0) | 7.117e+09(0) | (0, 0, 0) | 0 | 6.854e+10 | 9.030e+10 | 1.0 |
| C08 | 2.610e+04(0) | 1.504e+08(0) | 3.490e+06(0) | (0, 0, 0) | 0 | 1.690e+07 | 3.667e+07 | 1.0 |
| C09 | 2.752e+13(0) | 2.327e+15(1) | 4.526e+13(1) | (0, 1, 0) | −2.950e+01 | 3.607e+14 | 6.492e+14 | 0.12 |
| C10 | 7.187e+12(1) | 8.964e+14(1) | 3.631e+13(1) | (0, 0, 1) | −6.560e+02 | 1.805e+14 | 3.008e+14 | 0.0 |
| C11 | −5.486e+01(1) | −3.766e+01(1) | −4.441e+01(1) | (0, 0, 1) | 2.183e+12 | −4.561e+01 | 4.870e+00 | 0.0 |
| C12 | −5.327e+03(2) | 1.704e+03(2) | −3.904e+03(2) | (0, 0, 2) | 1.501e+12 | −3.008e+03 | 2.173e+03 | 0.0 |
| C13 | −3.702e+02(2) | −1.721e+02(2) | −2.804e+02(2) | (0, 0, 1) | 5.853e+02 | −2.823e+02 | 5.116e+01 | 0.0 |
| C14 | 2.141e+14(0) | 4.448e+15(0) | 1.304e+15(0) | (0, 0, 0) | 0 | 1.555e+15 | 1.018e+15 | 1.0 |
| C15 | 9.018e+12(0) | 1.041e+15(0) | 4.145e+14(0) | (0, 0, 0) | 0 | 4.449e+14 | 2.936e+14 | 1.0 |
| C16 | 4.900e−01(0) | 1.100e+00(2) | 6.500e−01(4) | (0, 1, 3) | −4.585e+06 | 6.200e−01 | 3.000e−01 | 0.16 |
| C17 | 1.310e+00(0) | 4.958e+02(1) | 1.784e+01(1) | (0, 1, 0) | −8.500e+05 | 6.456e+01 | 1.025e+02 | 0.4 |
| C18 | 3.365e+01(1) | 3.542e+03(1) | 5.894e+02(0) | (0, 0, 0) | 0 | 1.478e+03 | 2.594e+03 | 0.88 |

**Table 4** Numerical results (rank based on CEC rules) obtained from 25 independent of MAP-Elites (Configuration 2) on 10D problems

| Function | Best | Worst | Median | c | $\bar{v}$ | Mean | Std | FR |
|---|---|---|---|---|---|---|---|---|
| C01 | −6.700e−01(0) | −4.300e−01(0) | −4.800e−01(0) | (0, 0, 0) | 0 | −5.100e−01 | 6.000e−02 | 1.0 |
| C02 | 1.700e+00(0) | 4.980e+00(1) | 2.500e+00(2) | (1, 1, 0) | −1.046e+01 | 2.230e+00 | 1.930e+00 | 0.36 |
| C03 | 4.516e+13(1) | 7.934e+14(1) | 4.101e+14(1) | (0, 0, 1) | 2.642e+06 | 3.827e+14 | 1.928e+14 | 0.0 |
| C04 | −1.654e+01(4) | 3.941e+01(4) | 1.958e+01(4) | (1, 0, 3) | 3.074e+03 | 1.224e+01 | 1.789e+01 | 0.0 |
| C05 | −3.516e+02(2) | 5.255e+02(2) | −9.369e+01(2) | (1, 1, 0) | −6.720e+01 | −3.351e+01 | 2.121e+02 | 0.0 |
| C06 | −2.519e+02(2) | 5.891e+02(2) | −1.068e+02(2) | (0, 1, 1) | 1.434e+01 | −2.237e+02 | 2.178e+02 | 0.0 |
| C07 | 4.191e+08(0) | 2.371e+11(0) | 4.222e+09(0) | (0, 0, 0) | 0 | 2.836e+10 | 5.798e+10 | 1.0 |
| C08 | 8.302e+05(0) | 2.420e+08(0) | 1.664e+07(0) | (0, 0, 0) | 0 | 4.818e+07 | 6.160e+07 | 1.0 |
| C09 | 8.585e+11(1) | 7.835e+13(1) | 1.933e+13(1) | (1, 0, 0) | −1.388e+02 | 2.804e+13 | 2.302e+13 | 0.0 |
| C10 | 5.666e+12(0) | 3.540e+14(1) | 1.854e+13(1) | (1, 0, 0) | −2.489e+02 | 5.046e+13 | 7.657e+13 | 0.16 |
| C11 | −5.366e+01(1) | −3.885e+01(1) | −4.573e+01(1) | (0, 0, 1) | 1.894e+12 | −4.589e+01 | 3.750e+00 | 0.0 |
| C12 | −6.849e+03(2) | 5.433e+02(2) | −4.272e+03(2) | (0, 0, 2) | 7.650e+11 | −3.843e+03 | 1.779e+03 | 0.0 |
| C13 | −3.849e+02(2) | −4.087e+01(2) | −3.048e+02(2) | (1, 0, 1) | 4.854e+02 | −2.916e+02 | 7.552e+01 | 0.0 |
| C14 | 1.004e+07(0) | 9.089e+13(0) | 8.365e+08(0) | (0, 0, 0) | 0 | 3.681e+12 | 1.780e+13 | 1.0 |
| C15 | 3.640e+06(0) | 1.594e+14(0) | 4.011e+10(0) | (0, 0, 0) | 0 | 7.150e+12 | 3.112e+13 | 1.0 |
| C16 | 1.060e+00(0) | 1.080e+00(4) | 9.900e−01(4) | (1, 1, 2) | −5.208e+07 | 8.700e−01 | 2.300e−01 | 0.08 |
| C17 | 9.251e+01(0) | 6.959e+02(3) | 3.041e+01(3) | (0, 2, 1) | −3.098e+08 | 2.108e+02 | 2.661e+02 | 0.36 |
| C18 | 2.253e+02(0) | 2.242e+04(1) | 8.008e+03(0) | (0, 0, 0) | 0 | 9.665e+03 | 8.152e+03 | 0.8 |

**Table 5** Numerical results (rank based on CEC rules) obtained from 25 independent of MAP-Elites (Configuration 3) on 10D problems

| Function | Best | Worst | Median | $c$ | $\bar{v}$ | Mean | Std | FR |
|---|---|---|---|---|---|---|---|---|
| C01 | −8.300e−01(0) | −4.500e−01(0) | −5.500e−01(0) | (0, 0, 0) | 0 | −5.700e−01 | 1.000e−01 | 1.0 |
| C02 | 8.000e−01(0) | 3.050e+00(0) | 2.870e+00(2) | (1, 1, 0) | −8.750e+00 | 1.850e+00 | 2.230e+00 | 0.44 |
| C03 | 1.154e+14(1) | 9.250e+14(1) | 4.362e+14(1) | (0, 0, 1) | 1.310e+06 | 5.005e+14 | 2.213e+14 | 0.0 |
| C04 | −4.441e+01(4) | 5.003e+01(4) | 3.800e−01(4) | (0, 2, 2) | 9.570e+01 | 1.270e+00 | 2.920e+01 | 0.0 |
| C05 | −4.801e+02(2) | 1.527e+02(1) | −2.316e+02(2) | (0, 1, 1) | 1.188e+01 | −2.181e+02 | 1.804e+02 | 0.0 |
| C06 | −4.275e+02(2) | 4.210e+02(2) | −1.927e+02(1) | (0, 0, 1) | −1.397e+02 | −1.363e+02 | 2.086e+02 | 0.0 |
| C07 | 6.152e+03(0) | 5.908e+07(0) | 3.171e+05(0) | (0, 0, 0) | 0 | 4.070e+06 | 1.251e+07 | 1.0 |
| C08 | 4.550e+03(0) | 1.391e+06(0) | 1.518e+05(0) | (0, 0, 0) | 0 | 4.183e+05 | 4.949e+05 | 1.0 |
| C09 | 1.690e+07(0) | 1.600e+15(1) | 8.227e+07(1) | (0, 1, 0) | −6.704e+01 | 1.612e+14 | 4.344e+14 | 0.2 |
| C10 | 9.235e+06(0) | 7.494e+14(1) | 7.169e+07(1) | (1, 0, 0) | −3.527e+01 | 3.833e+13 | 1.490e+14 | 0.44 |
| C11 | −5.242e+01(1) | −3.835e+01(1) | −4.532e+01(1) | (0, 0, 1) | 3.883e+12 | −4.474e+01 | 3.120e+00 | 0.0 |
| C12 | −7.742e+03(2) | 3.236e+02(2) | −5.923e+03(2) | (1, 0, 1) | 1.573e+12 | −5.795e+03 | 1.603e+03 | 0.0 |
| C13 | −4.474e+02(2) | −2.925e+02(2) | −3.952e+02(2) | (0, 1, 1) | 8.816e+02 | −3.900e+02 | 3.742e+01 | 0.0 |
| C14 | 1.033e+05(0) | 3.280e+14(0) | 1.724e+09(0) | (0, 0, 0) | 0 | 3.958e+13 | 8.132e+13 | 1.0 |
| C15 | 4.980e+03(0) | 4.089e+11(0) | 9.060e+07(0) | (0, 0, 0) | 0 | 1.863e+10 | 8.010e+10 | 1.0 |
| C16 | 1.020e+00(0) | 1.100e+00(4) | 2.900e−01(3) | (2, 1, 0) | −3.723e+05 | 5.100e−01 | 3.800e−01 | 0.12 |
| C17 | 3.610e+00(0) | 1.560e+01(3) | 1.123e+01(3) | (2, 1, 0) | −8.454e+05 | 6.054e+01 | 1.196e+02 | 0.52 |
| C18 | 7.190e+00(2) | 1.480e+01(2) | 5.091e+01(0) | (0, 0, 0) | 0 | 1.078e+02 | 1.165e+02 | 0.92 |

**Table 6** Numerical results (rank based on CEC rules) obtained from 25 independent of MAP-Elites (Configuration 1) on 30D problems

| Function | Best | Worst | Median | c | $\bar{v}$ | Mean | Std | FR |
|---|---|---|---|---|---|---|---|---|
| C01 | −2.800e−01(0) | −2.000e−01(0) | −2.200e−01(0) | (0, 0, 0) | 0 | −2.300e−01 | 2.000e−02 | 1.0 |
| C02 | 3.880e+00(0) | 4.320e+00(2) | 3.760e+00(1) | (1, 0, 0) | −2.570e+00 | 3.210e+00 | 1.480e+00 | 0.36 |
| C03 | 6.664e+15(1) | 1.887e+16(1) | 1.472e+16(1) | (0, 0, 1) | 1.798e+07 | 1.463e+16 | 3.100e+15 | 0.0 |
| C04 | 1.869e+01(4) | 4.507e+01(4) | 3.239e+01(4) | (1, 1, 2) | 1.527e+06 | 3.235e+01 | 5.850e+00 | 0.0 |
| C05 | 1.739e+02(2) | 4.920e+02(1) | 2.784e+02(1) | (0, 1, 0) | −1.224e+01 | 3.036e+02 | 8.686e+01 | 0.0 |
| C06 | 2.237e+02(2) | 5.782e+02(1) | 3.076e+02(2) | (0, 1, 1) | 5.920e+00 | 3.545e+02 | 1.043e+02 | 0.0 |
| C07 | 3.226e+10(0) | 6.284e+12(0) | 9.281e+11(0) | (0, 0, 0) | 0 | 1.328e+12 | 1.386e+12 | 1.0 |
| C08 | 6.346e+07(0) | 6.746e+10(0) | 1.811e+09(0) | (0, 0, 0) | 0 | 7.092e+09 | 1.399e+10 | 1.0 |
| C09 | 7.159e+14(0) | 3.818e+15(1) | 8.511e+14(1) | (0, 0, 1) | −1.028e+03 | 1.323e+15 | 9.556e+14 | 0.2 |
| C10 | 6.189e+14(0) | 3.359e+15(1) | 7.702e+14(1) | (1, 0, 0) | −1.984e+02 | 9.456e+14 | 7.034e+14 | 0.16 |
| C11 | −3.289e+01(1) | −2.454e+01(1) | −2.802e+01(1) | (0, 0, 1) | 5.839e+12 | −2.833e+01 | 2.180e+00 | 0.0 |
| C12 | −1.363e+04(2) | 2.778e+03(2) | −8.056e+03(2) | (0, 1, 1) | 3.562e+12 | −7.862e+03 | 3.439e+03 | 0.0 |
| C13 | −3.129e+02(2) | −1.248e+02(2) | −2.373e+02(2) | (0, 1, 1) | 5.074e+02 | −2.253e+02 | 5.712e+01 | 0.0 |
| C14 | 6.971e+15(0) | 2.152e+16(0) | 1.160e+16(0) | (0, 0, 0) | 0 | 1.203e+16 | 3.312e+15 | 1.0 |
| C15 | 1.754e+15(0) | 2.898e+16(0) | 6.434e+15(0) | (0, 0, 0) | 0 | 7.380e+15 | 5.067e+15 | 1.0 |
| C16 | 9.100e−01(0) | 1.050e+00(3) | 8.500e−01(4) | (1, 2, 1) | −1.920e+19 | 8.200e−01 | 1.700e−01 | 0.08 |
| C17 | 6.822e+01(0) | 1.316e+03(2) | 8.772e+02(0) | (0, 0, 0) | 0 | 9.761e+02 | 8.256e+02 | 0.64 |
| C18 | 1.228e+03(0) | 1.276e+04(0) | 3.849e+03(0) | (0, 0, 0) | 0 | 4.380e+03 | 2.752e+03 | 1.0 |

**Table 7** Numerical results (rank based on CEC rules) obtained from 25 independent of MAP-Elites (Configuration 2) on 30D problems

| Function | Best | Worst | Median | $c$ | $\bar{v}$ | Mean | Std | FR |
|---|---|---|---|---|---|---|---|---|
| C01 | −2.600e−01(0) | −2.100e−01(0) | −2.300e−01(0) | (0, 0, 0) | 0 | −2.400e−01 | 1.000e−02 | 1.0 |
| C02 | 3.510e+00(0) | 4.940e+00(0) | 3.760e+00(2) | (0, 2, 0) | −5.310e+00 | 3.440e+00 | 1.030e+00 | 0.24 |
| C03 | 6.155e+15(1) | 1.970e+16(1) | 1.396e+16(1) | (0, 0, 1) | 1.316e+07 | 1.378e+16 | 3.582e+15 | 0.0 |
| C04 | 1.892e+01(4) | 4.564e+01(4) | 3.525e+01(4) | (0, 1, 3) | 1.317e+06 | 3.429e+01 | 6.280e+00 | 0.0 |
| C05 | −2.140e+00(2) | 5.772e+02(1) | 2.484e+02(2) | (0, 1, 1) | −2.131e+01 | 2.570e+02 | 1.174e+02 | 0.0 |
| C06 | 1.318e+02(2) | 5.961e+02(1) | 2.776e+02(1) | (0, 0, 1) | −1.296e+02 | 3.380e+02 | 1.436e+02 | 0.0 |
| C07 | 6.860e+10(0) | 6.957e+12(0) | 2.374e+12(0) | (0, 0, 0) | 0 | 2.560e+12 | 1.740e+12 | 1.0 |
| C08 | 4.992e+08(0) | 1.523e+12(0) | 6.357e+10(0) | (0, 0, 0) | 0 | 1.370e+11 | 2.923e+11 | 1.0 |
| C09 | 1.715e+15(0) | 1.520e+15(1) | 7.103e+14(1) | (1, 0, 0) | 4.026e+02 | 7.634e+14 | 4.300e+14 | 0.08 |
| C10 | 1.375e+14(1) | 1.511e+15(1) | 7.229e+14(1) | (0, 0, 1) | −5.762e+02 | 7.349e+14 | 3.095e+14 | 0.0 |
| C11 | −3.339e+01(1) | −2.650e+01(1) | −2.925e+01(1) | (0, 0, 1) | 5.776e+12 | −2.991e+01 | 2.060e+00 | 0.0 |
| C12 | −1.146e+04(2) | −4.734e+01(2) | −7.854e+03(2) | (0, 1, 1) | 3.687e+12 | −6.797e+03 | 3.248e+03 | 0.0 |
| C13 | −3.302e+02(2) | 2.444e+01(2) | −2.143e+02(2) | (1, 0, 1) | 5.036e+02 | −1.960e+02 | 8.269e+01 | 0.0 |
| C14 | 2.752e+10(0) | 4.214e+13(0) | 3.568e+12(0) | (0, 0, 0) | 0 | 7.098e+12 | 1.001e+13 | 1.0 |
| C15 | 1.921e+09(0) | 2.006e+15(0) | 1.110e+13(0) | (0, 0, 0) | 0 | 1.083e+14 | 3.901e+14 | 1.0 |
| C16 | 1.030e+00(4) | 1.300e+00(2) | 1.110e+00(4) | (0, 2, 2) | −6.197e+17 | 1.130e+00 | 8.000e−02 | 0.0 |
| C17 | 1.110e+03(0) | 2.864e+03(3) | 1.417e+03(0) | (0, 0, 0) | 0 | 1.539e+03 | 7.542e+02 | 0.44 |
| C18 | 5.037e+03(0) | 4.064e+04(0) | 3.043e+04(0) | (0, 0, 0) | 0 | 2.847e+04 | 9.539e+03 | 1.0 |

**Table 8** Numerical results (rank based on CEC rules) obtained from 25 independent of MAP-Elites (Configuration 3) on 30D problems

| Function | Best | Worst | Median | $c$ | $\bar{v}$ | Mean | Std | FR |
|---|---|---|---|---|---|---|---|---|
| C01 | −3.100e−01(0) | −2.300e−01(0) | −2.600e−01(0) | (0, 0, 0) | 0 | −2.600e−01 | 2.000e−02 | 1.0 |
| C02 | 4.590e+00(0) | 5.160e+00(2) | 3.150e+00(1) | (1, 0, 0) | −1.940e+00 | 2.610e+00 | 2.260e+00 | 0.32 |
| C03 | 1.033e+16(1) | 2.039e+16(1) | 1.521e+16(1) | (0, 0, 1) | 9.516e+06 | 1.534e−16 | 2.809e+15 | 0.0 |
| C04 | −1.859e+01(4) | 4.163e+01(4) | 1.518e+01(4) | (0, 2, 2) | 3.534e+05 | 1.215e+01 | 1.744e+01 | 0.0 |
| C05 | −2.304e+02(1) | 5.599e+02(2) | −7.310e+01(1) | (0, 0, 1) | −6.470e+00 | 9.960e+00 | 1.769e+02 | 0.0 |
| C06 | −1.678e+02(1) | 5.736e+02(1) | −1.477e+01(1) | (0, 0, 1) | −8.221e+01 | 7.093e+01 | 2.096e+02 | 0.0 |
| C07 | 7.226e+05(0) | 2.288e+09(0) | 2.058e+07(0) | (0, 0, 0) | 0 | 1.503e+08 | 4.498e+08 | 1.0 |
| C08 | 6.173e+05(0) | 3.216e+07(0) | 2.674e+06(0) | (0, 0, 0) | 0 | 6.223e+06 | 7.284e+06 | 1.0 |
| C09 | 5.770e+09(0) | 6.548e+10(1) | 1.463e+10(0) | (0, 0, 0) | 0 | 2.184e+13 | 1.050e+14 | 0.48 |
| C10 | 1.301e+08(0) | 1.153e+11(1) | 4.820e+09(1) | (1, 0, 0) | −2.706e+01 | 4.189e+13 | 1.970e+14 | 0.48 |
| C11 | −3.475e+01(1) | −2.439e+01(1) | −2.780e+01(1) | (0, 0, 1) | 5.744e+12 | −2.825e+00 | 2.340e+00 | 0.0 |
| C12 | −1.818e+04(2) | 2.716e+03(2) | −1.583e+04(2) | (0, 0, 2) | 3.737e+12 | −1.369e+04 | 5.767e+03 | 0.0 |
| C13 | −3.659e+02(2) | −2.891e+02(2) | −3.336e+02(2) | (1, 0, 1) | 6.148e+02 | −3.291e+02 | 2.011e+01 | 0.0 |
| C14 | 2.949e+06(0) | 9.116e+15(0) | 3.783e+08(0) | (0, 0, 0) | 0 | 6.363e−14 | 2.134e−15 | 1.0 |
| C15 | 9.592e+05(0) | 4.604e+09(0) | 1.375e+07(0) | (0, 0, 0) | 0 | 2.319e+08 | 8.981e+08 | 1.0 |
| C16 | 9.600e−01(0) | 1.040e+00(4) | 8.100e−01(4) | (1, 0, 3) | −1.020e+20 | 8.300e−01 | 1.400e−01 | 0.04 |
| C17 | 6.458e+01(0) | 1.234e+02(3) | 8.320e+01(3) | (1, 1, 1) | −1.291e+20 | 5.433e+02 | 7.307e+02 | 0.48 |
| C18 | 1.570e+02(0) | 2.793e−04(0) | 4.208e+02(0) | (0, 0, 0) | 0 | 1.896e+03 | 5.424e+03 | 1.0 |

**Table 9** Numerical results (rank based on CEC rules) obtained from 25 independent of $\varepsilon$DEag on 10D problems (taken from [28])

| Function | Best | Worst | Median | $c$ | $\bar{v}$ | Mean | Std | FR |
|---|---|---|---|---|---|---|---|---|
| C01 | −7.473104e−01(0) | −7.473104e−01(0) | −7.405572e−01(0) | 0, 0, 0 | 0.00e+00 | −7.47e−01 | 1.32e−03 | 1.0 |
| C02 | −2.277702e+00(0) | −2.269502e+00(0) | −2.174499e+00(0) | 0, 0, 0 | 0.00e+00 | −2.26e+00 | 2.39e−02 | 1.0 |
| C03 | 0.000000e+00(0) | 0.000000e+00(0) | 0.000000e+00(0) | 0, 0, 0 | 0.00e+00 | 0.00e+00 | 0.00e+00 | 1.0 |
| C04 | −9.992345e−06(0) | −9.977276e−06(0) | −9.282295e−06(0) | 0, 0, 0 | 0.00e+00 | −9.92e−06 | 1.55e−07 | 1.0 |
| C05 | −4.836106e+02(0) | −4.836106e+02(0) | −4.836106e+02(0) | 0, 0, 0 | 0.00e+00 | −4.84e+02 | 3.89e−13 | 1.0 |
| C06 | −5.786581e+02(0) | −5.786533e+02(0) | −5.786448e+02(0) | 0, 0, 0 | 0.00e+00 | −5.79e+02 | 3.63e−03 | 1.0 |
| C07 | 0.000000e+00(0) | 0.000000e+00(0) | 0.000000e+00(0) | 0, 0, 0 | 0.00e+00 | 0.00e+00 | 0.00e+00 | 1.0 |
| C08 | 0.000000e+00(0) | 1.094154e+01(0) | 1.537535e−01(0) | 0, 0, 0 | 0.00e+00 | 6.73e+00 | 5.56e+00 | 1.0 |
| C09 | 0.000000e+00(0) | 0.000000e+00(0) | 0.000000e+00(0) | 0, 0, 0 | 0.00e+00 | 0.00e+00 | 0.00e+00 | 1.0 |
| C10 | 0.000000e+00(0) | 0.000000e+00(0) | 0.000000e+00(0) | 0, 0, 0 | 0.00e+00 | 0.00e+00 | 0.00e+00 | 1.0 |
| C11 | −1.522713e−03(0) | −1.522713e−03(0) | −1.522713e−03(0) | 0, 0, 0 | 0.00e+00 | −1.52e−03 | 6.34e−11 | 1.0 |
| C12 | −5.700899e+02(0) | −4.231332e+02(0) | −1.989129e−01(0) | 0, 0, 0 | 0.00e+00 | −3.37e+02 | 1.78e+02 | 1.0 |
| C13 | −6.842937e+01(0) | −6.842936e+01(0) | −6.842936e+01(0) | 0, 0, 0 | 0.00e+00 | −6.84e+01 | 1.03e−06 | 1.0 |
| C14 | 0.000000e+00(0) | 0.000000e+00(0) | 0.000000e+00(0) | 0, 0, 0 | 0.00e+00 | 0.00e+00 | 0.00e+00 | 1.0 |
| C15 | 0.000000e+00(0) | 0.000000e+00(0) | 4.497445e+00(0) | 0, 0, 0 | 0.00e+00 | 1.80e−01 | 8.81e−01 | 1.0 |
| C16 | 0.000000e+00(0) | 2.819841e−01(0) | 1.018265e+00(0) | 0, 0, 0 | 0.00e+00 | 3.70e−01 | 3.71e−01 | 1.0 |
| C17 | 1.463180e−17(0) | 5.653326e−03(0) | 7.301765e−01(0) | 0, 0, 0 | 0.00e+00 | 1.25e−01 | 1.94e−01 | 1.0 |
| C18 | 3.731439e−20(0) | 4.097909e−19(0) | 9.227027e−18(0) | 0, 0, 0 | 0.00e+00 | 9.68e−19 | 1.81e−18 | 1.0 |

**Table 10** Numerical results (rank based on CEC rules) obtained from 25 independent of $\varepsilon$DEag on 30D problems (taken from [28])

| Function | Best | Worst | Median | $c$ | $\bar{v}$ | Mean | Std | FR |
|---|---|---|---|---|---|---|---|---|
| C01 | -8.218255e-01(0) | -8.206172e-01(0) | -8.195466e-01(0) | 0, 0, 0 | 0.00e+00 | -8.21e-01 | 7.10e-04 | 1.0 |
| C02 | -2.169248e+00(0) | -2.152145e+00(0) | -2.117096e+00(0) | 0, 0, 0 | 0.00e+00 | -2.15e+00 | 1.20e-02 | 1.0 |
| C03 | 2.867347e+01(0) | 2.867347e+01(0) | 3.278014e+01(0) | 0, 0, 0 | 0.00e+00 | 2.88e+01 | 8.05e-01 | 1.0 |
| C04 | 4.698111e-03(0) | 6.947614e-03(0) | 1.777889e-02(0) | 0, 0, 0 | 0.00e+00 | 8.16e-03 | 3.07e-03 | 1.0 |
| C05 | -4.531307e+02(0) | -4.500404e+02(0) | -4.421590e+02(0) | 0, 0, 0 | 0.00e+00 | -4.50e+02 | 2.90e+00 | 1.0 |
| C06 | -5.285750e+02(0) | -5.280407e+02(0) | -5.264539e+02(0) | 0, 0, 0 | 0.00e+00 | -5.28e+02 | 4.75e-01 | 1.0 |
| C07 | 1.147112e-15(0) | 2.114429e-15(0) | 5.481915e-15(0) | 0, 0, 0 | 0.00e+00 | 2.60e-15 | 1.23e-15 | 1.0 |
| C08 | 2.518693e-14(0) | 6.511508e-14(0) | 2.578112e-13(0) | 0, 0, 0 | 0.00e+00 | 7.83e-14 | 4.86e-14 | 1.0 |
| C09 | 2.770665e-16(0) | 1.124608e-08(0) | 1.052759e+02(0) | 0, 0, 0 | 0.00e+00 | 1.07e+01 | 2.82e+01 | 1.0 |
| C10 | 3.252002e+01(0) | 3.328903e+01(0) | 3.463243e+01(0) | 0, 0, 0 | 0.00e+00 | 3.33e+01 | 4.55e-01 | 1.0 |
| C11 | -3.268462e-04(0) | -2.843296e-04(0) | -2.236338e-04(0) | 0, 0, 0 | 0.00e+00 | -2.86e-04 | 2.71e-05 | 1.0 |
| C12 | -1.991453e-01(0) | 5.337125e+02(1) | 5.461723e+02(1) | 0, 1, 0 | 3.24e-01 | 3.56e+02 | 2.89e+02 | 0.12 |
| C13 | -6.642473e+01(0) | -6.531507e+01(0) | -6.429690e+01(0) | 0, 0, 0 | 0.00e+00 | -6.54e+01 | 5.73e-01 | 1.0 |
| C14 | 5.015863e-14(0) | 1.359306e-13(0) | 2.923513e-12(0) | 0, 0, 0 | 0.00e+00 | 3.09e-13 | 5.61e-13 | 1.0 |
| C15 | 2.160345e+01(0) | 2.160375e+01(0) | 2.160403e+01(0) | 0, 0, 0 | 0.00e+00 | 2.16e+01 | 1.10e-04 | 1.0 |
| C16 | 0.000000e+00(0) | 0.000000e+00(0) | 5.421011e-20(0) | 0, 0, 0 | 0.00e+00 | 2.17e-21 | 1.06e-20 | 1.0 |
| C17 | 2.165719e-01(0) | 5.315949e+00(0) | 1.889064e+01(0) | 0, 0, 0 | 0.00e+00 | 6.33e+00 | 4.99e+00 | 1.0 |
| C18 | 1.226054e+00(0) | 2.679497e+01(0) | 7.375363e+02(0) | 0, 0, 0 | 0.00e+00 | 8.75e+01 | 1.66e+02 | 1.0 |

inequality constraints only (except C13, that is, however, the only function with inequality constraints only whose volume of the feasible region is approximately zero); in 30D, it also finds feasible solutions on C18 in 100% of the runs in all configurations (92% in 10D for Configuration 3). The peculiarity of this function is that despite it has one equality constraint, the volume of its feasible region is non-zero. The only other functions on which a non-zero FR is obtained, although not in all configurations and dimensionalities, are C2, C9, C10, C16, C17. Except C10 that has 1 "rotated" equality constraint, all other functions have only separable constraints, which could explain why in some cases even by Gaussian mutation only (which acts independently on each variable) it is possible to reach the feasible region. Overall, among the 3 configurations Configuration 3 has the highest FR across all the tests, while it results that an excessively high value of $\sigma$ in Gaussian mutations (as in Configuration 2) is detrimental.

From these observations, we can conclude that the basic MAP-Elites algorithm we have used in our experimentation is not able to solve efficiently either problems with non-separable equality constraints, or with an approximately zero-volume feasible region. This is not surprising though, as the algorithm is only based on simple genetic operators (Gaussian mutations and uniform crossover in our case) that do not use any information about the constraints. In contrast, $\varepsilon$DEag [28] encapsulates highly efficient CHTs and uses gradient information about constraints that allows the algorithm to reach a 100% FR on all functions in 10D and 30D (except, respectively, for the two algorithms, C12 in 30D, and C11-C12 in both 10D and 30D), as reported in the original papers.

This comparison encourages though the idea to explore in the future the possibility to include into the MAP-Elites scheme at least one dedicated technique for better handling equality constraints, such as the $\varepsilon$ constrained method, initially introduced in [27] and since then used in most of the state-of-the-art algorithms for constrained optimization. Notably, the strength of this method is that it guides the search by allowing $\varepsilon$ level comparisons with a progressively shrinking relaxation (defined by the $\varepsilon$ parameter) of the constraint boundaries.

Considering the objective values, similar considerations can be drawn: limiting the analysis on the functions with 100% FR, it results that MAP-Elites is less efficient at finding optimal values than $\varepsilon$DEag [28]. In all cases MAP-Elites is several orders of magnitude worse than $\varepsilon$DEag, except C1 in 10D where instead the configuration with crossover finds a better optimal value.[4] Once again, this conclusion is not surprising and is also in line with what was observed by Runarsson and Yao [22], who identified the reason for the sometimes poor results obtained by multi-objective approaches (such as [10, 26, 33, 35]): in fact, when applied to constrained optimization, the Pareto ranking leads to a "bias-free" search that is not able to properly guide the search towards (and within) the feasible region. In other words, allowing the search to spend too many evaluations in the infeasible region makes it harder to find feasible

---

[4] We refer the interested reader to the Supplementary Material online, where we show a detailed report of the MAP-Elites results focused on a fitness-based rank, rather than the rank based on the sorting criteria described in the text. These results are omitted here for brevity.

solutions, but also to find feasible solutions with optimal values of the objective function. This might be the case also of MAP-Elites, where some form of bias (such as the $\varepsilon$ constrained method) might be needed.

## 6 Conclusions

In this paper, we have explored the use of MAP-Elites for solving constrained continuous optimization problems. In the proposed approach, each feature in the feature space explored by MAP-Elites corresponds, quite straightforwardly, to the violation of each constraint, discretized according to user-defined steps (tolerance levels). In this way, the algorithm allows to "illuminate" the search space and thus uncover possible correlations between the constraints and the objective. The visualization of MAP-Elites also gives users the possibility to focus on different solutions characterized by different values of constraint violations. We have tested this approach on a large number of benchmark problems in 10 and 30 dimensions, characterized by up to four equality/inequality constraints. Our numerical results showed that while MAP-Elites obtains results that are not particularly competitive with the state-of-the-art on all problems (especially those with equality constraints), it is still able to provide new valuable, easy-to-understand information that can be of great interest for practitioners. Additionally, the algorithm can be easily implemented and applied without any specific tuning to various real-world problems, for instance, in engineering design, where different tolerance levels can be defined depending on the specific constraints.

Since our goal was to evaluate the performance of the basic MAP-Elites on constrained optimization, the proposed approach is purposely quite simplistic, but clearly it can be extended in various ways. First of all, the basic MAP-Elites algorithm we used in this work (as shown in Algorithm 1 can be replaced with some more advanced variants recently proposed in the literature. In particular, the version of MAP-Elites based on centroidal Voronoi tessellation (CVT-MAP-Elites) [31] can be used instead of the basic one in order to scale the algorithm to a larger number of constraints. In order to better handle the unbounded feature spaces (thus avoiding the need for an explicit "upper" bin, (1.0, inf) in our case), the "expansive" MAP-Elites variants introduced in [30] can be employed instead, which are able to expand their bounds (in the feature space) based on the discovered solutions. Other possibilities will be to use the "directional variation" operator introduced in [32], that exploits inter-species (or inter-elites) correlations to accelerate the search, add, most of all, specific constraint handling techniques (especially for handling equality constraints, which as we have seen is the main weakness of this approach) [7, 14]. It is also worth considering the use of surrogate models, such as in [34], in order to further speed up the search and guide it towards the feasible region, still allowing the algorithm to keep infeasible solutions as part of the map. Another improvement can be obtained by using an explicit Quality Diversity measure [1, 4, 20, 21], so to enforce at the same time a better coverage of the feature space and a further improvement in terms of optimization results. It is also possible to hybridize the basic MAP-Elites algorithm with local search techniques (such that MAP-Elites explores the feature space and local search is applied within one bin to further refine the search), or to devise memetic computing approaches based on a combination of MAP-Elites and other metaheuristics, such as CMA-ES, that has been recently applied successfully also to constrained optimization [8, 11, 16].

Finally, on the application side, it will be interesting to evaluate the applicability of this approach on combinatorial constrained optimization problems, which can be obtained by simply modifying the variation operators, or multi-objective constrained optimization, which can be obtained by adding a Pareto-dominance check, as recently shown in the context of robotic experiments [23].

# References

1. Auerbach, J.E., Iacca, G., Floreano, D.: Gaining insight into quality diversity. In: Proceedings of the Genetic and Evolutionary Computation Conference (GECCO) Companion, pp. 1061–1064. ACM, New York, NY, USA (2016)
2. Bagheri, S., Konen, W., Bäck, T.: How to solve the dilemma of margin-based equality handling methods. In: Hoffmann, F., Hüllermeier, E., Mikut, R. (eds.) Proceedings of the Workshop Computational Intelligence, pp. 257–270. KIT Scientific Publishing, Karlsruhe (2018)
3. Cully, A., Clune, J., Tarapore, D., Mouret, J.B.: Robots that can adapt like animals. Nature 521(7553), 503 (2015)
4. Cully, A., Demiris, Y.: Quality and diversity optimization: a unifying modular framework. IEEE Trans. Evol. Comput. 22(2), 245–259 (2018)
5. Deb, K., Datta, R.: A hybrid bi-objective evolutionary-penalty approach for computationally fast and accurate constrained optimization. Technical report, KanGAL (2010)
6. Dolson, E., Lalejini, A., Ofria, C.: Exploring genetic programming systems with map-elites. PeerJ Preprints 6, 1–18 (2018)
7. Gong, W., Cai, Z., Liang, D.: Adaptive ranking mutation operator based differential evolution for constrained optimization. IEEE Trans. Cybern. 45(4), 716–727 (2015)
8. Hellwig, M., Beyer, H.G.: A matrix adaptation evolution strategy for constrained real-parameter optimization. In: Congress on Evolutionary Computation (CEC), pp. 1–8. IEEE, New York (2018)
9. Hellwig, M., Beyer, H.G.: Benchmarking evolutionary algorithms for single objective real-valued constrained optimization-a critical review. Swarm Evol. Comput. 44, 927–944 (2019)
10. Hernáandez Aguirre, A., Botello Rionda, S., Lizáarraga Lizáarraga, G., Coello Coello, C.: IS-PAES: multiobjective optimization with efficient constraint handling. In: Burczyński, T., Osyczka, A. (eds.) IUTAM Symposium on Evolutionary Methods in Mechanics, pp. 111–120. Springer, Dordrecht (2004)
11. Jamal, M.B., Ming, F., Zhengang, J.: Solving constrained optimization problems by using covariance matrix adaptation evolutionary strategy with constraint handling methods. In: Proceedings of the International Conference on Innovation in Artificial Intelligence, pp. 6–15. ACM, New York, NY, USA (2018)
12. Lehman, J., Stanley, K.O.: Exploiting open-endedness to solve problems through the search for novelty. In: International conference on Artificial Life (Alife XI), pp. 329–336. MIT Press, Cambridge, MA (2008)

13. Maesani, A., Iacca, G., Floreano, D.: Memetic viability evolution for constrained optimization. IEEE Trans. Evol. Comput. **20**(1), 125–144 (2016)
14. Mallipeddi, R., Suganthan, P.N.: Differential evolution with ensemble of constraint handling techniques for solving CEC 2010 benchmark problems. In: Congress on Evolutionary Computation (CEC), pp. 1–8. IEEE, New York (2010)
15. Mallipeddi, R., Suganthan, P.N.: Problem definitions and evaluation criteria for the CEC 2010 competition on constrained real-parameter optimization. Technical report, NTU, Singapore (2010)
16. de Melo, V.V., Iacca, G.: A modified covariance matrix adaptation evolution strategy with adaptive penalty function and restart for constrained optimization. Expert Syst. Appl. **41**(16), 7077–7094 (2014)
17. Michalewicz, Z., Schoenauer, M.: Evolutionary algorithms for constrained parameter optimization problems. Evol. Comput. **4**(1), 1–32 (1996)
18. Mouret, J.B.: Novelty-based multiobjectivization. In: Doncieux, S., Bredèche, N., Mouret, J.B. (eds.) New Horizons in Evolutionary Robotics, pp. 139–154. Springer, Berlin, Heidelberg (2011)
19. Mouret, J.B., Clune, J.: Illuminating search spaces by mapping elites. CoRR 1–15 (2015). arXiv:1504.04909
20. Pugh, J.K., Soros, L.B., Stanley, K.O.: Quality diversity: a new frontier for evolutionary computation. Front. Robot. AI **3**, 40 (2016)
21. Pugh, J.K., Soros, L.B., Szerlip, P.A., Stanley, K.O.: Confronting the challenge of quality diversity. In: Proceedings of the Genetic and Evolutionary Computation Conference (GECCO), pp. 967–974. ACM, New York, NY, USA (2015)
22. Runarsson, T.P., Yao, X.: Search biases in constrained evolutionary optimization. IEEE Trans. Syst. Man, Cybern. Part C (Appl. Rev.) **35**(2), 233–243 (2005)
23. Samuelsen, E., Glette, K.: Multi-objective analysis of map-elites performance. CoRR 1–8 (2018). arXiv:1803.05174
24. Shahrzad, H., Fink, D., Miikkulainen, R.: Enhanced optimization with composite objectives and novelty selection. CoRR 1–7 (2018). arXiv:1803.03744
25. Singh, H.K., Asafuddoula, M., Ray, T.: Solving problems with a mix of hard and soft constraints using modified infeasibility driven evolutionary algorithm (IDEA-M). In: Congress on Evolutionary Computation (CEC), pp. 983–990. IEEE, New York (2014)
26. Surry, P.D., Radcliffe, N.J.: The COMOGA method: constrained optimisation by multi-objective genetic algorithms. Control Cybern. **26**, 391–412 (1997)
27. Takahama, T., Sakai, S.: Constrained optimization by $\varepsilon$ constrained particle swarm optimizer with $\varepsilon$-level control. In: Soft Computing as Transdisciplinary Science and Technology, pp. 1019–1029. Springer, Berlin, Heidelberg (2005)
28. Takahama, T., Sakai, S.: Constrained optimization by the $\varepsilon$ constrained differential evolution with an archive and gradient-based mutation. In: Congress on Evolutionary Computation (CEC), pp. 1–9. IEEE, New York (2010)
29. Urquhart, N., Hart, E.: Optimisation and illumination of a real-world workforce scheduling and routing application via map-elites. CoRR, 1–13 (2018). arXiv:1805.11555
30. Vassiliades, V., Chatzilygeroudis, K., Mouret, J.B.: A comparison of illumination algorithms in unbounded spaces. In: Proceedings of the Genetic and Evolutionary Computation Conference (GECCO) Companion, pp. 1578–1581. ACM, New York, NY, USA (2017)
31. Vassiliades, V., Chatzilygeroudis, K., Mouret, J.B.: Using centroidal voronoi tessellations to scale up the multidimensional archive of phenotypic elites algorithm. IEEE Trans. Evol. Comput. **22**(4), 623–630 (2018)
32. Vassiliades, V., Mouret, J.B.: Discovering the elite hypervolume by leveraging interspecies correlation. In: Proceedings of the Genetic and Evolutionary Computation Conference (GECCO), pp. 149–156. ACM, New York, NY, USA (2018)

33. Venkatraman, S., Yen, G.G.: A generic framework for constrained optimization using genetic algorithms. IEEE Trans. Evol. Comput. **9**(4), 424–435 (2005)
34. Wang, Y., Yin, D.Q., Yang, S., Sun, G.: Global and local surrogate-assisted differential evolution for expensive constrained optimization problems with inequality constraints. IEEE Trans. Cybern. **1**(1), 1–15 (2018)
35. Zhou, Y., Li, Y., He, J., Kang, L.: Multi-objective and MGG evolutionary algorithm for constrained optimization. In: Congress on Evolutionary Computation (CEC), pp. 1–5. IEEE, New York (2003)

# Optimization of Fresh Food Distribution Route Using Genetic Algorithm with the Best Selection Technique

Douiri Lamiae, Abdelouahhab Jabri, Abdellah El Barkany, and A.-Moumen Darcherif

**Abstract** All along the food supply chain, managers face the challenge of making important cost-optimized decisions relating to transportation and storage conditions. An efficient management of food products requires the consideration of their perishable nature to solve the safety problem, and logistics efforts in minimizing the total cost while maintaining the quality of food products above acceptable levels. This paper addresses a methodology to resolve a capacitated model for food supply chain (FSC). The model is a constrained mixed integer nonlinear programming problem (MINLP) that computes the cost of quality internally for a three echelon FSC, to minimize the total cost under overall quality level constraints, due to the perishable nature of the products, and the other constraints of demand, capacity, flow balance, and cost. The practical application of the model is demonstrated using two approaches: an exact method based on the Branch & Bound technique, and a Genetic Algorithm solution method. Then, we propose a comparison of different GA selection strategies, such as Tournament, Stochastic Sampling without Replacement, and Boltzmann tournament selection; the performance of each selection method is studied using paired "T-Test" of statistical analysis. The results of computational testing are reported and discussed, and implications and insights for managers are provided by studying instances of practical and realistic size. It was evident that the tournament selection is more likely to produce a better performance than the other selection strategies.

**Keywords** Food supply chain · MINLP · CPLEX software · Genetic algorithm · Simulated annealing · T-test

D. Lamiae (✉) · A. Jabri · A. El Barkany
Mechanical Engineering Laboratory, Faculty of Science and Techniques, Sidi Mohammed Ben Abdellah University, B.P. 2202, Route d'Imouzzer, FEZ, Morocco
e-mail: lamiae_iad@yahoo.fr

A.-M. Darcherif
ECAM-EPMI, 13 Boulevard de l'Hautil, 95092 Cergy-Pontoise cedex, France

© The Author(s), under exclusive license to Springer Nature Singapore Pte Ltd. 2021
A. J. Kulkarni et al. (eds.), *Constraint Handling in Metaheuristics and Applications*,
https://doi.org/10.1007/978-981-33-6710-4_8

## List of Acronyms

B&B      Branch and Bound
COQ      Cost of Quality
CPU      Central Processing Unit
FSC      Food Supply Cain
GA       Genetic Algorithm
MILP     Mixed Integer Linear Programming
MINLP    Mixed Integer Non-Linear Programming
PAF      Prevention-Appraisal-Failure
PSO      Particle Swarm Optimization
QL       Quality Level
RBF      Radial Basis Function
SA       Simulated Annealing
SC       Supply Chain

## 1  Introduction

This paper addresses the problem of supply chain performance measurement in a food supply chain (FSC) design. The term food supply chain (FSC) has been coined to describe the activities dedicated to manufacturing food products from production to distribution of final products to the consumer [1]. As any other supply chain, the FSC is formed by a network of entities such as producers, processors, and distributors dedicated to deliver superior consumer value at less cost. But there are relevant characteristics that make the FSC different and more complex to manage, especially the wide range of decisions that must be taken in food supply chain, related to quality and safety, given the perishable nature of food products. A general production–distribution network comprises three echelons: production, distribution, and customer. The FSC network design problem consists of selecting the facilities to open in order to minimize total costs and then, maximize customer service and overall quality levels. Product quality degradation is often incorporated in the modeling of supply [2]. Therefore, we find in literature a number of studies that deal with the optimization of specifics which has a practical value to prevent food quality degradation, such as cost, time, and storage conditions of fresh food during transportation. For example, de Keizer et al. [2] presents a MILP model to design a logistics network for distribution of perishable products, taking into consideration product quality decay and its heterogeneity. The quality decay was based on a time–temperature model, and also other environmental conditions like humidity. The model was tested for various small and large instances for cut flowers logistics network. Li and O'Brien [3] proposed a model to improve supply chain efficiency and effectiveness based on profit, lead-time performance, and waste elimination. Zhao and Lv [4] proposed a multi-echelon and multi-product agri-food supply chain network to minimize the production and transportation costs. The model is a mixed integer programming (MIP) problem that was solved through a particle swarm optimization (PSO) approach. Tsao [5] proposed a fresh food supply chain network. He developed a model to determine the optimal service for agricultural product marketing corporations. The author proposed an algorithm to solve the nonlinear problem, provides numerical analysis to illustrate the proposed solution procedure, and discusses the effects of various system parameters on the decisions and total profits. A real case of an agricultural product supply chain in Taiwan is used to verify the model. Wang et al. [6] developed a multi-objective vehicle routing model for perishable food distribution route dealing with the minimization of total cost and maximization of the freshness state of delivered products. The author used fuzzy method for determining a quantified food safety risk indicator to support operational decisions and to be valuable as a component of a HACCP system. Nakandala [7] developed a methodology to assess fresh food supply chain to optimize the total cost while maintaining a required quality level. The application of the proposed total cost model was demonstrated using three approaches of Genetic Algorithm, Fuzzy Genetic Algorithm, and Simulated Annealing. The author suggested that all three approaches are adoptable but the FGA provided a better performance than the other two approaches of GA and SA based on the performance evaluation. Zhang and Chen [8] developed a model for the vehicle routing problem in the cold chain logistics for a multi-food product. The objective was to minimize the delivery cost that includes transportation cost, cooling cost, and penalty cost under the constraints of time and loading weight and volume. A genetic algorithm (GA) method was proposed to resolve the model, and the experiments showed that the GA method can provide sound solutions with good robustness and convergence characteristics in a reasonable time span. Aramyan et al. [9] presented a conceptual model for agri-food supply chain including four performance measuring: efficiency, flexibility, responsiveness, and food quality and applied it in the tomato supply chain. Bag and Anand [10] employed ISM methodology to design sustainable network for food processing network in India. This methodology aims at including the environmental and social aspects in the supply chain of food sector. Boudahri et al. [11] developed a mathematical model to solve the location-allocation problem for a chicken meat distribution center, while minimizing the total logistics costs and respecting vehicle and slaughterhouses capacity. LINGO optimization solver version12.0 has been used to get the solution to the problem. Doganis [12] presented a complete framework that can be used for developing nonlinear time series for short shelf-life food products and applied it successfully to sales data of fresh milk. The method is a combination of two artificial intelligence technologies: the radial basis function (RBF) and a specially designed genetic algorithm (GA). Entrup [13] developed a mixed integer linear programming approach, considering the shelf life in production planning, applied to yogurt production. To solve the model, the author used Branch and Bound procedures that were implemented using ILOG's OPL Studio 3.6.1 as a modeling environment and its incorporated standard optimization software CPLEX 8.1. Ekşioğlu [14] developed a production and distribution planning problem in a dynamic two-stage supply chain considering the perishability and limited shelf life of the products. But this model does not consider transportation costs. The aim of this work presented in the paper is to develop a GA-based procedure to resolve the FSC-COQ model, and to quantify the effect of different parameters on total cost and further compare it with solution found by CPLEX solver. The remainder of this paper is organized as follows: in the next section, the capacitated FSC-COQ problem is formulated and discussed. Next, in Sect. 3, we provide a comprehensive explanation of the proposed GA methodology for optimizing the total cost through the FSC followed by a presentation of the computational results obtained to show the performance of the GA using actual data versus results obtained with CPLEX software. In Sect. 4, we propose a comparison between different selection techniques for the GA algorithm. Section 5 presents a conclusion and the paper finishes with suggestions for further research.

## 2  The Capacitated FSC-COQ Model

The food supply chain design problem discussed in this paper is an integrated multi-echelon single product system. This generic supply chain is based on production facilities, storage facilities, transportation, and retailers. Our modeling approach consists of one objective function that attempts to minimize the total cost. The design task involves the choice of facilities production sites and storage facilities to be opened, and transportation equipment to be chosen to satisfy the customer demand with minimum cost. The following sets are defined: $P$, set of production sites ($p \in P$); $D$, set of storage facilities ($d \in D$); $Y$, set of transportation vehicles ($y \in Y$); $R$, set of retailers ($r \in R$). The model constants are $dem_{r,t}$, demand at retailer $r$ in period $t$; $a_{p,t}$, $p$ roduction capacity for production sites; $b_{d,t}$, storage capacity for storage facilities; $g_d^1$, cooling cost for storage facility d per period; $g_d^2$, cost for storing one product for one period in storage facility $d$; $P_{p,t}$, cost for producing one product unit in production site ($p \in P$) in period $t$, $f_{d,r,y}$, Cost for transporting one product per period from storage facility to retailer $r$; $\Delta q_d$, Quality degradation in one period for products stored in storage facility $d$; $\Delta q_{d,y}$, quality degradation for products transported from storage facility $d$ to retailer $r$ by transportation vehicle $y$. It is worth noting that quality degradation of food products changes according to product characteristics and storage conditions. In general, degradation of food products is dependent on storage time, storage temperature, and various constants such as activation energy, and gas constant. The model variables are $W_{p,d,t}$, flow quantities from production sites $p$ to storage facility $d$ in period $t$; $W_{d,r,t}$, flow quantities from storage facility $d$ to retailer $r$ in period $t$; $\alpha_p$, fraction defective at storage facility; $\beta_d$, inspection error rate after storage process; $Z_{p,t}$, binary variable which equals 1 if production site $p$ is selected, 0 otherwise; $O_{d,r,t}$, binary variable which equals 1 if transportation equipment is selected in period $t$, 0 otherwise; $X_{d,t}$, binary variable which equals 1 if storage facility $d$ is selected in period $t$, 0 otherwise. The objective function consists of minimizing the total cost of the food supply chain, where the total cost is the sum of operational costs, the food quality costs, and the transportation costs.

According to the classical PAF model, the cost of quality calculated through a logistics chain is the sum of the costs of prevention, evaluation, internal and external failure costs, as shown in (1):

$$COFQ = C_P + C_A + C_{IF} + C_{EF} \tag{1}$$

Table 1 shows the expressions of the different cost categories calculated for our FSC model.

The optimization model of the FSC-COQ is formulated as follows:

**Min**

$$\sum_t \sum_p P_{p,t} W_{p,d,t} Z_{p,t} + \sum_t \sum_d g_{d,k}^2 W_{p,d,t} X_{d,t} + \sum_t \sum_p f_{d,r} W_{d,r,t} O_{d,r,t}$$
$$+ COFQ(W_{p,d,t}, W_{d,r,t}, Z_{p,t}, X_{d,t}, O_{d,r,t}, \alpha_p, \beta_d) \tag{9}$$

**Subject to:**

$$\sum_t W_{d,r,t} \geq dem_{r,t}; \quad \forall r \in R, \forall t \in T \tag{10}$$

$$\sum_t W_{p,d,t} \geq \sum_t W_{d,r,t}; \quad \forall d \in D \tag{11}$$

$$\sum_t W_{p,d,t} \leq \sum_p a_{p,t} Z_{p,t} \tag{12}$$

$$\sum_t W_{d,r,t} \leq \sum_d b_{d,t} X_{d,t} \tag{13}$$

$$FQ_L \geq lR_t \tag{14}$$

$$Z_{k,t} \in \{0.1\} \quad \forall t \in T \tag{15}$$

$$O_{d,r,t} \in \{0.1\} \quad \forall t \in T \tag{16}$$

$$X_{d,t} \in \{0.1\} \quad \forall t \in T \tag{17}$$

$$0 \leq \alpha_p \leq 1 \quad \forall p \in P \tag{18}$$

$$0 \leq \beta_d \leq 1 \quad \forall d \in D \tag{19}$$

**Table 1** Different cost categories for the FSC-COQ model

| Cost category | Expression | |
|---|---|---|
| Production cost $C_{pr}$ | $\sum_t \sum_p P_{p,t} W_{p,d,t} Z_{p,t}$ | (2) |
| Storage cost $C_{st}$ | $\sum_t \sum_d g_d^2 W_{d,r,t} X_{d,t}$ | (3) |
| Transportation cost $C_{tr}$ | $\sum_t \sum_d \sum_y f_{d,r,y} W_{d,r,t} O_{d,r,t}$ | (4) |
| Prevention cost $C_P$ | $\sum_d g_d^1 X_{d,t} + \sum_d \sum_p \sum_t \dfrac{C_{pvar}(1-\alpha_p)}{W_{p,d,t}(1-\Delta q_d)} Z_{p,t}$ | (5) |
| Appraisal cost $C_A$ | $\sum_p C_{Afix} Z_{p,t} + \sum_d \sum_t \sum_p \dfrac{C_{Avar}(1-\beta_d)}{W_{p,d,t}} Z_{p,t} X_{d,t}$ | (6) |
| Internal failure cost $C_{IF}$ | $\sum_p C_{IFfix} Z_{p,t} + \sum_p \sum_t \sum_d \dfrac{g_d^2(1-\beta_d)}{W_{p,d,t}}\left[(1-\alpha_p)\Delta q_d\right] X_{d,t} Z_{p,t} +$ $\sum_p \sum_d \sum_t (g_d^2 + C_s) W_{p,d,t}(1-\beta_d)\alpha_p X_{d,t} Z_{p,t}$ | (7) |
| External failure cost $C_{EF}$ | $\sum_d \sum_p \sum_t \sum_y \overline{C_{EF}}\, W_{d,r,t}\left[\Delta q_{d,y}(1-\alpha_p)(1-\Delta q_d) + \beta_d\left[(1-\alpha_p)\Delta q_d + \alpha_p\right]\right] X_{d,t} O_{d,r,t}$ | (8) |

In the formulated model (9)–(19), the objective function (9) aims to minimize the total costs. The first term of (9) is the total production costs, the second term is the storage cost and the sum of these two terms is the operational costs. The third term is the transportation cost. The fourth term represents the total COFQ for the network. The parameters for the COFQ function are shown in the Appendix. Constraint (10) enforces that demand at retailer is not exceeded, constraint (11) makes sure that the number of food products going through storage facility equals the number of food products transported to retailers, constraint (12) reflect that capacity at producers is not exceeded, and constraint (13) enforces that capacity at storage facilities is not exceeded. Constraint (14) is the quality level constraint; thus, the quality of the food products delivered at each retailer must meet the minimum required quality level calculated by

$$
FQ_L = \frac{\sum_d \sum_p \sum_y (1 - \Delta q_{d,y}) W_{d,r,t} (1 - \alpha_p)(1 - \Delta q_d) Z_{p,t} X_{d,t} O_{d,r,t}}{\sum_d W_{d,r,t} X_{d,t}} \tag{20}
$$

which represents the proportion of good products relative to all products transported to final customers.

Constraints (15), (16), and (17) are non-negativity constraints for decision variables. The remaining constraints (18) and (19) are feasible ranges of model variables.

## 3 Solution Procedures

The formulated FSC-COQ model (9)–(19) is a constrained mixed integer nonlinear programming problem (MINLP). The complexity of the solution is NP-hard, because of the difficulty of nonlinear problems (NLP), and the combinatorial nature of mixed integer programming (MIP). We propose two procedures for solving the FSC-COQ model. One is the exact algorithm which includes the branch and bound method. For this method, we use the standard optimization software CPLEX 12.6 to solve a FSC-COQ with a small number of production sites, storage sites, and transportation vehicles. In the second solution procedure, we use an evolutionary algorithm for larger test instance. Existing literature that compares different algorithms for Supply Chain Network Design has demonstrated that genetic algorithm (GA) performs other metaheuristics as it can handle greater problems with less computational time. The GA proposed for this problem was coded in Python 3.4.1, and all test runs were performed on a 1.70 GHz Intel® core™ i5 PC with 4 GB RAM.

## 3.1   Genetic Algorithm-Based Optimization

GA simulates the survival of the fittest among individuals over consecutive generations throughout the solution of a problem [15]. Thus, only individuals who are able
to adapt to constraints generated by the natural environment can survive and generate
offspring to ensure the sustainability of the species. On the other hand, unsuitable
individuals are automatically discarded. Selection makes it possible to have individuals more and more adapted during the generations. In a similar way, in optimization
problems, genetic algorithms are based on the principle of selecting the most appropriate individual or candidate solution, which is represented by a chromosome or
string of characters. Each individual represents a point in the search space. The pseudocode of the genetic algorithm developed for our FSC-COQ model is given in the
following Algorithm 1. We consider the following notations for genetic operators:
Pop_size: population size, Max_iter: maximum iterations, Pc: Crossover probability,
Pm: mutation probability.

---

**Algorithm 1. Genetic Algorithm for FSC-COQ**

---

**Step 1**. Set the initial values of parameters used to generate GA instances, Pop_size,
Pc, Pm and Max_iter.
**Step 2**. Generate an initial population of chromosomes which represents a feasible
solution
**Step 3**. Calculate the fitness function 'COFQ' to evaluate each chromosome of the
current population
**Step 4**. Select the best fitness of the decision variables of the current population via
stochastic remainder selection without replacement method.
**Step 5**. *While* Max_iter is not reached, start the reproduction
**Step 6**. *For* j < Pop_size:
Randomly select a sequence of P (N)
 if random_value1 < Pc :
 Apply the standard crossover and produce two children
 *if* random_value2 < Pm :
 Apply the mutation
Copy the obtained ascending into the new population
Evaluation of the objective functions of all the chromosomes of the new population,
return to step 3.
**Step 7**. Increment j until meeting the stopping criteria: Max_iter and maximum
number of iterations without improvement.
**Step 8**. End **While**

---

## 3.2   Taguchi Calibration for GA Parameters

The effectiveness of optimization approaches and the quality of the convergence
process in GA depends on the specific choices and combinations of parameter values

**Table 2** Considered levels for each factor of AG

| Factors | Level (1) | Level (2) | Level (3) |
|---------|-----------|-----------|-----------|
| Popsize | 20 | 50 | 100 |
| Max_iter | 100 | 150 | 200 |
| Pc | 0.2 | 0.4 | 0.7 |
| Pm | 0.1 | 0.4 | 0.7 |
| Penalty | 5000 | 6500 | 8000 |

of genetic operators. The Taguchi method is a powerful tool for the design of high-quality systems. It is firstly advanced by a Japanese quality control expert named Genechi Taguchi in the 1960s. This approach is based on optimal designs that provide the maximum amount of information with minimal testing, exploring factors that influence mean and variance. In addition, known by its robustness, allows to minimize the impact of the noise factor that cannot be controlled by the designers and to find the best level of the controllable factors, and this, by using a lot of notions whose controlled factor, orthogonal table, noise factor, signal-to-noise (S/N) ratio, etc. The main idea of this methodology is to find factors and levels and to get the appropriate combination of these factors and levels by the method of design of experiments. To solve this problem, the Taguchi method uses a quality loss function to calculate the deviation between experimental value and expected value. This loss function can be transformed into a S/N ratio. Taguchi recommends the use of the S/N ratio to measure the quality characteristics deviating from the desired values. Usually, there are three categories of quality characteristics in the analysis of the S/N ratio, i.e., the lower-the-better, the-higher-the-better, and the-nominal-the better. The S/N ratio for each level of process parameters is computed based on the S/N analysis. Regardless of the category of the quality characteristic, a greater S/N ratio corresponds to a better quality characteristic. In this order of ideas, 27 sets of experiments were performed based on the distribution of the orthogonal table, and three levels were assigned to each factor, as shown in Table 2.

Figure 1 shows a graph of the main effects created on "Minitab 19" by plotting the average of the characteristics for each factor level.

Moreover, as regards the calibration of the optimization approach in question, the four factors considered above must be adjusted such as shown in Table 3.

## 3.3 Experimental Study

### 3.3.1 Test Instances

In this section, we illustrate this modeling approach by use of a small instance with three production sites, three storage sites, and three transport vehicles. The resulting problem instance has 30 constraints and 18 decision variables. As described in solution procedure section, we use both the standard solver CPLEX 12.6 and the

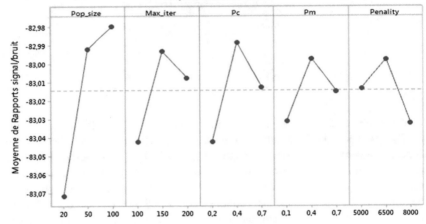

Fig. 1 Main effects graph for S/N ratio for AG

**Table 3** Optimal parameters of AG

| Methodology | Pop_size | Max_iter | Pc | Pm | Penality |
|---|---|---|---|---|---|
| AG | 100 | 150 | 0.4 | 0.4 | 6500 |

AG algorithm to solve this small instance. Moreover, we study a larger test instance with a variety of network sizes as shown in Table 4, to demonstrate that the developed AG procedure is able to handle the trade-offs between all cost categories. We note that the larger the population size, the longer computation time it takes. For solving the small instance, we fix a limit of 4 h for the solution time for the CPLEX solver.

**Table 4** Test problems

| Test problem | Problem size $|P| \times |D| \times |Y|$ | Number of constraints | Number of decision variables |
|---|---|---|---|
| 1 | $3 \times 3 \times 3$ | 30 | 18 |
| 2 | $5 \times 5 \times 5$ | 44 | 28 |
| 3 | $10 \times 10 \times 10$ | 79 | 53 |
| 4 | $15 \times 15 \times 15$ | 114 | 78 |
| 5 | $20 \times 20 \times 20$ | 149 | 103 |
| 6 | $25 \times 25 \times 25$ | 184 | 128 |

### 3.3.2 Performance of the AG

We demonstrate in this section how the Genetic Algorithm performs CPLEX. We measure the performance by solution quality and computational time in CPU seconds. Table 5 reports the computational time required for each solution procedure over different test problems.

We observe that GA performs better, even for small problem size. The CPU time calculated for GA procedure is for 500 iterations, which means that solutions are generated extremely fast. On the other hand, CPLEX requires important computational time even for only one iteration, as it can no longer solve large-size problems. We conclude that managers should use GA for real size problems that requires less computational time. Next, we compare the quality solutions of GA-based procedure and CPLEX. For this purpose, we study the average gaps between GA and CPLEX for two scenarios:

**Scenario 1**

We investigate the impact of the quality degradation for products in storage facility $\Delta q_d$ on total cost curve, by changing this parameter randomly, while keeping the parameters $\alpha_p$ (which is internal decision variable in our model) and $\Delta q_{d_y}$ constant. Table 6 give computational results for two levels of $\Delta q_d$. The results show that the greater the $\Delta q_d$, the slower is the total quality level which minimizes the total cost. This means that we need to invest more in prevention activities especially, the variable cost for prevention of poor quality after a failed storage process, and internal failure costs generated by bad products. We need also to maintain initial product quality, for example:

- by performing regular analyses to assess quality and strict food safety practices,
- by respecting soil conditions, harvest time, and also by avoiding process losses, contamination in process causing loss of quality, and
- by considering fraction defective when selecting storage facilities, in order to avoid food waste and decisions based solely on higher initial quality product that leads to higher purchasing costs, to meet the quality requirements at minimum cost.

**Table 5** Solution time of CPLEX and GA

| Test problem | Solution time in CPU (s) | |
|---|---|---|
| | CPLEX | AG (150 iterations) |
| 1 | 1.75 | 78.26 |
| 2 | 17.34 | 110.20 |
| 3 | 10 min 26 | 221.90 |
| 4 | 3 h 28 min 58 | 282.43 |
| 5 | – | 387.46 |
| 6 | – | 532.39 |

**Table 6** Performance of AG against CPLEX for different $\Delta q_d$

| QL | $\Delta q_{d,y}$ | CPLEX | AG | Gap (%) |
|---|---|---|---|---|
| *Total cost (\$) for* $\Delta q_d = 0.03$ | | | | |
| 0.948 | 0.005 | 29622.84 | **29821.75** | 0.67 |
| 0.94 | 0.01 | 16993.48 | **17043.50** | 0.29 |
| 0.93 | 0.02 | 14031.70 | **14092.56** | 0.43 |
| 0.92 | 0.04 | 17453.86 | **17455.56** | 0.009 |
| 0.87 | 0.08 | 24298.18 | **24594.23** | 1.21 |
| 0.85 | 0.1 | 27720.34 | **28694.64** | 3.5 |
| 0.76 | 0.2 | 44831.14 | **45791.02** | 2.14 |
| 0.67 | 0.3 | 61941.94 | 63619.92 | 2.7 |
| 0.57 | 0.4 | 79052.74 | **79496.63** | 0.56 |
| 0.48 | 0.5 | 96163.54 | **96635.91** | 0.49 |
| 0.38 | 0.6 | 113274.34 | **113790.58** | 0.45 |
| 0.28 | 0.7 | 130385.14 | **135132.38** | 3.64 |
| 0.192 | 0.8 | 147495.94 | **150076.03** | 1.74 |
| | | | **Avg GAP** | **1.37** |
| *Total cost (\$) for* $\Delta q_d = 0.3$ | | | | |
| 0.7 | 0.005 | 26663.47 | **27460.85** | 2.99 |
| 0.68 | 0.01 | 17680.65 | **18017.55** | 1.9 |
| 0.66 | 0.03 | 14230.12 | **14338.44** | 0.76 |
| 0.65 | 0.04 | 14464.921 | **15496.10** | 0.2 |
| 0.63 | 0.08 | 20404.12 | **21132.55** | 3.5 |
| 0.61 | 0.1 | 22873.72 | **23233.76** | 1.57 |
| 0.54 | 0.2 | 35221.72 | **36807.02** | 4.5 |
| 0.48 | 0.3 | 47569.72 | **48258.12** | 1.44 |
| 0.41 | 0.4 | 59917.72 | 61951.41 | 3.39 |
| 0.34 | 0.5 | 72265.72 | **74686.52** | 3.34 |
| 0.27 | 0.6 | 84613.72 | **86271.18** | 1.95 |
| 0.2 | 0.7 | 96961.72 | **100009.11** | 3.14 |
| 0.137 | 0.8 | 109309.72 | **116371.18** | 6.46 |
| | | | **Avg GAP** | **2.71** |

From this scenario 1, the average gaps between GA and CPLEX are 1.37 and 2.71% for $\Delta q_d = 0.03$ and $\Delta q_d = 0.3$. It means that GA performs better when $\Delta q_d$ is smaller. And according to results obtained in Table 4, which demonstrate that solutions are generated extremely fast with GA, we conclude that GA is therefore suitable for integration in IT-based decision support systems.

**Table 7** The relative transportation cost for different temperature levels

| Temperature (°C) | 2 | 4 | 6 | 8 | 10 |
|---|---|---|---|---|---|
| Shelf life (days) | 34 | 29 | 24 | 19 | 14 |
| Quality degradation $\Delta q$ | 11 | 13 | 16 | 20 | 27 |
| Relative transportation cost ($rc_y$) | 1 | 0.88 | 0.77 | 0.65 | 0.54 |

## Scenario 2

We study in this second scenario, the impact of transportation cost $f_{d,r,y}$ on total cost curve and the quality level achieved. We consider $f_{d,r,y} = \gamma \times rc_y$ where we vary the parameter $\gamma$ between 0, 15, and 1, and $rc_y$ is the relative transportation cost given by the values reported in Table 7. The results are reported in Table 8.

For the second scenario, the average gap changes proportionally over the transportation cost factor $\gamma$. The lower is $\gamma$, the smaller is the average GAP too. This is due to the reduction of the total cost for lower transportation cost factor, while maintaining the quality level achieved by the food supply chain. We conclude that GA performs better when $\gamma$ is smaller.

**Table 8** Performance of AG against CPLEX for different $\gamma$

| $\gamma$ | T (°C) | Total cost ($) | | Gap (%) |
|---|---|---|---|---|
| | | CPLEX | AG | |
| 1 | 2 | 29551.42 | 30313.76 | 2.58 |
| | 4 | 32829.58 | 33637.76 | 2.46 |
| | 6 | 37830.82 | 37974.13 | 0.38 |
| | 8 | 44531.14 | 45445.24 | 2.05 |
| | 10 | 56376.70 | 58544.32 | 3.84 |
| **Avg** | | | | **2.26** |
| 0.6 | 2 | 29071.42 | 30078.02 | 3.46 |
| | 4 | 32407.18 | 32791.60 | 1.19 |
| | 6 | 37461.22 | 37691.86 | 0.62 |
| | 8 | 44219.14 | 44636.62 | 0.94 |
| | 10 | 56117.50 | 57310.13 | 2.13 |
| **Avg** | | | | **1.77** |
| 0.15 | 2 | 28531.42 | 29057.65 | 1.84 |
| | 4 | 31931.98 | 32344.87 | 1.29 |
| | 6 | 37045.42 | 37093.51 | 0.13 |
| | 8 | 43868.14 | 44616.83 | 1.71 |
| | 10 | 55825.90 | 57075.80 | 2.24 |
| **Avg** | | | | **1.5** |

### 3.3.3   Evaluation of AG Selection Strategies

In this section, we investigate the performance of GA with two selection strategies for solving the FSC-COQ model, in which our objective is to find the optimal quality level that minimizes the total cost. In the above, we used stochastic technique in the selection phase. Choosing a right selection technique is a very critical step in GA, since if not chosen correctly, it may lead to convergence of the solution to a local optimum. There are various selection strategies for AG algorithm presented in Goldberg (1989) [15]: deterministic sampling, remainder stochastic sampling without replacement, remainder stochastic sampling with replacement, stochastic sampling without replacement, stochastic sampling with replacement, and stochastic tournament. In our study, we compare tournament selection and stochastic sampling without replacement. We present first a remainder of the operating mode of each technique:

i.   **Stochastic sampling without replacement**

The stochastic selection without replacement is based on the concept that the fitness of each chromosome should be reflected in the incidence of this chromosome in the reproductive pool. The stochastic rest selection technique involves calculating the relative physical form of a chromosome, which is the ratio between the physical form associated with an individual and the average physical form, say $m_i = f_i/f_{moy}$. The integer part of $m_i$ is used to select a parent deterministically, for the rest the algorithm applies the roulette selection. For example, if the $m_i$ value of an individual is 2.3, then individual is selected twice as a parent because the whole party is 2. The rest of the parents are chosen stochastically with a probability proportional to the party fractional of its scaled value. If the stochastic selection of the remainder is performed without replacement, the fractional parts of the expected occurrence value are treated as probabilities. A fractional part $e_i$ is calculated as follows:

$$e_i = \frac{f(s)}{\sum f(s)/popsize} \tag{21}$$

Fractions of $e_i$ are treated as probabilities that each string has copies in subsequent generations. According to Goldberg (1988), the strategy of stochastic sampling without replacement is an improved method of the roulette wheel selection.

ii.   **Tournament selection**

The principle of this technique is to make a meeting between several pairs of individuals randomly selected from the population, then choose among these pairs the individual who has the best quality of adaptation who will be the winner of the tournament and will be reproduced in the new population. The procedure is iterated until the new population is complete. This tournament selection technique allows individuals of lower quality to participate in the improvement of the population. There are different ways to implement this technique: An individual can participate in multiple tournaments, multiple tournaments can be created with multiple participants, and so on.

---

**Algorithm 2. Tournament selection algorithm**

---

Set the values of parameters Pop_size, Pc, Pm and Max_iter.
 Set tournament size t=15
*Step 4 of algorithm 1*
While j<t
Begin Pick t random individuals from Pop_size
Calculate fitness value of these individuals
Select the best and store in new vectors
Loop until all spots are filled

---

The calibration of the tournament selection parameters is performed following the Taguchi technique with the same parameters of the Genetic Algorithm presented previously. Similarly, 27 sets of experiments were performed based on the distribution of the orthogonal table, and three levels were assigned to each factor, as shown in Table 9.

Figure 2 shows the graph of the main effects created on Minitab by plotting the average of the characteristics for each factor level.

Table 10 summarizes the selected levels of each factor to control and optimizes the tournament selection technique.

We report in Table 11 the best results obtained for 13 quality levels, and for 3, 5, and 10 problem size, run with the two selection techniques.

It is clearly shown that, for small size problems, GA with stochastic selection gives the highest solution quality, (i.e., minimum total cost) for most quality levels tested. For large problem size, the tournament technique is able to achieve optimal solution for all quality levels tested. However, as the size of test problems increases, the quality of solution reduces. According to Fig. 3, we can see that the percentage of deviation from the optimal solution is less than 1% for small size, and 2% for large size for stochastic selection. While tournament selection does not give any deviation from the optimal solution for small instances (0%), and less than 1% for large size problems. We conclude that tournament selection gives better results than stochastic selection for all sizes of problems.

The performance of both optimization selection method is also related to other main factors, namely convergence curve. Figure 4 corresponds to the convergence

| Factors | Level (1) | Level (2) | Level (3) |
|---------|-----------|-----------|-----------|
| Pop_size | 20 | 50 | 100 |
| Max_iter | 100 | 150 | 200 |
| Pc | 0.2 | 0.4 | 0.7 |
| Pm | 0.1 | 0.4 | 0.7 |
| Penalty | 5000 | 6500 | 8000 |
| T | 5 | 10 | 15 |
| S | 2 | 5 | 10 |

**Table 9** Considered levels for each factor of tournament selection

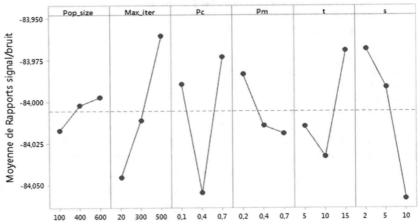

Fig. 2 Main effects graph for S/N ratio for tournament

**Table 10** Optimal Parameters of tournament selection

| Methodology | Pop_size | Max_iter | Pc | Pm | t | s |
|---|---|---|---|---|---|---|
| Tournament | 600 | 500 | 0.7 | 0.2 | 15 | 2 |

**Table 11** Results of the best solution for 3, 5, and 10 problem size

| | Size = 3 | | Size = 5 | | Size = 10 | |
|---|---|---|---|---|---|---|
| Ql | S | T | S | T | S | T |
| 0.96 | 27,546 | 27,936 | 29,151 | 29,073 | 30,131 | 29,105 |
| 0.94 | 20,266 | 20,428 | 20,356 | 18,546 | 20,665 | 20,275 |
| 0.93 | **15,754** | **15,741** | **15,804** | **15,741** | **16,089** | **15,746** |
| 0.92 | 17,485 | 17,470 | 17,520 | 17,470 | 17,835 | 17,470 |
| 0.88 | 24,384 | 24,384 | 24,465 | 24,384 | 25,364 | 24,392 |
| 0.86 | 24,477 | 27,842 | 27,929 | 27,847 | 28,645 | 27,842 |
| 0.76 | 34,480 | 45,129 | 45,317 | 45,163 | 47,465 | 45,129 |
| 0.67 | 54,481 | 62,416 | 62,502 | 62,416 | 64,494 | 62,416 |
| 0.57 | 81,261 | 79,703 | 79,724 | 79,703 | 84,800 | 79,703 |
| 0.48 | 98,461 | 96,991 | 97,056 | 96,991 | 102,564 | 96,991 |
| 0.38 | 115,754 | 114,278 | 114,371 | 114,278 | 117,431 | 114,312 |
| 0.28 | 133,085 | 131,565 | 131,677 | 131,567 | 139,481 | 131,565 |
| 0,19 | 150,331 | 148,852 | 148,873 | 148,852 | 155,309 | 148,852 |

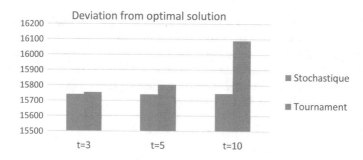

**Fig. 3** Deviation from optimal solution

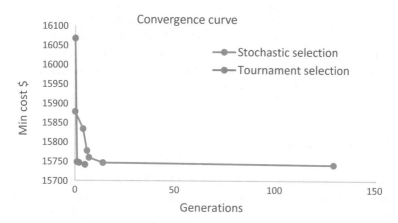

**Fig. 4** Convergence curves

curve of the proposed GA selection methods: tournament and stochastic. It is clearly shown that as the computation time and the number of generations increase, the convergence curve of the tournament selection algorithm reaches the lowest level (at the 5th iteration) compared to Stochastic selection.

### iii. **Boltzmann tournament selection**

The notion of Boltzmann distribution is borrowed from Simulated Annealing and adapted to Genetic Algorithm practice to escape the problem of premature convergence, asymptotic or otherwise. Specifically, a Boltzmann tournament selection procedure is implemented in order to give stable distributions within a population of structures that are near Boltzmann. This selection technique selection has been proposed as an alternative to obtain niching-like behavior in a tournament scheme [15]. It was primarily used in simulated annealing (SA) algorithm to accept or reject solutions after the process of neighborhood exploration. Kirkpatrick [16] and Cerny [17] firstly introduced SA optimization approach; in Boltzmann selection, the temperature that controls the rate of selection is decreased gradually and the selection pressure increases because it is inversely proportional to temperature. Initially, the temperature is high and selection pressure is low. The temperature is decreased gradually and the selection pressure increases. This results in a reduction of the research space with maintaining the diversity in population. Boltzmann tournament is aiming to find a global optimum. The selection of an individual is done with Boltzmann probability that is given by (22)

$$P(\Delta E) = e^{-(s_i - s_j)/T} \tag{22}$$

where $\Delta E$ is the difference of energy between two values ($S_i$ and $S_j$) of the objectives function (fitness in GA) and the level of the current temperature, $P(\Delta E)$ is the acceptance probability, and T is the temperature level, and the minus sign in the exponent is necessary because a minimization is performed. Similarly, to the thermodynamics properties where the possibility to change between two energy states is very high at high temperatures. With Boltzmann selection method, the acceptance of bad solutions at high temperature can be taken as number close to one. Nevertheless, as the system is cooled, this probability decreases to zero [18].

## 4   Results of Statistical Tests

The performance was tested for ten runs of COFQ for each of the selection techniques. For this experiment, the crossover and mutation probabilities are 0.7 and 0.2, respectively. We use the size 2 for Tournament selection, the population size is 120, 200, and 300. Penalty cost is fixed at 5000 $, temperature is 1750, and cool rate is 0.9. The maximum number of iterations is fixed at 100. The result is to find the optimal total cost of food quality. In order to determine which selection technique provides the best solution, we perform a statistical analysis using paired sample T-Test. We use the statistical software package IBM SPSS statistics version 23 to analyze the data. The following Table 12 shows the obtained experiment results of the three selection strategies presented above. As the population size of the algorithm has a considerable impact on the solution quality, we propose to investigate different "Pop size". Hence, paired T-test will be conducted to analyse these collected data.

The paired-Samples T-test procedure is based on the verification of two contrary hypothesis. The first one is called the null hypothesis ($H_0$) that could be stated as the mean of each observation is the same. As for the second one, the alternative hypothesis ($H_A$) is simply the opposite of H0. We define the hypothesis to work with as follows.

### Hypothesis
$H_0$: $\mu_1 = \mu_2 = \mu_3$ (i.e., all the selection strategies give the same results)
$\quad H_A$: at least one selection strategy is different from the above.
*Significance is set at 0.05 (95% confidence).*

**Table 12** Examples of obtained results with different costs

|  | Solutions | | | | |
|---|---|---|---|---|---|
| O | [1, 0, 0] | [0, 1, 0] | [0, 1, 0] | [0, 1, 0] | [1, 0, 0] |
| X | [1, 0, 0] | [0, 1, 0] | [0, 1, 0] | [0, 0, 1] | [1, 0, 0] |
| Z | [1, 0, 0] | [0, 1, 0] | [0, 1, 0] | [0, 0, 1] | [1, 0, 0] |
| $W_{pd}$ | [1300, 0, 0] | [0, 1300, 0] | [0, 1302, 0] | [0, 0, 1303] | [1301, 0, 0] |
| $W_{dr}$ | [1200, 0, 0] | [0, 1201, 0] | [0, 1201, 0] | [0, 0, 1200] | [1200, 0, 0] |
| Ql | 93% | 93% | 93% | 93% | 93% |
| Min cost ($) | 14031.56 | 14037.61 | 14048.01 | 14048.96 | 14045.76 |
| CEF ($) | 3422.16 | 3186.8 | 3425.01 | 3422.16 | 3422.16 |
| CIF ($) | 216.95 | 216.95 | 217.28 | 217.44 | 217.11 |
| CA ($) | 3186.8 | 3092.65 | 3191.7 | 3194.15 | 3189.25 |
| CP ($) | 3092.65 | 3092.65 | 3097.4 | 3099.78 | 3095.03 |
| Operational cost ($) | 273.0 | 273.0 | 273.42 | 273.63 | 273.21 |
|  | 2760.0 | 2762.3 | 2762.3 | 2760.0 | 2760.0 |
| Transport cost ($) | 1080.0 | 1080.9 | 1080.9 | 1081.8 | 1089.0 |

Measurements are taken from the same program with different selection strategies; we have to verify if these techniques give the same results. In this case, if the average difference between the measurements is equal to 0, then the null hypothesis holds. On the other hand, if their impact on the obtained results is similar, the average difference is not 0 and the null hypothesis is rejected and the alternative $H_a$ is accepted. To conduct the T-Test analysis, we need to calculate some descriptive statistics for each observation such as mean, standard error, and standard deviation. Table 13 shows all data needed in the next step.

Table 14 gives the mean, standard deviation, and mean standard error of each selection technique.

Subsequently, we have to pair the selection techniques together in three combinations:

- Pair 1: Tournament and stochastic selection
- Pair 2: Tournament and Boltzmann selection
- Pair 3: Boltzmann and stochastic selection.

Based on these data, we can now conduct the analysis to verify the difference between each observation. Consequently, for each pair, we evaluate the value of "t" by the following (23):

$$t = \frac{\bar{x} - \mu}{s/\sqrt{n}} \tag{23}$$

where $\bar{x}$: the mean, N: the number of samples that are taken into consideration "N = 10", $\mu$: the standard error mean, and s: the standard deviation.

**Table 13** Statistical results of the three selection techniques

| Test N° | | Boltzmann | Tournament | Stochastic |
|---|---|---|---|---|
| Popsize 120 | 1 | 14036.71 | 14031.56 | 14088.77 |
| | 2 | 14067.02 | 14052.36 | 14077.42 |
| | 3 | 14036.76 | 14156.19 | 14106.83 |
| | 4 | 14059.26 | 14037.66 | 14093.87 |
| | 5 | 14059.91 | 14046.86 | 14118.93 |
| | 6 | 14057.56 | 14071.96 | 14095.67 |
| | 7 | 14042.81 | 14052.36 | 14085.02 |
| | 8 | 14048.01 | 14075.47 | 14148.13 |
| | 9 | 14065.51 | 14052.36 | 14058.41 |
| | 10 | 14031.56 | 14078.36 | 14089.33 |
| Popsize 200 | 1 | 14054.92 | 14036.76 | 14109.57 |
| | 2 | 14065.86 | 14031.56 | 14073.07 |
| | 3 | 14064.47 | 14052.36 | 14069.67 |
| | 4 | 14041.96 | 14052.36 | 14107.47 |
| | 5 | 14031.56 | 14031.56 | 14134.33 |
| | 6 | 14031.56 | 14031.56 | 14107.77 |
| | 7 | 14036.76 | 14031.56 | 14042.81 |
| | 8 | 14034.26 | 14031.56 | 14058.41 |
| | 9 | 14049.71 | 14031.56 | 14039.81 |
| | 10 | 14041.96 | 14031.56 | 14057.12 |
| Popsize 300 | 1 | 14048.06 | 14031.56 | 14123.44 |
| | 2 | 14031.56 | 14031.56 | 14048.06 |
| | 3 | 14042.81 | 14031.56 | 14101.87 |
| | 4 | 14041.96 | 14031.56 | 14074.21 |
| | 5 | 14043.66 | 14031.56 | 14102.47 |
| | 6 | 14039.66 | 14052.36 | 14060.97 |
| | 7 | 14031.56 | 14042.81 | 14042.26 |
| | 8 | 14036.76 | 14049.61 | 14059.26 |
| | 9 | 14045.01 | 14031.56 | 14152.03 |
| | 10 | 14043.66 | 14031.56 | 14116.13 |

The following tables recapitulate the output of the paired T-test analysis for each population size of GA program. This table provides relevant information about the studied measurement presented above; in other words, it contains the mean, standard deviation, and standard error mean differences of the pairs, for instance, the value of the first row (pair 1: Boltzmann Tournament − Tournament) is 14050.51 − 14065.51 = −15. The additional columns help us to make a decision about the measurements (with degrees of freedom 9), using these outputs we will be able to

**Table 14** Paired sample T-test

| | | N | Mean | | Std. deviation |
|---|---|---|---|---|---|
| | | | Statistic | Std. error | |
| Popsize 120 | Boltzmann | 10 | 14050.51 | 4.10 | 12.98 |
| | Tournament | 10 | 14065.51 | 11.22 | 35.47 |
| | Stochastic | 10 | 14096.24 | 7.70 | 24.36 |
| Popsize 200 | Boltzmann | 10 | 14045.30 | 4.08 | 12.90 |
| | Tournament | 10 | 14036.24 | 2.74 | 8.65 |
| | Stochastic | 10 | 14080.00 | 10.27 | 32.48 |
| Popsize 300 | Boltzmann | 10 | 14040.47 | 1.76 | 5.57 |
| | Tournament | 10 | 14036.57 | 2.65 | 8.39 |
| | Stochastic | 10 | 14088.07 | 11.54 | 36.49 |

choose the best selection technique. As an example, for the population size of 200, the average optimum solution given by tournament selection is less than the one given by stochastic selection: 43.76 \$ $-31.88$ SD. Moreover, there is a strong evidence that tournament is better than stochastic, since the obtained $P$-value is 0.002 which is less than our chosen significance level $\alpha = 0.05$. Therefore, we can conclude that the average optimal tournament selection and stochastic costs are significantly different. The last column of this table shows that the probability to get the result by chance is 0.7% for pair 1 of Popsize 300. From the above tables, we can aggregate in Fig. 5 the given $P$-values in order to discuss the results of this statistical analysis. We report in Table 15 the paired T-Test for various population sizes.

As we can see in Fig. 5, larger population sizes give lesser $p$-values. For pop size 120, there no significant difference between Boltzmann and Tournament selection, so the null hypothesis is true. However, for the second and the third pairs, $p$ is less than 5% which means that the $H_0$ is rejected and there is a significant difference between stochastic selection method and the other ones. When pop size is fixed at 200, we can see that tournament performs better than Boltzmann and stochastic selection techniques. The same conclusion can be made for pop size fixed at 300. Moreover,

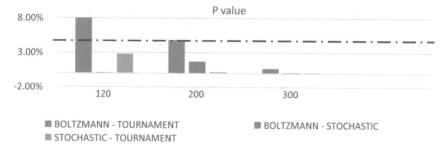

**Fig. 5** Aggregated $P$-values

**Table 15** Paired T-test for various Pop size

| | | Paired differences | | | | | | t | df | Sig. (2-tailed) |
|---|---|---|---|---|---|---|---|---|---|---|
| | | Mean | Std. deviation | Std. error mean | 95% confidence interval of the difference | | | | | |
| | | | | | Lower | Upper | | | | |
| Popsize 120 | Pair 1 | Boltzmann-Tournament | −15.00 | 42.56 | 13.46 | −45.45 | 15.44 | −1.11 | 9 | 0.294 |
| | Pair 2 | Boltzmann-Stochastic | −45.72 | 30.08 | 9.51 | −67.25 | −24.20 | −4.80 | 9 | 0.001 |
| | Pair 3 | Tournament-Stochastic | −30.72 | 36.99 | 11.69 | −57.18 | −4.26 | −2.62 | 9 | 0.028 |
| Popsize 200 | Pair 1 | Boltzmann-Tournament | 9.06 | 12.52 | 3.96 | 0.10 | 18.02 | 2.29 | 9 | 0.048 |
| | Pair 2 | Boltzmann-Stochastic | −34.70 | 37.47 | 11.85 | −61.50 | −7.90 | −2.93 | 9 | 0.017 |
| | Pair 3 | Tournament-Stochastic | −43.76 | 31.88 | 10.08 | −66.57 | −20.96 | −4.34 | 9 | 0.002 |
| Popsize 300 | Pair 1 | Boltzmann-Tournament | 6.90 | 6.36 | 2.01 | 2.35 | 11.45 | 3.43 | 9 | 0.007 |
| | Pair 2 | Boltzmann-Stochastic | −47.60 | 31.73 | 10.04 | −70.30 | −24.90 | −4.74 | 9 | 0.001 |
| | Pair 3 | Tournament-Stochastic | −54.50 | 36.72 | 11.61 | −80.76 | −28.24 | −4.69 | 9 | 0.001 |

in this case, the P-values are smaller and variability decreased, for example, the average optimal cost at pop size 200 pair 1 (Boltzmann–Tournament) decreased from 9.06$–12.52SD to 6.9$–6.36SD.

# 5  Conclusion

In this paper, we presented a model for optimizing the food supply chain, especially for fresh products. The model is a MINLP problem that demonstrates the importance of managing FSC to maintain the quality and safety of food products, and provides managers with a practical tool for making cost-optimized decisions. The proposed resolution procedure based on genetic algorithms proved more efficient in terms of speed of execution as well as the quality of the results compared to the solver. The calculation of the difference between the two approaches gave a GAP of 1.37% and of 2.71% for a quality degradation equal to 3% and 30%, respectively, which means that the performances of developed AG are better compared to the solver, especially for low values of quality degradation. We also addressed the issue of selection strategy to be chosen in Genetic Algorithm optimization approach. We demonstrate that tournament selection is the best selection technique compared to stochastic sampling without replacement and Boltzmann tournament using the results obtained by T-test statistical analysis. Future research can be conducted to investigate hybrid algorithms that combine two metaheuristics.

# Appendix

See Tables 16 and 17.

**Table 16**  Parameters for the COFQ function

| | |
|---|---|
| $C_{Afix}$ | Fixed costs for inspection after storage |
| $C_{Avar}$ | Variable costs for inspection at the end of storage |
| $C_{pvar}$ | Variable costs for prevention activities |
| $C_{IFfix}$ | Fixed internal failure costs |
| $C_s$ | Loss generated by food products shipped from production sites to meet quality requirements |
| $\overline{C_{EF}}$ | Cost per defective products intended to waste disposal for treatment |

**Table 17** Food supply chain network parameters

| Parameter | Value | Parameter | Value |
|---|---|---|---|
| $P_{p,t}$ | 0.21 | $C_{IFfix}$ | 2.5 |
| $g_d^2$ | 2.3 | $\Delta q_d$ | 0.03 |
| $f_{d,r,y}$ | 0.9 | $\Delta q_{d,y}$ | 0.02 |
| $g_d^1$ | 3.2 | $C_s$ | 2.7 |
| $C_{pvar}$ | 2.5 | $\overline{C_{EF}}$ | 150 |
| $\alpha_p$ | 0.2 | $dem_{r,t}$ | 500 |
| $C_{Afix}$ | 1.8 | $a_{p,t}$ | 1450 |
| $C_{Avar}$ | 2.5 | $b_{d,t}$ | 1450 |
| $\beta_d$ | 0.2 | $lR_t$ | [1–98%] |
| t | 1 | p | 3 |
| d | 3 | y | 3 |
| r | 3 | $C_{IFfix}$ | 2.5 |

# References

1. Manzini, R., Accori, R.: The new conceptual framework for food supply chain assessment. J. Food Eng. **115**, 251–263 (2013)
2. de Keizer, M., Akkerman, R., Grunow, M., Bloemhof, J.M., Haijema, R., van der Vorst, J.G.A.J.: Logistics network design for perishable products with heterogeneous quality decay. Eur. J. Oper. Res. **262**, 535–549 (2017)
3. Li, D., O'Brien, C.: Integrated decision modelling of supply chain efficiency. Int. J. Prod. Econ. **59**, 147–157 (1999)
4. Zhao, X., Lv, Q.: Optimal design of agri-food chain network: an improved particle swarm optimization approach. In: 2011 International Conference on Management and Service Science (2011). https://doi.org/10.1109/icmss.2011.5998308
5. Tsao, Y.-C.: Designing a fresh food supply chain network: an application of nonlinear programming. J. Appl. Math. 1–8 (2013). https://doi.org/10.1155/2013/506531
6. Wang, X., Wang, M., Ruan, J., Zhan, H.: The multi-objective optimization for perishable food distribution route considering temporal-spatial distance. Procedia Comput. Sci. **96**, 1211–1220 (2016)
7. Nakandala, D.: Cost-optimization modelling for fresh food quality and transportation. Ind. Manag. Data Syst. **116**(3), 564–583 (2016)
8. Zhang, Y., Chen, X.D.: An optimization model for the vehicle routing problem in multi-product Frozen food delivery. J. Appl. Res. Technol. **12**(2), 239–250 (2014). https://doi.org/10.1016/s1665-6423(14)72340-5
9. Aramyan, H., Oude Lansink, A.G.J.M., van der Vorst, J.G.A.J., van Kooten, O.: Performance measurement in agri food supply chain, a case study. Supply Chain Manag. Int. J. **12**(4), 304–315 (2007)
10. Bag, S., Anand, N.: Modelling barriers of sustainable supply chain network design using interpretive structural modelling: an insight from food processing sector in India. Int. J. Autom. Logist. **1**(3), 234 (2015). https://doi.org/10.1504/ijal.2015.071722
11. Boudahri, F., Sari, Z., Maliki, F., Bennekrouf, M.: Design and optimization of the supply chain of agri-foods: application distribution network of chicken meat. In: 2011 International Conference on Communications, Computing and Control Applications (CCCA) (2011).https://doi.org/10.1109/ccca.2011.6031424.

12. Van der Vorst, J.G.A.J.: Effective food supply chains. Generating, modelling and evaluating supply chain scenarios, Ph.D. Thesis, Wageningen University, Wageningen (2000)
13. Lütke entrup, M., Günther, H.-O., Van Beek, P., Grunow, M., Seiler, T.: Mixed-integer linear programming approaches to shelf-life-integrated planning and scheduling in yoghurt production. Int. J. Prod. Res. **43**(23), 5071–5100 (2005). https://doi.org/10.1080/002075405001 61068
14. Ekşioğlu, S.D., Jin, M.: Cross-facility production and transportation planning problem with perishable inventory. Lect. Notes Comput. Sci. 708–717 (2006). https://doi.org/10.1007/117 51595_75
15. Goldberg, D.E., Deb, K.: A comparative analysis of selection schemes used in genetic algorithms. In: Rawlins, G.J.E. (ed.) Foundations of Genetic Algorithms, pp. 69–93. Morgan Kaufmann, Los Altos (1991)
16. Kirkpatrick, S., Gelatt, S., Vecchi, C.: Optimization by simulated annealing. Science **220**, 671–680 (1983)
17. Cerny, V.: Thermodynamical approach to the traveling salesman problem: an efficient simulation algorithm. J. Optim. Theory Appl. **45**(1), 41–51 (1985)
18. AARTS, Emile et KORST, Jan. Simulated annealing and Boltzmann Machines (1988)

# Optimal Cutting Parameters Selection of Multi-Pass Face Milling Using Evolutionary Algorithms

Ghita Lebbar, Abdelouahhab Jabri, Abdellah El Barkany, Ikram El Abbassi, and Moumen Darcherif

**Abstract** Over decades, particular attention was devoted to the optimal selection of the cutting conditions associated with the material removal processes, particularly for the multi-pass face milling operations considered as a highly complex problem both theoretically and practically. The machining conditions in the milling operations consist commonly of cutting speed, depths of cut, and feed rate. Researches in this field are of huge interest due to their considerable importance for the Computer-Aided Process Planning (CAPP) on one hand and the large impact of those variables on the quality of machining products, the operational costs, and the machining efficiency on the other hand. In this paper, various evolutionary optimization techniques are proposed to minimize the unit production cost of multi-pass face milling operations while considering technological constraints. The proposed optimization tools are based initially on the Genetic Algorithm (GA) with two different selection strategies, namely stochastic and tournament selections. And secondly, on the Hybrid Simulated Annealing Genetic Algorithm (SAGA). The integration target of the simulated annealing (SA) based local search strategy with the genetic search is to prevent accurately the trap of GAs in premature convergence. Parameters of these three optimization approaches are then calibrated using Taguchi design of experiment (DEO) $L_{27}$ orthogonal array method. Finally, different case studies are considered in order to adequately show the effectiveness of the proposed mechanisms. The comparison of the obtained results with the suggested literature approaches; show the effectiveness of the proposed SAGA for selecting optimal cutting parameters of the multi-pass face milling operation.

**Keywords** Multi-face milling process · Optimization · Simulated annealing · Genetic algorithm · Taguchi method

G. Lebbar (✉) · I. E. Abbassi · M. Darcherif
ECAM-EPMI, 13 Boulevard de l'Hautil, 95092 Cergy-Pontoise cedex, France
e-mail: g.lebbar@ecam-epmi.com

A. Jabri · A. E. Barkany
Mechanical Engineering Laboratory, Faculty of Science and Techniques, Sidi Mohammed Ben Abdellah University, B.P. 2202, Route d'Imouzzer, FEZ, Morocco

© The Author(s), under exclusive license to Springer Nature Singapore Pte Ltd. 2021
A. J. Kulkarni et al. (eds.), *Constraint Handling in Metaheuristics and Applications*,
https://doi.org/10.1007/978-981-33-6710-4_9

201

## Nomenclature

| | |
|---|---|
| ap | Approach distance |
| D | Tool diameter (mm) |
| B | Width of cut |
| L | Length of the workpiece (mm) |
| Z | Number of machining teeth of the tool |
| e | Arbitrary distance to avoid possible accidents and damages |
| h1 and h2 | Constants related to tool travel and approach/depart time |
| $C_{total}$ | Unit production cost |
| $C_m$ | Machining cost |
| $C_l$ | Machine idle cost |
| $C_r$ | Replacement cost |
| $C_t$ | Tool cost |
| $C_{ri}$ | Unit cost for the $i$th rough pass ($/part) |
| $C_s$ | Unit cost for the finish pass ($/part) |
| $k_0$ | Overhead costs per unit time ($/min) |
| $k_t$ | Cost of tool material ($/edge) |
| $L_r$ | Length of the travel of the cutter |
| $L_f$ | Length of machining travel measured from the first to the last contact between the tool and the workpiece |
| n | Number of rough passes |
| $V_{ri}$ and $V_s$ | Machining speed (m/min) of rough and finish passes, respectively |
| $f_{ri}$ and $f_s$ | Feed per tooth (mm/tooth) of rough and finish passes, respectively |
| $d_{ri}$ and $d_s$ | Cutting depth (mm) of rough and finish passes, respectively |
| $T_m$ | Machining time |
| $T_l$ | Machine idling time |
| $T_r$ | Tool changing time |
| $T_p$ | Loading and unloading time (min/part) |
| $T_i$ | Idle tool motion time (min) |
| $T_r$ | Tool exchange time (min) |
| $T_{tc}$ | Tool changing time required for each edge (min/edge) |
| $T_{ri}$ | Tool life for the $i$th rough pass (min) |
| $T_s$ | Tool life for the final finish pass |
| $F_{ri}$ and $F_s$ | Machining force of rough and finish passes, respectively (N) |
| $P_{ri}$ and $P_s$ | Machining power of rough and finish passes, respectively (kW) |
| $R_s$ | Surface roughness ($\mu$m) |

## 1 Introduction

Material removal process cost represents a very huge part of the global component manufacturing charges. Thereby, an important concern associated with the reduction of that overall cost consists of the control and optimization of cutting conditions in the machining systems. Among the material removal process that has been investigated to a certain extent, the milling operation is considered as the most crucial given its operational significance introduced by its ability to generate complex forms surface with a desirable precision and a good surface quality.

Multi-pass milling operation consists of several rough passes and one finish pass, it is commonly used to remove that cannot be subtracted in a single pass. Each pass can be affected by various parameters, as illustrated in Fig. 1, ranging from the cutting speed (V) and feed rate (f) to the depth of cut (d) condition. An appropriate selecting of those cutting parameters is indispensable to assure necessities concerning machine parts quality, machining cost as well as machining systems productivity.

Despite the efforts conducted by researches to determine the optimal cutting parameters, the literature in this area is still limited. However, a large part of the published works in this direction is devoted to the turning processes. The slow progress in developing constrained optimization for milling systems is explained by the high complexity of their cutting mechanisms on one hand and the consistency of the applied constraint on the other. Due to this lack of investigation, this paper introduces various meta-heuristic algorithms basing on the genetic search to minimize the unit production cost subject to practical constraint while finding the optimal cutting conditions is the main target. The first technique is lying on the GAs procedure with different selection implementations. Whereas in the second, a hybrid simulated annealing genetic algorithm is suggested. Moreover, to calibrate the algorithms input parameters, a design of experiment technique is used in this research study.

**Fig. 1** Example of Face milling operation

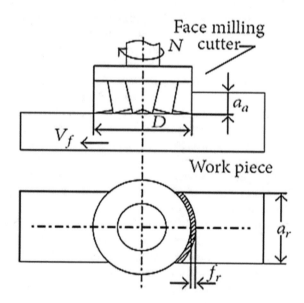

The remainder of this paper is organized as follows: the next section reports the main published works related to the considered issue. A mathematical model is then given in Sect. 3. Proposed solution algorithms approaches are described in Sect. 4. Parameters calibration of the suggested tools is then studied in Sect. 5. Section 6 reports case studies and obtained results. Finally, conclusion and some possible future research areas are carried out.

## 2 Literature Review of the Machining Operations Optimization

It is worth emphasizing that currently, different optimization approaches have been proposed to solve the multi-pass machining operations starting from geometric programming and graphical tools to exact and meta-heuristic techniques. With this in mind, some contemporary works linked to the optimization of the multi-pass machining operations are holistically presented. In earlier studies, Petropoulos [1] developed a new non-linear programming technique, namely geometric programming with the aim to optimize the unit cost in turning operations subject to select the optimal machining conditions. Later, Shin and Joo [2] proposed a conventional differential and a dynamic programming method to determine the optimum of machining turning conditions with the consideration of practical constraints and variation of machining idle time. Gupta et al. [3] developed an integer programming to find the optimal subdivision of depth of cut in multi-pass turning by adding the minimum costs of individual rough passes and the finish pass. Tolouei-Rad and Bidhendi [4] used a new approach based on feasible directions method to find optimum machining parameters for milling operations. It starts with an initially feasible vector and via an iterative scheme; the intermediate vectors improving the objective function are produced. Sönmez et al. [5] proposed a dynamic programming method to select the optimum machining parameters for multi-pass milling operations. Based on the maximum production rate criterion, the authors funded the optimal values through a geometric programming technique. A modified GA was presented by [6] to optimize milling process parameters. Genetic algorithm was similarly proposed by [7] to optimize the machining parameters for multi-pass milling operations. Regarding single pass machining operations, Wang et al. [8] introduced deterministic graphical programming based on the criteria typified by the minimum production time. With the progress of intelligent computing, more bright algorithms have been introduced to machining parameters such as the Artificial Neural Network (ANN) that was suggested by [9] for modeling and simulating the milling process. An and Chen [10] proposed an integer programming method for optimal selection of machining parameters in multi-pass face milling process. Based on genetic algorithm (GA) and simulated annealing (SA), a hybrid approach GSA was developed by [11] to select the optimal machining parameters for plain milling process. Correspondingly, Wang et al. [12] suggested a parallel genetic simulated annealing to minimize machining cost of multi-pass milling operations. Multi-pass milling operations such as plain milling and face milling was also considered by [13] using an optimization technique based on tribes to select the optimum machining parameters. Moreover, Saha et al. [14] proposed a binary-coded GA to minimize the cost of unit production to obtain adequate machining parameters of multi-pass face milling. António et al. [15] proposed a binary-coded and integer-coded genetic algorithm by substituting the depth of cut (d) with a sequence of depths of cut. The solution was obtained based on an elitist procedure. Krimpenis and Vosniakos [16] used a GA to optimize rough milling operations for sculptured surfaces and select process parameters such as feed rate, cutting speed, width of cut, raster pattern angle, spindle speed, and number of machining slices of variable thickness.

Several studies had been focused on the optimization of important cutting process parameters using hybrid techniques. For instance, Öktem [17] studied the surface roughness using ANN and GA in order to provide the best combinations of cutting parameters. Mahdavinejad et al. [18] also studied the effect of milling parameters on the surface roughness of Ti-6Al-4V using a hybrid optimization approach by combining the immune algorithm with ANN. Rao and Pawar [19] used an Artificial Bee Colony (ABC) algorithm to minimize production time of a multi-pass milling process. Yang [20] proposed a fuzzy particle swarm optimization (FPSO) combined with a methodology for distribution of the total stock removal. Zhou et al. [21] also applied fuzzy particle swarm optimization algorithm (PSO) to select the machining parameters for milling operations. Zarei et al. [22] proposed a Harmony Search (HS) algorithm to predict optimal cutting parameters for a multi-pass face milling operation. Yang et al. [23] developed an efficient fuzzy global and personal best-mechanism-based multi-objective particle swarm optimization (F-MOPSO) to minimize production time and cost and maximize profit rate of multi-pass face milling problem. Fratila and Caizar [24] applied the Taguchi method for the design of experiments (DOE), to optimize the cutting parameters in face milling when machining AlMg3 (EN AW 5754) with HSS (high speed steel) tool. A novel hybrid optimization approach based on differential evolution algorithm and receptor editing property of immune system was proposed by Yildiz et al. [25] to optimize machining parameters in milling operations. Qu et al. [26] suggested a non-dominated sorting genetic algorithm (NSGAII) to determine and validate the optimum machining parameters in milling thin-walled plates. Recently, Shaik et al. [27] developed a multi-objective approach to minimize simultaneously vibration amplitudes all through the operations as well as surface roughness average values of the workpiece. A new population-based meta-heuristic algorithm called chaotic imperialist competitive algorithm (CICA) is proposed by [28] to select machining parameters optimization of multi-pass face milling processes. Chen et al. [29] proposed a multi-objective integrated optimization model for minimizing energy footprint and production time and select optimal via a multi-objective Cuckoo Search algorithm.

Even though some excellent improvements have been attained, these problems still require in-depth investigations owing to the complexity of the issue and the inconsistency between parameters and targets.

## 3   Mathematical Formulation of Face Milling Process

The mathematical model of face milling operation considered in this paper is based on the research works of [14, 15, 20]. As illustrated in Fig. 2, in the multi-pass milling operations, the total stock removal is divided into multiple rough passes "n" and one finish pass. The length of the travel of the cutter is calculated as:

$$L_r = L + a_p + e \tag{1}$$

Given that B is the width of cut and D represents the diameter of the tool, the approach distance for symmetrical milling and the length of cutting in finish machining are defined as:

$$a_p = \frac{D}{2} - \sqrt{\left(\frac{D}{2}\right)^2 - \left(\frac{B}{2}\right)^2} \tag{2}$$

$$L_f = L + D + e \tag{3}$$

### 3.1   Objective Function

The minimum total production cost is used as objective function which is based on machining time, machine idling time, and tool changing time. The actual machining time (min) is estimated using:

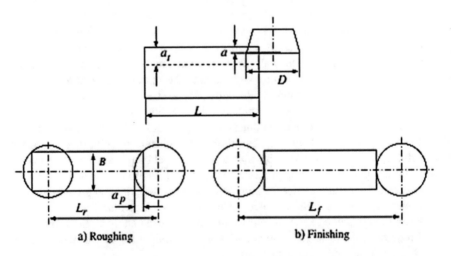

a) Roughing                              b) Finishing

**Fig. 2**  Multi-pass milling scheme according to [15]

$$T_m = \sum_{i=1}^{n} t_{mri} + t_{ms} \tag{4}$$

where, $t_{mri}$ and $t_{ms}$ denote the machining time of one rough pass "i" and a final finish pass, that is:

$$t_{mri} = \frac{\pi D L_r}{1000 V_{ri} f_{ri} Z} \tag{5}$$

$$t_{ms} = \frac{\pi D L_f}{1000 V_s f_s Z} \tag{6}$$

where, $V_s$ and $V_{ri}$ represent the machining speed (m/min) for roughing and finishing passes, $f_{ri}$ and $f_s$ denote the feed per tooth (mm/tooth) for rough and finish passes. And $Z$ is the number of machining teeth of the tool.

Considering that $h_1$ and $h_2$ are related to tool travel and depart time. Machine idling time is calculated by summing the loading and unloading operations $T_p$ and idle tool motion time $T_i$. It can be expressed as:

$$T_l = T_p + T_i = T_p + (h_1 L_f + h_2) + n(h_1 L_r + h_2) \tag{7}$$

During the machining operation, the tool life is constantly consumed. A tool exchange is needed and the corresponding time of this operation is given by the following equation:

$$T_r = Z \left( T_{tc} \sum_{i=1}^{n} \frac{t_{mri}}{t_{ri}} + T_{tc} \frac{t_{ms}}{t_s} \right) \tag{8}$$

where:

$$t_{ri} = \left( \frac{C_v K_v D^{q_v}}{V_{ri} d_{ri}^{x_v} f_{ri}^{x_v} W^{s_v} Z^{p_v}} \right)^{1/l} \tag{9}$$

$$t_s = \left( \frac{C_v K_v D^{q_v}}{V_s d_s^{x_v} f_s^{x_v} W^{s_v} Z^{p_v}} \right)^{1/l} \tag{10}$$

where, $T_{tc}$ denotes the tool changing time required for each edge; $t_{ri}$ and $t_s$ are the tool life for rough and finish pass. $C_v$, $K_v$, $x_v$, $y_v$, $s_v$, $q_v$, $p_v$, represent the constants associated with the tool and the material of the workpiece. The tool cost is then calculated as follows:

$$C_t = Z \left( K_t \sum_{i=1}^{n} \frac{t_{mri}}{t_{ri}} + K_t \frac{t_{ms}}{t_s} \right) \tag{11}$$

Finally, the total production cost of face milling operation represents the sum of machining cost $C_m$, machine idle cost $C_l$, tool replacement cost $C_r$ and tool cost $C_t$, as a result, the unit production cost can be calculated as:

$$C_{\text{total}} = C_m + C_l + C_r + C_t \tag{12}$$

where:

$$C_m = K_0 T_m \tag{13}$$

$$C_l = K_0 T_l \tag{14}$$

$$C_r = K_0 T_r \tag{15}$$

The total cost can be written as follows:

$$C_{\text{total}} = \sum_{i=1}^{n} C_{ri} + C_s + K_0 T_p \tag{16}$$

where:

$$C_{ri} = t_{mri} \left( K_0 + Z \frac{K_0 T_{tc}}{t_{ri}} + Z \frac{K_t}{t_{ri}} \right) + K_0 (h_1 L_r + h_2) \tag{17}$$

$$C_s = t_{mri} \left( K_0 + Z \frac{K_0 T_{tc}}{t_s} + Z \frac{K_t}{t_s} \right) + K_0 (h_1 L_s + h_2) \tag{18}$$

where $C_{ri}$ ($/part) is the unit cost for the $i$th rough pass and $C_s$ ($/part) is the unit cost for the finish pass.

## 3.2  Machining Constraints

For each parameter of this machining operation, there is permissible range of values to be selected. These limitations are expressed by Eqs. 19–24:

$$V_{\min} \le V_{ri} \le V_{\max} \tag{19}$$

$$V_{\min} \le V_s \le V_{\max} \tag{20}$$

$$f_{\min} \le f_{ri} \le f_{\max} \tag{21}$$

$$f_{min} \leq f_s \leq f_{max} \tag{22}$$

$$d_{min} \leq d_{ri} \leq d_{max} \tag{23}$$

$$d_{min} \leq d_s \leq d_{max} \tag{24}$$

where, $V_{min}$ and $V_{max}$ are the minimum and maximum machining speed, $f_{min}$ and $f_{max}$ denote the minimum and maximum feed rate and $d_{min}$ and $d_{max}$ denote the minimum and the maximum depth of cut. Furthermore, as multi-pass operations consist of multiple passes of rough passes and a final finish pass. Obviously, the total stock to remove ($d_t$) should be the sum of the cutting depths of rough passes and the final depth of cut. This constraint must be expressed by the following equation:

$$d_t = \sum_{i=1}^{n} d_{ri} + d_s \tag{25}$$

In order to prevent damage to the cutting tool produced by repeated impact that causes dimensional errors. Machining force must be lower than the maximum machining force permitted by the machine and the cutting tool. Using the model proposed by [30], the constraints imposed on cutting force for rough and final passes are expressed as:

$$F_{ri} = \frac{C_u K_u W^{S_u} Z^{P_u} d_{ri}^{x_u} f_{ri}^{y_u}}{D^{q_u}} \leq F_{max} \tag{26}$$

$$F_s = \frac{C_u K_u W^{S_u} Z^{P_u} d_s^{x_u} f_s^{y_u}}{D^{q_u}} \leq F_{max} \tag{27}$$

where, $C_u$, $K_u$, $P_u$, $q_u$, $S_u$, $x_u$, $y_u$ are constants; $F_{ri}$ is the machining force in the $i$th rough operation; $F_s$ represents the machining force for the finish pass; and $F_{max}$ represents the maximum available machining force.

The cutting power should not exceed the maximum power allowed for machine tool spindle during machining process. The power consumed in face milling for both rough and finish passes are given by the following equations:

$$P_{ri} = \frac{C_\lambda K_\lambda W^{s_\lambda} Z^{P_\lambda} d_{ri}^{x_\lambda} f_{ri}^{y_\lambda}}{D^{q_\lambda}} \leq P_{max} \tag{28}$$

$$P_s = \frac{C_\lambda K_\lambda W^{s_\lambda} Z^{P_\lambda} d_s^{x_\lambda} f_s^{y_\lambda}}{D^{q_\lambda}} \leq P_{max} \tag{29}$$

where, $C_\lambda$, $K_\lambda$, $P_\lambda$, $Q_\lambda$, $S_\lambda$, $x_\lambda$ and $y_\lambda$ are constants; $F_{ri}$ is the machining power in the ith rough operation; $P_s$ is the machining power for the finish pass; $P_{max}$ represents the maximum available machining.

Another important constraint is related to surface roughness that represents the product quality. The surface roughness depends essentially on feed rate and it must meet the roughness requirement, the following equations give the actual surface roughness in the ith finish operation and the actual surface roughness for the final finish pass $R_s$:

$$R_{ri} = 0.0321 \frac{f_s^2}{r_e} \leq R_{rmax} \tag{30}$$

$$R_s = 0.0321 \frac{f_s^2}{r_e} \leq R_{smax} \tag{31}$$

where, $r_e$ is the nose radius of the machining edge, $R_{rmax}$ and $R_{smax}$ are the surface roughness requirement for rough and final finish passes, respectively.

The tool life should not be shorter than the tool replacement life. In the model proposed by [30], tool life is inversely and exponentially dependent on cutting speed, depth of cut and feed, is adopted. Since, $T_R$ denotes the tool life requirement pre-specified by user, the tool life for the ith rough pass $t_{ri}$ and the tool life for the final finish pass $t_s$ are calculated as follow:

$$t_{ri} = \left( \frac{C_v K_v D^{qv}}{V_{ri} d_{ri}^{x_v} f_{ri}^{x_v} W^{s_v} Z^{p_v}} \right)^{1/l} \geq T_R \tag{32}$$

$$t_s = \left( \frac{C_v K_v D^{qv}}{V_s d_s^{x_v} f_s^{x_v} W^{s_v} Z^{p_v}} \right)^{1/l} \geq T_R \tag{33}$$

## 4   Proposed Solution Algorithms

- **Genetic algorithms based research**

Initiated in the 1970s, the genetic algorithms (AGs) represent stochastic search tools that solve a wide range of combinatorial optimization problems. They are inspired by the biological mechanism of reproduction and natural selection. Based on the durability of the most promising intuition, AGs allow to efficiently finding a new generation with a better cost function. Starting from the basic principle of AGs, the evolutionary process is simulated through a population of solutions representing individuals; each individual is endowed with a genotype consisting of one or more chromosomes. These are made up of a set of elements, which can take on several values. In these so-called evolutionary algorithms, three fundamental operators are involved, ranging from selection, which eradicates the least promising solutions, to crossover and mutation, which gives rise to new competing solutions by exploring the state space. In addition, to implement a genetic algorithm, four pieces of data are needed that corresponds practically to the size of the population (Pop_Size), the probability of crossover (Pc), the probability of mutation (Pm), and the total number of generations (Max_iter).

The genetic algorithm starts with a step called genesis in which an initial population of Pop_Size of individuals is generated. For each generated individual, a cost function is computed in order to define the adaptation score of the individuals during the selection process. These individuals evolve through the application of the crossover according to a probability Pc. Subsequently, the children obtained undergo an inversion at the gene level with a probability of mutation Pm. These three phases of evolution allow with a high chance of producing a new population better than the previous generation. With each new generation, the new populations become stronger and a loop is made until the assessment considers that the solution is not yet optimal. The main operators will be presented, in detail, in the following sections.

*Selection operator*: it represents a decisive heuristic in the implementation of the genetic mechanism to direct evolution and guide research by determining the candidate solutions supposed to be the most promising for the generation of offspring. Different selection techniques have emerged to eliminate chromosomes leading to poor quality children. In our research work, the stochastic Universal Sampling and the Tournament selection strategies are adopted.

- **Stochastic Universal Sampling selection**: It is used to improve the fitness proportionate by reducing selection Bias and embedding the gene pool to be conquered by only a small amount of fittest members of the population.
- **Tournament selection**: It consists of randomly selecting two individuals and selecting the best in terms of performance by comparing their adaptation function. Individuals who participate in a tournament are handed over or removed from the population, depending on the user's choice. This technique has the advantage of allowing the user to create tournaments with many participants or to highlight those who win the tournaments, which will promote the durability of their genes.

*Crossover operator*: It tends to increase the strength of the current population by randomly dividing individuals into hermaphrodite pairs. It is used to create new combinations of component parameters to form two descendants with characteristics from both parents. This operator is executed with a probability Pc, called crossover probability. Several crossing techniques have been used in the literature, in our research work; the Two Cut-Points Crossover is used and implemented as follows: two cut-off points are randomly selected for each ascendant. The genes at the ends are inherited directly from parents to children. The missing part is then completed by copying from left to right and in the same order of genes, the residual genes from string 2 to new string 1 and from string 1 to new string 2, respectively.

*Mutation operator*: Mutation is the third operator used in the process of genetic research; it is applied to preserve genetic diversity from one generation to the next by allowing the generation of points in regions that are of no interest. The mutation is carried out with a probability Pm, called the probability of mutation. Several mutation techniques have been introduced in the literature. In this research work, we use a reciprocal exchange mutation in which a gene at one position is removed and implanted at the position of the second gene chosen with a probability of mutation.

The final step of the GAs lies in the termination condition that plays an important role in assessing the quality of the individuals obtained. In this study, the execution stops when the algorithm makes n iterations without improvement in the best solution. The GAs procedure is shown in Algorithm 1.

---

**Algorithm 1 : The GA based search algorithm**

- Initialize : Pop_Size, Pc, Pm, Max_iter;
- Generate randomly the initial population Pop_Size (n);
- Calculate the fitness function for each chromosome of the current population;
- Select by tournament or stochastic selection strategies the best fitness of the current population;
- As long as the maximum number of iterations has not been reached:
- i=1
- For i  < Pop_Size :
  - Select randomly two individuals of P(N)
  - if $\alpha \leq$Pc :
    - Apply standard two-cut point's crossover and produce two children's
    - if $\beta \leq$ Pm :
    - Apply mutation
  - Copying the ascendants obtained in the new population P(n+1)
  - Evaluate the fitness of all chromosomes of the new fi(P(n+1)) population
- i=i+1
- **Until** i = Max-iter ( until maximum number of iteration is reached without improvement)
- **End while**

---

# 5  Simulated Annealing Genetic Algorithm (SAGA) Procedure

Although GA has several advantages over traditional techniques, the successful application of this algorithm depends on the population size and the diversity of individual solutions in the search space. If its diversity cannot be maintained before the global optimum is reached, it may converge to a local optimum that avoids the global targeted convergence. This genetic phenomenon called premature convergence is due to the fact that a good individual dominates the space of several generations, afterwards it risks invading the whole population and preventing a stable and balanced evolution.

To achieve effective research, genetic diversity must be preserved by maintaining a balance between exploiting the best solutions and exploring areas of interest. Typically, in order to achieve global convergence, there is a trend toward hybrid algorithms combining the strengths of the local and the global methods.

Simulated annealing (SA) is an efficient local approach inspired by the analogy with the annealing of metals in metallurgy. It consists of performing a movement according to a distribution probability that depends on the quality of the different neighbors based on the Metropolis algorithm. In the original Metropolis scheme, the probability of a physical system possessing energy E, when thermodynamic equilibrium is reached at a temperature T, is proportional to the Boltzmann–Gibbs factor. Under these conditions, at the implementation level of the process, a primary solution S is randomly generated, the transition from the current solution S to an adjacent solution S* depends on the probability P of the abovementioned Boltzman-Gibbs distribution. Indeed, this transition is accepted if the following condition is fulfilled:

$$P = \min\left\{1, e^{\frac{-f}{T}}\right\} \geq \Omega \qquad (34)$$

where, $f$ denotes the difference between the objective function of the two states, $\Omega$ is a random number generated in the interval [0, 1] and $T$ is the current control parameter in the process.

$$f = f_i - f_{worst} \qquad (35)$$

A hybrid method can be effective or ineffective depending on the selection of its components. In order to design a successful hybrid method, the strengths and weaknesses of each candidate method must be mastered. In this research work, a hybrid simulated annealing genetic algorithm is proposed, the choice of sequential hybridization of genetic algorithms and simulated annealing is dictated by a great mastery of the assets and defects of each of these optimization tools. Thus, this integration not only prevents premature convergence, but also overcomes the main drawback of SA consisting of its stopping when no neighboring point can improve the current solution. Indeed, this disadvantage can only be overcome if the process can be repeated from several randomly generated starting points through a commonly used measurement, or if the starting solutions are provided by a genetic algorithm.

In the last decade, several researchers attempted to combine GA and SA to provide a more robust optimization method that has both good convergence control and efficient stability. In their paper, Chen and Flann [31] proved that the hybrid of GA and SA can perform better for ten difficult optimization problems than either GA or SA separately. Sirag and Weisser [32] proposed a unified thermodynamic genetic operator to solve ordering problems. The unified operator is applied to the conventional GA operation of crossover and mutation to yield offspring. This operator can ensure greater population diversity at high temperature and less population diversity at low temperature. Mahfoud and Goldberg [33] also introduced a GA and SA hybrid. Their hybrid runs SA procedures in parallel, which uses mutation as the SA neighborhood operator and incorporates crossover to reconcile solutions across the processors. Similarly, hybrid method of GA and SA was also used by Varanelli and Cohoon [34]. In addition, Chen et al. [35] also proposed a hybrid method, which maintains one solution per Processing Element (PE). Each PE accepts a visiting solution from other PEs for crossover and mutation. For the selection process, the

SA cooling schedule and system temperature were used to decide whether the newly generated individual was accepted or not. In this method, they used the local selection of SA to replace the conventional selection process of GA. Hiroyasu et al. [36] proposed an algorithm involving several processes. In each process SA is employed. The genetic crossover is used to exchange information between individuals at fixed intervals.

The SAGA procedure is established in two stages: In the first one, a set of descendants is produced by applying different genetic research operators. In other words, after crossover and mutation for a couple of individuals, a new population is created. While in the second one, the obtained new population is accepted or rejected to pass to the next generation according to the Metropolis criterion formulated previously. Individuals with higher fitness values have a greater probability of surviving into the next generation. Those with less fitness values are not necessarily discarded. Four parameters $(f_{best}, f_{worst}, T_n, f_i)$ are involved to describe this selection process, $f_{best}$ represents the best fitness value of two parents, $f_{worst}$ is the worst fitness value of two parents, $f_i$ denotes the fitness value of one offspring $(i = 1, 2)$ and $T_n$ is the controlling temperature. Starting from a high temperature value, a state search is performed at each level with a number of iterations designated as Markov chain length. This length represents the number of movements performed for a fixed value of temperature that decreases logarithmically with time. When the algorithm reaches very low temperatures, the most probable states are in principle excellent solutions to the optimization problem. Since $0 < Cool\_rate < 1$ is the cooling rate parameter, the annealing program is therefore defined as follows:

$$T_{n+1} = Cool\_rate * T_n \tag{36}$$

Without omitting to point out that the replacement of the old population takes place if $f_{best}$ is better than $f_{worst}$ through Boltzmann's probability. This process is repeated iteratively until the maximum number of iterations is reached and the overall optimal value is obtained. The specific steps of the proposed SAGA cooperative approach are illustrated in the Algorithm 2 and Fig. 3.

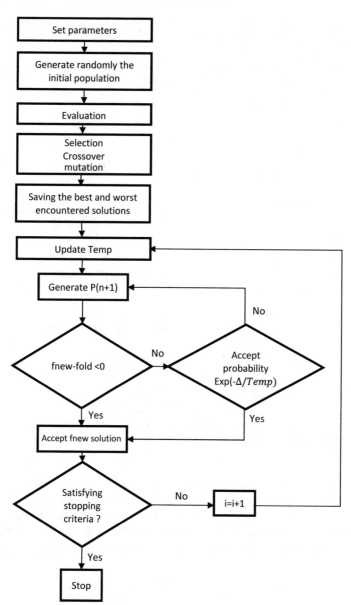

**Fig. 3** Flowchart of the proposed SAGA algorithm

| Algorithm 2 : The proposed SAGA algorithm |
| --- |

- Initialize : Pop_Size, Pc, Pm, Max_iter, Cool_rate, Temp=T0 ;
- Generate randomly the initial population Pop_Size (n);
- Calculate the fitness function for each chromosome of the current population;
- Select by tournament or stochastic selection strategies the best fitness of the current population;
- As long as the maximum number of iterations has not been reached:
- i=1
- For i < Pop_Size :
  - Select randomly two individuals of P(N)
  - if $\alpha \leq Pc$:
  - Apply standard two-cut point's crossover and produce two children's
  - If $\beta \leq Pm$ :
  - Apply mutation
  - Copying the ascendants obtained in the new population P(n+1)
  - Evaluate the fitness of all chromosomes of the new fi(P(n+1)) population
  - Saving the best and worst encountered solutions encountered:
    f_newbest = min [f]
    f_newworst= max [f]
  - if _newbest < f_oldbest:
    Population_Temp = New_Population [f_newbest:]
  - else if f_newbest == f_oldbest:
    i=i+1
  - else if f_newbest > f_oldbest:
    Calculate X= exp [- (f_newbest - f_oldbest) / Temp]
    Si $X \geq \Omega$ :
    Population_Temp=New_Population [f_newbest:]
    New_Population [: f_newworst ]= Population_Temp
    P (n+1) = New_Population
    Slow cooling of the temperature:
    Temp = Cool_rate *Temp
- If the stop criterion is met, stop the SAGA procedure and return the best configuration of the unit production cost found, otherwise go to step 5.
- **End while**

For both the procedures, the penalty function is adopted to penalize the chromosomes who violate the constraints. The more the constraints are violated, the heavier the penalty will be done. Consequently, the objective function will be small and the unfeasible results have more chance to be eradicated from the solution space. Basing on the objective function the probability of selection is calculated as follows:

$$Penalty = \frac{f_{worst} - f_j}{\sum_{j=1}^{Pop\_size} (f_{worst} - f_j)/Pop\_size} \tag{38}$$

where, $f_{worst}$ is the maximum value of objective function and $f_j$ is the value of objective function of the current element.

# 6 Algorithms Parameters Calibration: Design of Experiment of Taguchi

The strength and effectiveness of optimization approaches depend largely on the proper selection of parameters. With this in mind, a selection methodology called "design of experiments" has been used to optimize the parameters of the genetic proposed processes. Generally, this technique aims to identify and establish the links between input variables of the process and a quantity of interest, called the response. Several types of experimental designs have been developed to parameterize and manage the impact of variables on the response variation; they differ in their corresponding levels and the interference between them. The Taguchi approach represents one of the most adopted design-of-experiment methods in the literature. Developed by [37] in the 1960s, this approach aims to minimize the impact of the noise factor, that cannot be controlled by designers, and find the best level of controllable factors by using a number of concepts. Including the signal-to-noise (SNR) ratio, which is an indicator of variation used to measure the existing variation in order to identify control factors that reduce variability. There are three categories of performance characteristics for studying the signal-to-noise ratio, namely: smaller better, nominal better, and larger better. Since this study addresses a minimization problem, the signal-to-noise ratio should be measured using the lowest value, as shown below:

$$\frac{S}{N} = -10\log_{10}\left(\frac{1}{n}\sum_{i=1}^{n}\frac{1}{y_i^2}\right) \tag{39}$$

where, $n$ and $y$ represent the number of observations and the response variable, respectively.

Execution of the experimental trials was based on $L_{27}$ Taguchi's orthogonal array, and three levels were assigned to each factor whether for the SAGA or the GA with tournament and stochastic selection strategies as illustrated in Tables 1 and 2.

Regarding the GA tool, four factors, including population size, maximum number of iterations, probability of crossover, and probability of mutation are defined as critical control factors. It is worth mentioning that tournament and stochastic selections

**Table 1** Considered levels of GAs factors (tournament and stochastic selection)

| Considered levels of GAs factors (tournament and stochastic selection) | | | |
|---|---|---|---|
| Factors | Level 1 | Level 2 | Level 3 |
| Pc | 0.5 | 0.7 | 0.9 |
| Pm | 0.2 | 0.4 | 0.6 |
| Max_iter | 50 | 100 | 200 |
| Pop_size | 100 | 200 | 400 |
| Penalty | 1 | 5 | 10 |

**Table 2** Considered levels of SAGA factors

| Considered levels of SAGA factors | | | |
|---|---|---|---|
| Factors | Level 1 | Level 2 | Level 3 |
| Pc | 0.7 | 0.75 | 0.8 |
| Pm | 0.1 | 0.15 | 0.20 |
| Max_iter | 50 | 100 | 200 |
| Pop_size | 100 | 200 | 400 |
| Temp | 500 | 750 | 900 |
| Cool_rate | 0.7 | 0.8 | 0.9 |
| Penalty | 1 | 5 | 10 |

are used separately, the obtained results based on Taguchi method for these two traditional GAs are presented in Figs. 4 and 5. Results show that both cases need high population size and high maximum iterations to reach the optimization. Results of GA with stochastic selection method are altered with high levels of mutation and crossovers probabilities as well as penalty cost. In contrary GA with tournament selection method needs higher values of penalty cost, crossover, and mutation probabilities.

Concerning the proposed SAGA method, it involves various parameters namely: population size, cooling rate, initial temperature, mutation, crossover probabilities, and finally the maximum iterations without any best solution improvement. Basing on the orthogonal array of Taguchi, the obtained results shown in Fig. 6, illustrate that a higher population size and lower penalty costs are more appropriate to obtain

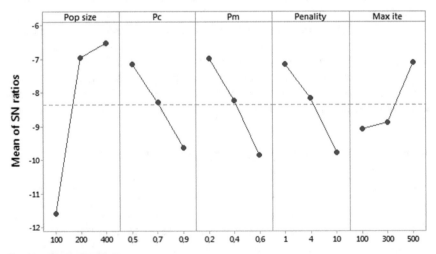

Signal-to-noise: Smaller is better

**Fig. 4** Mean effect plot for SN ratio of GA with tournament selection parameters

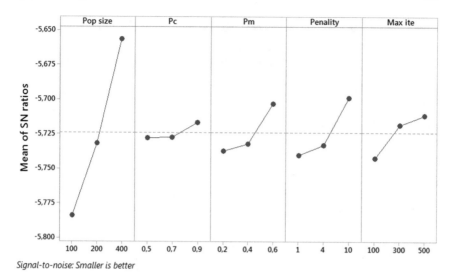

**Fig. 5** Mean effect plot for SN ratio of GA with stochastic selection parameters

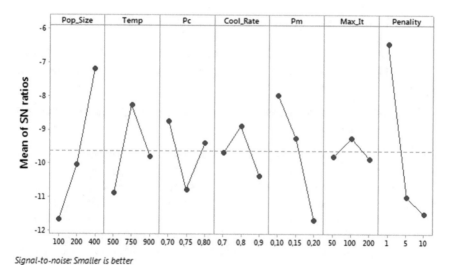

**Fig. 6** Mean effect plot for SN ratio of SAGA parameters

the best results. In contrast, respecting the mutation probability, it has to be fixed at the lower level and the crossover probability must not exceed 70%. Finally, the other factors are to be fixed at the medium levels. Table 3 regroups the selected level of each factor for the three optimization approaches.

**Table 3** Optimal parameters of the proposed optimization approaches

Optimal parameters of the proposed optimization approaches

| Factors | GA with tournament selection | GA with stochastic selection | SAGA |
|---------|------------------------------|------------------------------|------|
| Pc | 0.9 | 0.5 | 0.7 |
| Pm | 0.6 | 0.2 | 0.1 |
| Max_iter | 500 | 500 | 100 |
| Pop_size | 400 | 400 | 400 |
| Temp | – | – | 750 |
| Cool_rate | – | – | 0.8 |
| Penalty | 10 | 1 | 1 |

## 7  Numerical Results and Discussion

To evaluate the effectiveness of the traditional GAs with tournament and stochastic strategies and the hybrid SAGA for machining parameters optimization of multi-pass face milling, two cases of multi-pass face milling operations are studied. Data given in Table 4 summarizes workpiece features as well as cutting tool and machining characteristics. The considered data is the same adopted by [15, 30].

The GA and the hybrid SAGA are written in python 3.5 and runs on Intel(R) Core(TM) i5-2430M CPU @2.4 GHz, with 4 GB of RAM. A set of parameters are involved in the proposed algorithm.

- **Comparison of the proposed SAGA with GA (tournament and stochastic selection strategies)**

From the following Tables 5 and 6, it can be seen that for the first case $d_t = 8$ mm, where the number of passes is considered between 2 to 3 passes, GA technique combined to SA optimization tool is more powerful than both GA approach with Stochastic and Tournament selections. Particularly, for the second case $d_t = 15$ mm, where the number of passes is considered between 4 and 7 passes. The performance of the proposed algorithm is more proved. Since stock removal is too much large, more rough passes are needed which means that more decision variables are taken into consideration in the program.

The results show that optimal costs of SAGA technique are less than the traditional GA technique, moreover: GA with stochastic selection method is not enough powerful to give better results. For number of passes of 6 and 7, GA approach with stochastic selection cannot give satisfactory results.

The following figures report the convergence curves of the proposed SAGA and traditional GA optimization techniques. Figures 7 and 8 illustrates that the program takes more than 550 iterations before satisfying constraint related to number of rough passes and finish pass to total stock removal. Moreover, initial temperature of 750 and cooling rate of 0.8 justify the acceptance of non-feasible solutions which represents

**Table 4** Parameters setting of face milling operations

| Item | Symbol | Value | Item | Symbol | Value |
|------|--------|-------|------|--------|-------|
| Workpiece | L | 240 (mm) | Constants and exponents | $C_v$ | 445 |
|  | W | 100 (mm) |  | $K_v$ | 1 |
| Tool | D | 160 (mm) |  | $x_v$ | 0.15 |
|  | $r_e$ | 1 (mm) |  | $y_v$ | 0.35 |
|  | Z | 16 |  | $p_v$ | 0 |
| Milling costs and constraints | $K_0$ | 0.5 ($/min) |  | $q_v$ | 0.2 |
|  | $K_t$ | 2.5 ($/cut edge) |  | $s_v$ | 0.2 |
|  | Ttc | 1.5 (min/cut edge) |  | m | 0.32 |
|  | $T_p$ | 0.75 (min/cut edge) |  | $C_u$ | 534.6 |
|  | h1 | 7 $10^{-4}$ (min/mm) |  | $K_u$ | 1 |
|  | h2 | 0.3 (min) |  | $x_u$ | 0.9 |
|  | $d_{r,min}$ | 2.0 (mm) |  | $y_u$ | 0.74 |
|  | $d_{r,max}$ | 4.0 (mm) |  | $p_u$ | 1 |
|  | $d_{s,min}$ | 0.5 (mm) |  | $q_u$ | 1.0 |
|  | $d_{s,max}$ | 2.0 (mm) |  | $t_u$ | 1.0 |
|  | $f_{min}$ | 0.1 (mm/touth) |  | $C_\lambda$ | 0.5346 |
|  | $f_{max}$ | 0.6 (mm/touth) |  | $K_\lambda Z$ | 1 |
|  | $V_{min}$ | 50 (m/min) |  | $x_\lambda$ | 0.9 |
|  | $V_{max}$ | 300 (m/min) |  | $y_\lambda$ | 0.74 |
|  | $T_r$ | 240 (min) |  | $p_\lambda$ | 0 |
|  | $r_{r,a}$ | 25.0 ($\mu$m) |  | $q_\lambda Z$ | 1.0 |
|  | $r_{s,a}$ | 2.5 |  | $t_\lambda$ | 1.0 |
|  | $F_{max}$ | 8000 |  |  |  |
|  | $P_{max}$ | 8 |  |  |  |

the main feature of this optimization technique. It can also be seen that some worst solutions are accepted at the 200 first generations as the temperature is high. With lower temperature best solutions are more likely to be accepted and the program stops search after some iteration without improvement of minimum unit cost. In contrast, GA traditional optimization tool performs the search in the space of solution by accepting merely best solution as shown in Fig. 9.

- **Comparison of the proposed SAGA with other procedures**

In order to test the performance of the proposed SAGA, comparison is carried out for the case of $d_t = 8$ mm. Several optimization techniques have been used in the same conditions, such as GA [7, 14], Genetic Search (GS) [15] and Fuzzy Particle Swarm Optimization Algorithm (FPSO) [20]. The comparison of SAGA with previous research works is presented in Table 7. The obtained results of the

**Table 5** Optimal machining parameters and total cost for total depth of cut $d_t$ = 8 mm

| $d_t$ = 8 mm | n | SAGA | | | | GA stochastic | | | | GA tournament | | | |
|---|---|---|---|---|---|---|---|---|---|---|---|---|---|
| | | V | f | A (mm) | Cost ($) | V | f | a | Cost ($) | V | f | A | Cost ($) |
| Rough | 2 | 93 | 0.41 | 3.19 | 1.59 | 99.25 | 0.31 | 3.9 | 1.66 | 79.52 | 0.46 | 2.72 | 1.673 |
| | | 93.96 | 0.46 | 2.94 | | 76.65 | 0.4 | 2.55 | | 86.28 | 0.29 | 3.6 | |
| Finition | | 92.4 | 0.59 | 1.87 | | 89.78 | 0.52 | 1.55 | | 91.33 | 0.49 | 1.64 | |
| Rough | 3 | 61.8 | 0.59 | 2.23 | 1.91 | 91 | 0.47 | 2 | 2.10 | 92.47 | 0.5 | 2.28 | 1.92 |
| | | 94.54 | 0.5 | 2.27 | | 91.84 | 0.31 | 2.16 | | 90.8 | 0.5 | 2.67 | |
| | | 90.05 | 0.47 | 2.92 | | 76.14 | 0.27 | 2.48 | | 93.36 | 0.5 | 2.54 | |
| Finition | | 108.36 | 0.6 | 0.55 | | 87.1 | 0.49 | 1.36 | | 111.6 | 0.5 | 0.1 | |

**Table 6** Optimal machining parameters and total cost for total depth of cut $d_t$ = 15 mm

| $d_t$ = 15 mm | n | SAGA | | | | GA stochastic | | | | GA tournament | | | |
|---|---|---|---|---|---|---|---|---|---|---|---|---|---|
| | | V | f | A | Cost ($) | V | f | a | Cost ($) | V | F | A | Cost ($) |
| Rough | 4 | 92.36 | 0.4968 | 2.73 | 2.40 | 139.75 | 0.15 | 2.73 | 2.93 | 111.23 | 0.21 | 3.9 | 2.71 |
| | | 93.37 | 0.3745 | 3.477 | | 101.5 | 0.29 | 3.81 | | 77.93 | 0.24 | 2.93 | |
| | | 95.36 | 0.4369 | 3.0779 | | 93 | 0.31 | 3.84 | | 111.8 | 0.23 | 3.89 | |
| | | 103.37 | 0.30 | 3.9259 | | 101.57 | 0.26 | 2.71 | | 97.95 | 0.34 | 2.47 | |
| Finition | | 94.64 | 0.554 | 1.7771 | | 146.5 | 0.13 | 1.94 | | 71.24 | 0.46 | 1.83 | |
| Rough | 5 | 107.55 | 0.25 | 3.8 | 2.755 | 91.44 | 0.43 | 2.56 | 3.11 | 88.89 | 0.48 | 2.54 | 2.78 |
| | | 92.35 | 0.5 | 2.62 | | 79 | 0.28 | 2.14 | | 88.17 | 0.42 | 2.54 | |
| | | 93.09 | 0.5 | 2.32 | | 113.24 | 0.24 | 3.71 | | 88.87 | 0.49 | 2.71 | |
| | | 94.42 | 0.44 | 3.1 | | 72.23 | 0.47 | 2.06 | | 61.45 | 0.43 | 2.68 | |
| | | 91.52 | 0.5 | 2.65 | | 76.5 | 0.18 | 2.56 | | 88.1 | 0.42 | 2.52 | |
| Finition | | 107.56 | 0.5 | 0.52 | | 69.48 | 0.47 | 1.93 | | 87.44 | 0.49 | 1.94 | |
| Rough | 6 | 91.32 | 0.5 | 2.088 | 3.105 | – | – | – | – | 61.76 | 0.5 | 2.37 | 3.11 |
| | | 92.3 | 0.5 | 2.022 | | – | – | – | | 115.65 | 0.25 | 2.02 | |
| | | 92.11 | 0.5 | 2.09 | | – | – | – | | 91.4 | 0.49 | 2.47 | |
| | | 126.1 | 0.23 | 2 | | – | – | – | | 91.22 | 0.5 | 2.78 | |
| | | 93.36 | 0.37 | 3.47 | | – | – | – | | 93.15 | 0.5 | 2.52 | |
| | | 87.45 | 0.48 | 2.8 | | – | – | – | | 91.13 | 0.5 | 2.27 | |
| Finition | | 108 | 0.6 | 0.514 | | – | – | – | | 109.41 | 0.5 | 0.54 | |
| Rough | 7 | 61.88 | 0.5 | 2.23 | 3.358 | – | – | – | – | 85.85 | 0.5 | 2.06 | 3.39 |
| | | 92.17 | 0.5 | 2 | | – | – | – | | 90.2 | 0.49 | 2.05 | |

(continued)

**Table 6** (continued)

| $d_t = 15$ mm | n | SAGA | | | | GA stochastic | | | | GA tournament | | | |
|---|---|---|---|---|---|---|---|---|---|---|---|---|---|
| | | V | f | A | Cost ($) | V | f | a | Cost ($) | V | F | A | Cost ($) |
| | | 91.64 | 0.5 | 2 | | – | – | – | | 84.37 | 0.5 | 2.05 | |
| | | 77.54 | 0.59 | 2 | | – | – | – | | 91.49 | 0.49 | 2 | |
| | | 92.67 | 0.5 | 2 | | – | – | – | | 82.95 | 0.5 | 2.08 | |
| | | 90.49 | 0.5 | 2 | | – | – | – | | 86.14 | 0.5 | 2.1 | |
| | | 92.04 | 0.5 | 2.16 | | – | – | – | | 86.87 | 0.5 | 2.1 | |
| Finition | | 109.7 | 0.6 | 0.51 | | – | – | – | | 98.1 | 0.47 | 0.5 | |

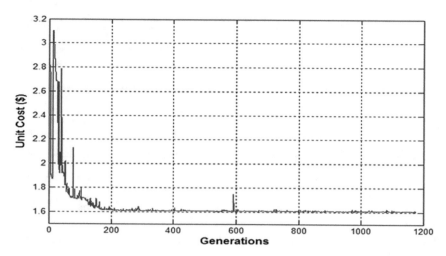

**Fig. 7** The convergence curve of the proposed SAGA approach for $d_t = 8$ mm

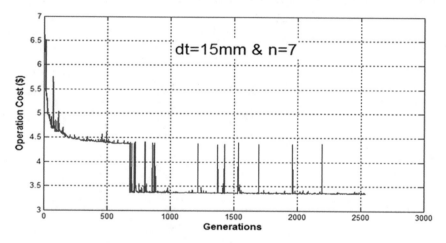

**Fig. 8** The convergence curve of the proposed SAGA approach for $d_t = 15$ mm

proposed approach are marked in bold. They show that the unit production cost ($1.59US) obtained by SAGA is smaller than those of the aforementioned comparison studies. Additionally, Fig. 10 gives a comparison of obtained optimal cost with our proposed optimization techniques and the optimal costs as presented by [20] and the two different GA approaches. Yang et al. [20] proposed a FPSO optimization technique where, a methodology to distribute the total stock removal is incorporated. This figure confirms that without a methodology to distribute the total stock removal SAGA is more powerful and gives satisfactory results for the machining parameters optimization of multi-pass face milling.

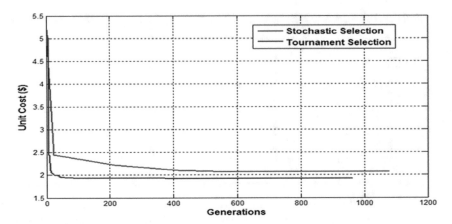

**Fig. 9** The convergence curve of the propos $d_t$ GA approach with different selection methods for $d_t = 8$ mm

**Table 7** Benchmark of optimal machining parameters and total cost for total depth of cut $d_t =$ 15 mm

| Methods | F (mm/tool) | V (m/min) | A (mm) | Cost ($) |
|---|---|---|---|---|
| GA [7] | 0.1 | 114.7 | 4 | 2.00 |
| | 0.30 | 115.3 | 2 | |
| | 0.27 | 119.2 | 2 | |
| GS [15] | 0.4 | 66.7 | 2.8 | 1.86 |
| | 0.33 | 66.7 | 3.2 | |
| | 0.6 | 66.7 | 2 | |
| GA [14] | 0.43 | 60 | 3.1 | 1.76 |
| | 0.43 | 60 | 3.1 | |
| | 0.27 | 124.46 | 2.8 | |
| FPSO [20] | 0.59 | 89 | 2.4 | 1.96 |
| | 0.59 | 89 | 2.4 | |
| | 0.59 | 91 | 2.6 | |
| Current SAGA | **0.41** | **93** | **3.19** | **1.59** |
| | **0.46** | **93.96** | **2.94** | |
| | **0.59** | **92.4** | **1.78** | |
| Current GA with tournament selection | 0.46 | 79.52 | 2.72 | 1.67 |
| | 0.29 | 86.28 | 3.6 | |
| | 0.49 | 91.33 | 1.64 | |
| Current GA with stochastic selection | 0.31 | 99.25 | 3.9 | 1.66 |
| | 0.4 | 76.65 | 2.55 | |
| | 0.52 | 89.78 | 1.55 | |

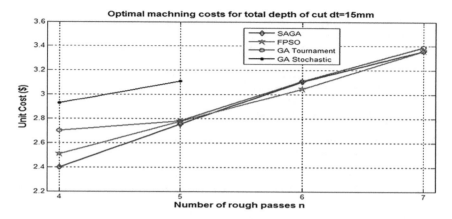

**Fig. 10** Benchmark of the unit cost improvement yielded by SAGA with FPSO

# 8 Conclusion and Perspectives

The face milling operation which involves various parameters is studied in this paper for the purpose of minimizing the unit production cost. Three different approaches are proposed and analyzed, namely the hybrid SAGA optimization approach and the traditional GA with tournament selection strategy and traditional GA with stochastic selection technique. The aim of the integration of the SA with the genetic search is to avoid the premature convergence of the genetic algorithm on one hand and overcome the main drawback of the SA on the other hand. Parameters of the different approaches are then calibrated with Taguchi method. Finally, results show that for the different considered depth of cut, the SAGA is more powerful compared to the proposed GA tools with different selection procedures and to some many others methodology of total stock distribution proposed in previous published papers.

In future works, a methodology to generate feasible solutions will be integrated into our optimization tool. Additionally, this study can be extended to optimize others objective functions such as production time and can be also extended for other material removal processes as turning and grinding operations.

# References

1. Petropoulos, P.G.: Optimal selection of machining rate variables by geometric programming. Int. J. Prod. Res. **11**, 305–314 (1973)
2. Shin, Y.C., Joo, Y.S.: Optimization of machining conditions with practical constraints. Int. J. Prod. Res. **30**, 2907–2919 (1992)
3. Gupta, R., Batra, J.L., Lal, G.K.: Determination of optimal subdivision of depth of cut in multipass turning with constraints. Int. J. Prod. Res. **33**, 2555–2565 (1995)
4. Tolouei-Rad, M., Bidhendi, I.M.: On the optimization of machining parameters for milling operations. Int. J. Mach. Tools Manuf. **37**, 1–16 (1997)

5. Sönmez, A.İ, Baykasoğlu, A., Dereli, T., Fılız, İH.: Dynamic optimization of multipass milling operations via geometric programming. Int. J. Mach. Tools Manuf. **39**, 297–320 (1999)
6. Liu, Y., Wang, C.: A modified genetic algorithm based optimisation of milling parameters. Int. J. Adv. Manuf. Technol. **15**, 796–799 (1999)
7. Shunmugam, M.S., Reddy, S.B., Narendran, T.T.: Selection of optimal conditions in multi-pass face-milling using a genetic algorithm. Int. J. Mach. Tools Manuf. **40**, 401–414 (2000)
8. Wang, J., Kuriyagawa, T., Wei, X.P., Guo, D.M.: Optimization of cutting conditions for single pass turning operations using a deterministic approach. Int. J. Mach. Tools Manuf. **42**, 1023–1033 (2002)
9. Briceno, J.F., El-Mounayri, H., Mukhopadhyay, S.: Selecting an artificial neural network for efficient modeling and accurate simulation of the milling process. Int. J. Mach. Tools Manuf. **42**, 663–674 (2002)
10. An, L., Chen, M.: On optimization of machining parameters. In: Proceedings of the 4th International Conference on Control and Automation, 2003. ICCA'03, pp. 839–843. IEEE (2003)
11. Wang, Z.G., Wong, Y.S., Rahman, M.: Optimisation of multi-pass milling using genetic algorithm and genetic simulated annealing. Int. J. Adv. Manuf. Technol. **24**, 727–732 (2004)
12. Wang, Z.G., Rahman, M., Wong, Y.S., Sun, J.: Optimization of multi-pass milling using parallel genetic algorithm and parallel genetic simulated annealing. Int. J. Mach. Tools Manuf. **45**, 1726–1734 (2005)
13. Onwubolu, G.C.: Performance-based optimization of multi-pass face milling operations using Tribes. Int. J. Mach. Tools Manuf. **46**, 717–727 (2006)
14. Saha, S.: Genetic algorithm based optimization and post optimality analysis of multi-pass face milling. arXiv preprint arXiv:0902.0763 (2009)
15. António, C.C., Castro, C.F., Davim, J.P.: Optimisation of multi-pass cutting parameters in face-milling based on genetic search. Int. J. Adv. Manuf. Technol. **44**, 1106–1115 (2009)
16. Krimpenis, A., Vosniakos, G.C.: Rough milling optimisation for parts with sculptured surfaces using genetic algorithms in a Stackelberg game. J. Intell. Manuf. **20**, 447–461 (2009)
17. Öktem, H.: An integrated study of surface roughness for modelling and optimization of cutting parameters during end milling operation. Int. J. Adv. Manuf. Technol. **43**, 852–861 (2009)
18. Mahdavinejad, R.A., Khani, N., Masoud, M., Fakhrabadi, S.: Optimization of milling parameters using artificial neural network and artificial immune system. J. Mech. Sci. Technol. **26**, 4097–4104 (2012)
19. Rao, R.V., Pawar, P.J.: Parameter optimization of a multi-pass milling process using non-traditional optimization algorithms. Appl. Soft Comput. **10**, 445–456 (2010)
20. Yang, W.A., Guo, Y., Liao, W.H.: Optimization of multi-pass face milling using a fuzzy particle swarm optimization algorithm. Int. J. Adv. Manuf. Technol. **54**, 45–57 (2011)
21. Zhou, A., Qu, B.Y., Li, H., Zhao, S.Z., Suganthan, P.N., Zhang, Q.: Multiobjective evolutionary algorithms: a survey of the state of the art. Swarm Evol. Comput. **1**, 32–49 (2011)
22. Zarei, O., Fesanghary, M., Farshi, B., Saffar, R.J., Razfar, M.R.: Optimization of multi-pass face-milling via harmony search algorithm. J. Mater. Process. Technol. **209**, 2386–2392 (2009)
23. Yang, W.A., Guo, Y., Liao, W.: Multi-objective optimization of multi-pass face milling using particle swarm intelligence. Int. J. Adv. Manuf. Technol. **56**, 429–443 (2011)
24. Fratila, D., Caizar, C.: Application of Taguchi method to selection of optimal lubrication and cutting conditions in face milling of AlMg3. J. Clean. Prod. **19**(6–7), 640–645 (2011)
25. Yildiz, A.R.: A new hybrid differential evolution algorithm for the selection of optimal machining parameters in milling operations. Appl. Soft Comput. **13**(3), 1561–1566 (2013)
26. Qu, S., Zhao, J., Wang, T.: Experimental study and machining parameter optimization in milling thin-walled plates based on NSGA-II. Int. J. Adv. Manuf. Technol. **89**, 2399–2409 (2017)
27. Shaik, J.H., Srinivas, J.: Optimal selection of operating parameters in end milling of Al-6061 work materials using multi-objective approach. Mech. Adv. Mater. Modern Processes **1**, 5 (2017)
28. Yang, Y.: Machining parameters optimization of multi-pass face milling using a chaotic imperialist competitive algorithm with an efficient constraint-handling mechanism. Comput. Model. Eng. Sci. **116**(3), 365–389 (2018)

29. Chen, X., et al.: Integrated optimization of cutting tool and cutting parameters in face milling for minimizing energy footprint and production time. Energy **175**, 1021–1037 (2019)
30. Nefedov, N., Osipov, K.: Typical examples and problems in metal cutting and tool design. MIR, Moscow (1987)
31. Chen, H., Flann, N.S.: Parallel simulated annealing and genetic algorithms: a space of hybrid methods. In: International Conference on Parallel Problem Solving from Nature, pp. 428–438. Springer, Berlin, Heidelberg (1994)
32. Sirag, D.J., Weisser, P.T.: Toward a unified thermodynamic genetic operator. In: Erlhaum, L. (ed.) Genetic Algorithms and Their Applications: Proceedings of the Second International Conference on Genetic Algorithms: Massachusetts Institute of Technology. Cambridge, MA. Hillsdale, NJ (1987)
33. Mahfoud, S.W., Goldberg, D.E.: Parallel recombinative simulated annealing: a genetic algorithm. Parallel Comput. **21**, 1–28 (1995)
34. Varanelli, J.V., Cohoon, J.C.: Population-oriented simulated annealing: a genetic/thermodynamic hybrid approach to optimization. In: Eshelman, L.J. (ed.) Proceedings of Sixth International Conference on Genetic Algorithms, pp. 174–181. M. Kaufman, San Francisco, Calif (1995)
35. Chen, H., Flann, N.S., Watson, D.W.: Parallel genetic simulated annealing: a massively parallel SIMD algorithm. IEEE Trans. Parallel Distrib. Syst. **9**, 126–136 (1998)
36. Hiroyasu, T., Miki, M., Ogura, M.: Parallel simulated annealing using genetic crossover. Sci. Eng. Rev.-Doshisha University **41**, 130–138 (2000)
37. Taguchi, G.G., Chowdhury, S., Taguchi, S.: Robust Engineering. McGraw-Hill, New York (2000)

# Role of Constrained Optimization Technique in the Hybrid Cooling of High Heat Generating IC Chips Using PCM-Based Mini-channels

V. K. Mathew, Naveen G. Patil, and Tapano Kumar Hotta

**Abstract** Optimization has gained enormous importance in the area of electronics to control the temperature of the high heat-generating electronic components. The present study emphasizes the transient numerical simulations on seven asymmetric IC chips kept next to the Left–Right-Bottom-Top (LRBT) and Left–Right-Bottom (LRB) mini-channels fabricated on the SMPS board using the three different phase change materials (PCMs); Suntech P116 (Tm: 49.5 °C), Paraffin wax (Tm: 44 °C), and n-eicosane (Tm: 40.5 °C). The objective is to minimize the maximum temperature of the configuration (arrangement of the seven IC chips) and to maintain it below the critical limit ($\leq$125 °C). The results suggest that the configuration maximum temperature has dropped by 15.33–22.2% using the n-eicosane (lower melting point PCM). A correlation is established for the dimensionless configuration maximum temperature excess ($\theta$) in terms of the IC chips heat input ($q_o$), PCM volume content ($v_o$), and IC chips size ($\delta$). The optimal temperature of the configuration is identified using the GA-based constrained optimization strategy. A sensitivity analysis is carried out to study the effect of the constraints on the optimal temperature of the IC chips. It confirms the dependency of the IC chips temperature on their size, heat input, and PCM volume content.

**Keywords** Constrained optimization · Genetic algorithm · IC chips · Mini-channel · Phase change material · Thermal management

V. K. Mathew
MIT School of Engineering, MIT-ADT University, Pune 412201, India

N. G. Patil
School of Engineering, Ajeenkya DY Patil University, Pune, Maharashtra 412102, India

T. K. Hotta (✉)
Department of Thermal and Energy Engineering, School of Mechanical Engineering, Vellore Institute of Technology (VIT), Vellore, Tamilnadu 632014, India
e-mail: tapano.hotta@vit.ac.in

# 1 Introduction

The temperature control of the electronic components is playing a very vital role in day-to-day practices to improve their reliability. This can be accomplished using various innovative cooling techniques. Reduction in the size of the electronic components has led to the shrinkage in the effective area available for their heat dissipation. The energy requirements have led to the search for alternative energy sources as well as other energy-saving ways that can fulfill the present energy demand. Since then; the focus on the phase change material (PCM)-based cooling has increased. The PCMs are substances that can absorb, store, and release a large amount of thermal energy. The exchange of energy plays a vital role, as several materials display this phenomenon in a temperature range near to the PCM melting point. The overview of different phase change materials, their typical applications, and the general limitations of this study were given by Humphires and Griggs [1]. The thermal energy storage is a key aspect of energy management with an emphasis on efficient energy conservation.

# 2 Background Study

The analytical and experimental analysis for the temperature control of electronics using the PCM-based heat sinks subjected to the cyclic heat load was carried by Saha and Dutta [2]. They considered two important factors; the working cycle and convective cooling of the chips using the PCM. They suggested that the modified design can effectively reduce the heat flow during the cooling cycle as compared to the heating one. Sabbah et al. [3] carried out the 3D numerical investigation on the cooling of the electronic devices using the micro-channel heat sinks (MCHS) filled with a mixture of PCM and slurry of water. They observed that the heat dissipation rate (100–500 W/cm$^2$) was strongly dependent on the channel inlet and outlet conditions, and melting temperature. Baby and Balaji [4] investigated experimentally the heat transfer performance of a PCM-based plate-fin heat sink matrix under the constant and intermittent heat loads (5–10 W) using the n-eicosane by varying their volume fraction between 0.5 and 1. They observed around 60% temperature in the heat sink baseline. Faraji et al. [5] carried out the numerical analysis to study the thermal performance of n-eicosane-based heat sink for the cooling of protruding electronic chips and observed the maximum temperature of the chip at 75 °C. They have proposed a correlation for the dimensionless temperature of the IC chips in terms of the PCM working time and PCM liquid fraction. The thermal management studies of the electronic devices using the PCM embedded in a pin–fin heat sink was carried out by Parlak and Etiz [6]. They suggested that the pin–fin heat sink has led to an increase in the conductivity and surface area of the devices. The 3D transient numerical simulations for the cooling of the portable electronic device were carried out by Wang and Yang [7] using the multi-fin heat sink filled with PCM. They varied the performance parameters such as fin number (0, 3, and 6), power input (2, 3, and 4 W) and orientation (vertical/horizontal/slanted). They showed that the use of PCM was able to control the operating temperature of the components for a longer melting time. Baby and Balaji [8] carried out the experimental and ANN-based optimization studies for the thermal performance of the PCM-based heat sinks. They observed the enhancement factor of 15 using the PCM and suggested that the PCM was not advisable at the low power level (1 kW/m$^2$). The experimental and numerical investigations of the hybrid PCM-air heat sink was carried out by Kozak et al. [9]. They found that the use of PCM was justified when the accumulation of the latent heat was very high. Ollier et al. [10] investigated experimentally and numerically the cooling of the electronic devices using the composite materials (CNT and PCM) integrated with silicon. These new materials can provide solutions for better thermal management at higher heat fluxes. The multi-objective geometric optimization for the PCM was carried out by Sridharan et al. [11] on the cylindrical heat sinks using the thermal conductivity enhancer. The ANN-based multi-objective optimization forward model has been used to get the optimal parameters. They observed a 13% improvement in the performance of the heat sinks. The numerical investigation to study the heat sink cooling in the motherboard chip of a computer with temperature-dependent thermal conductivity was carried out by Zaretabar et al. [12]. They observed that the increase in the airflow velocity has led to an increase in the Nusselt number and ultimately there was an enhancement in the heat transfer coefficient. Loganathan and Mani [13] studied the fuzzy-based hybrid multi decision-making methodology adopted for the different PCMs used for the cooling of the power electronics system. They found that RT-80 has shown better results for the thermal management of power electronics. Mathew and Hotta [14] carried the PCM-based cooling of discrete heat sources under the mixed convection heat transfer mode. They used the mini-channels filled with Paraffin wax at different melt fractions and found that the conjugate heat transfer between the IC chips and the PCMs can effectively drop the $T_{max}$ of the chips by 3–6 °C. Sharma et al. [15] conducted the numerical and optimization analysis for the size, shape of the composite heat sinks filled with n-eicosane (ECHS). They found that the optimal sized vertical ECHS has exhibited good results with the conventional designs. Liu et al. [16] carried out the experimental investigation using the composite PCM (n-eicosane and expanded graphite) for the electro-driven thermal energy storage. They achieved the heat storage efficiency of C20/EG15 at 65.7% PCM volume content. The numerical investigation for the transient cooling of the finned heat sinks embedded with PCM was carried out by Arshad et al. [17]. They proposed the correlation for the liquid fraction, corrected Nusselt number in terms of the corrected Fourier number, Stefan number, and Rayleigh number. Mathew and Hotta [18] studied the thermal management of multiple protruding IC chips mounted on the SMPS board under different positions of the mini-channels filled with different PCMs. They suggested that n-eicosane can reduce the $T_{max}$ of the configuration by 5.5%. The numerical simulation and hybrid optimization (ANN-GA) strategies used for the cooling of the IC chips was carried out by Patil and Hotta [19] under the forced convection. They suggested that the bigger size chips have led to the dissipation of more heat (better cooling), and essentially kept at the substrate bottom. Zeng et al.

[20] performed an experimental investigation on silver nano-wire composite PCM (1-Tetradecanol) at different aspect ratios of the silver nano-wire. They observed that the thermal conductivity of the composite PCM with nano-wire has increased to 1.45 W/mK and the enthalpy content has increased to 76.5 J/g. Usman et al. [21] carried out the experimental investigation on finned and un-finned heat sinks filled with PCM. They showed that RT-44 has exhibited better passive thermal management. Kannan and Kamatchi [22] conducted the experimental studies on the cooling of the electronic devices using the thermo-siphon assisted with different PCMs. They observed 98.9% of heat removal for acetone at 90 W.

A detailed review of the literature has been carried out on the experimental and numerical investigations for the cooling of heat sources using the PCM-based heat sinks under different heat transfer modes. From the intense review of the literature, it was noticed that most of the investigations were based on the cooling of the heat sources with direct contact, but very few studies were reported on the conjugated heat transfer analysis with indirect contact of the chips using the PCM. The cooling of the heat sources using the mini and micro-channels with constrained optimization strategies are very scarce in the literature. Hence, the present work emphasizes the numerical modeling for the cooling of the IC chips [14] using the PCM-filled mini-channels.

## 3 Methodology

Time-dependent numerical investigations are carried out on the 7 asymmetric IC chips placed on the substrate board (silicon) for the PCM-filled mini-channels under the forced convection. The computational model considered for the present study is shown in Fig. 1. The specifications of the substrate board, IC chips, mini-channels (LRBT and LRB cases), and the PCM were reported in Mathew and Hotta [14]. The simulations are carried out using the commercial software ANSYS fluent V16.0 using three different phase change materials with melting point (SunTech P116—49.5 °C, Paraffin Wax—44 °C, and n-eicosane—40.5 °C). The PCMs are then filled inside these mini-channels as shown in Fig. 2.

The simulations are carried out for the LRBT (left, right, top, and bottom) and LRB (left, right, and bottom) mini-channel cases. In each case of LRBT and LRB, three different PCMs under two constant heat fluxes (50 and 25 W/cm$^2$) and with two different PCM volume contents (100 and 50%) are considered for the analysis. This has led to the total 24 cases as shown in Table 1. In each case, the uniform velocity of 13 m/s is supplied at the inlet of the channel which forms the combined effect of forced convection and PCM leading to the hybrid cooling. The velocity is selected as per the actual rotation of the fan located in the CPU which can run at a maximum speed of 2500 rpm. The rpm is then converted to the axial velocity of 13 m/s (Fig. 3 and Table 2).

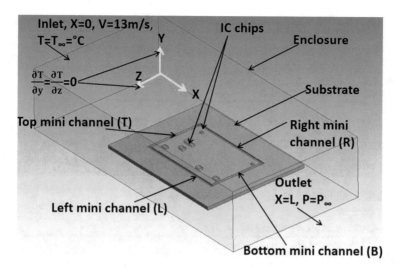

**Fig. 1** The computational model used for the present analysis

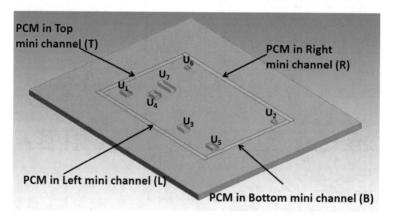

**Fig. 2** Position of the IC chips and the PCM-filled mini-channels

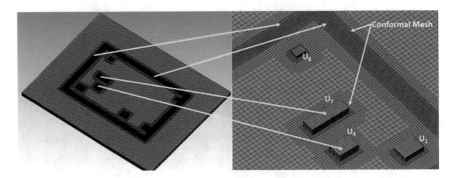

**Fig. 3** Mesh profile of the IC chips and the substrate board

**Table 1** Different cases considered for the simulation

| Volume variation | Constant heat flux W/cm² | Phase change material | | | | | |
|---|---|---|---|---|---|---|---|
| | | SunTech P116 (49.5 °C) | Paraffin Wax (44 °C) | n-eicosane (40.5 °C) | SunTech P116 (49.5 °C) | Paraffin Wax (44 °C) | n-eicosane (40.5 °C) |
| 100% PCM volume | 50 | LRBT | LRBT | LRBT | LRB | LRB | LRB |
| | 25 | LRBT | LRBT | LRBT | LRB | LRB | LRB |
| 50% PCM volume | 50 | LRBT | LRBT | LRBT | LRB | LRB | LRB |
| | 25 | LRBT | LRBT | LRBT | LRB | LRB | LRB |

**Table 2** Properties of PCM

| Sl.no | Property | Suntech P116 [24] | Paraffin wax [23] | n-eicosane [24] |
|---|---|---|---|---|
| 1 | Density (kg/m³) | 760 | 760 | 769 |
| 2 | Thermal conductivity (W/mK) | 0.24 | 0.24 | 0.21 |
| 3 | Dynamic viscosity (Ns/m²) | 1.90 | 1.90 | 0.00355 |
| 4 | Specific heat (J/kg K) | 2950 | 2950 | 2460 |
| 5 | Latent heat of fusion (kJ/kg) | 266 | 266 | 215 |
| 6 | Melting point (°C) | 49.5 | 44 | 40.5 |

## 3.1  Numerical Framework

Numerical simulations are carried out for all the cases mentioned in Table 1. The ANSYS V16 uses the Solidification and Melting model with meshy zone constant (C) = $10^5$. The detailed methodology, governing equations, validation of the numerical model can be found in Mathew and Hotta [14].

## 3.2  Boundary Conditions

The boundary conditions applied to the computational model are mentioned below.
    At X = 0 (inlet), T = T∞ = 25 °C, V = 13 m/s.
    At X = L (outlet), P = P∞ = 25 °C and the lateral boundary conditions are assumed to be adiabatic: $\frac{\partial T}{\partial y} = \frac{\partial T}{\partial z} = 0$.

## 3.3  Grid Sensitivity Study

The grid size of the computational domain plays a vital role in the numerical analysis and the accuracy of the solution is largely dependent on this. The grid size selected for the study decides the accuracy and the computational cost of the analysis. In the present study, the cut cell method is adopted in which all the cells of the components are Hexa dominant and the accuracy of the solution for this Hexa cell is more accurate and reliable in the computational domain. The mesh profile is shown in Fig. 3. A mesh sensitivity analysis is carried for the different elements to evaluate the changes in the IC chips temperature, and to choose the optimum mesh elements for which the solution is accurate and precise. Figure 4 shows the mesh sensitivity analysis for

**Fig. 4** Mesh sensitivity study

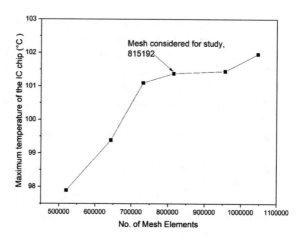

the different elements. This suggests that 815192 elements are deemed to be optimal and is selected for the present study.

The detailed methodology, governing equations, and validation of the numerical model can be found in Mathew and Hotta [14].

# 4   Results and Discussion

The transient analysis is conducted for a time period of 300 s under the different cases of constant heat flux (50 and 25 W/cm$^2$), PCM volume content (100 and 50%), and with three different phase change materials (Suntech P116—49.5 °C, Paraffin Wax—44 °C and n-eicosane—40.5 °C). The properties of the PCMs are mentioned in Table 2. The mini-channels are thrust inside the substrate board and around the periphery of the IC chips; so that, the actual wiring and traces of the electrical connections of the board are not disturbed. The mini-channels are filled with PCM which are embedded in the substrate board and forms a pool of PCM around the IC chips. Constant heat flux (50 and 25 W/cm$^2$) are provided to the IC chips which results in different volumetric heat generation. The simulations are performed using the hybrid cooling technique with PCM filled inside the mini-channels under the forced convection (uniform velocity of 13 m/s is supplied). The simulations are performed for all 24 cases as mentioned in Table 1. The goal is to minimize the temperature of the IC chips under the hybrid cooling technique using the different PCMs. The study is also extended to determine the optimum temperature of the configuration under the constraints of IC chip heat input, PCM volume content, and IC chip size.

## 4.1   IC Chip Temperature Variation Under Different Cooling Techniques

The transient analysis is conducted for different modes of heat transfer under the natural convection with PCM filled mini-channels (NC + PCM), forced convection (V = 13 m/s) without the PCM-filled mini-channels (FC) and forced convection (V = 13 m/s) with the PCM filled mini-channels (FC + PCM). Seven asymmetric IC chips are supplied with a constant heat flux of 50 W/cm$^2$. Figure 5 depicts the IC chip temperature variation under different cooling techniques and it is observed that under the natural convection with PCM filled mini-channels there is a huge rise in the temperature of the IC chips up to 254 °C. This temperature is reduced under the forced convection cooling technique without the use of PCM. Under the forced convection there is a huge reduction in the temperature of the IC chips by 130–146 °C which is further reduced using the PCM (Suntech P116) filled mini-channels under the forced convection cooling technique. The use of the hybrid cooling technique has reduced the temperature of the IC chips by 5–12.5 °C in comparison to the forced convection

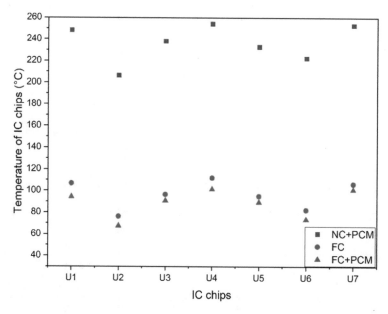

**Fig. 5** IC chip temperature variation under the different cooling techniques

cooling without PCM. It indicates that for the high heat-dissipating IC chips, their temperature is maintained below the critical value using the hybrid cooling technique.

The present study is further extended for the hybrid cooling technique using different PCMs, different heat fluxes, variable PCM volume content, and PCM-filled four mini-channels (LRBT) and three mini-channels (LRB) cases.

## 4.2  IC Chip Temperature Variation for Different PCMs

Figure 6 represents the IC chip temperature variation supplied with a constant heat flux of 50 W/cm$^2$ using different PCMs. Three different PCMs; Suntech P116, Paraffin wax, and n-eicosane are used having melting points of 49.5 °C, 44 °C, and 40.5 °C, respectively. The PCMs are filled inside the mini-channels in such a way that, it occupies 100% volume inside all the mini-channels embedded on the substrate board. It is observed that the IC chip temperature is lowered for the n-eicosane in comparison to Suntech P116 and Paraffin wax. It is significant to note that the PCM gets melted at the early stage due to its lower melting point; thereby the solid PCM turns to the mushy zone by the conjugate heat transfer from the IC chips. The same trend is observed for the PCMs filled inside the entire four channels (LBRT) and three channels (LBR) cases. The IC chip's temperature dropped by 15.33%–22.2% and 8.62%–12.89% using the n-eicosane and Paraffin wax, respectively, in comparison to Suntech P116.

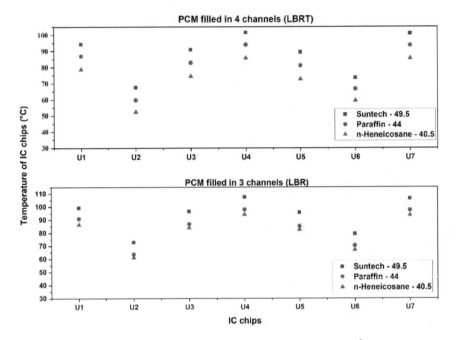

**Fig. 6** IC chip temperature variation using different PCMs with q = 50 W/cm²

When the PCM is filled in all the four channels (LBRT); it forms a pool that helps in storing a large amount of sensible heat during its melting and thereby significantly reduces the IC chip temperature. When the storage capacity of the PCM is reduced by the volume content from the four mini-channels to three mini-channels, there is a rise in the temperature of the IC chips. This is mainly due to the decrease in volume contents of the PCM and less PCM is available to absorb the IC chip temperature through the convective heat transfer mode and conduction through the substrate board. The same trend is observed when the constant heat flux of 25 W/cm² is supplied to the IC chips where the temperature of the IC chips is high for LRB case in comparison to the LBRT case, as shown in Fig. 7.

Figure 8 represents the IC chips temperature and the volume fraction of the n-eicosane filled inside the LRBT mini-channels. It is observed that the PCM in right, left and top mini-channels has reached 60% melting while at the same time; the PCM in the bottom mini-channel has reached 90% melting. The temperature distribution can be observed on the substrate board with a green patch which suggests that the heat is getting conducted from the substrate board to the PCM and also the convective heat transfer mode plays a vital role in the PCM melting, thereby reducing the IC chips temperature.

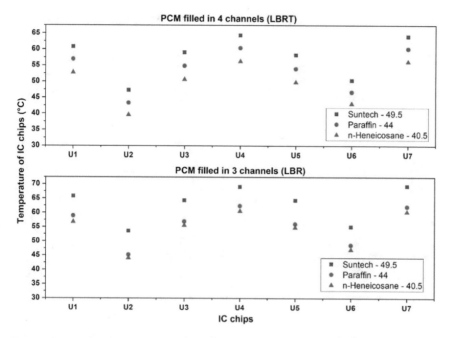

**Fig. 7** IC chip temperature variation using different PCM with q = 25 W/cm²

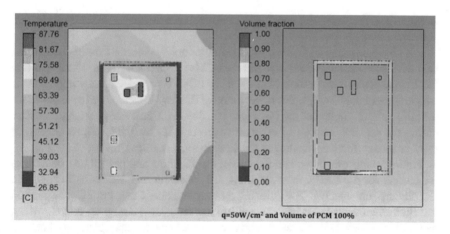

**Fig. 8** Temperature contour and volume fraction for n-eicosane with q = 50 W/cm²

## 4.3 Variation of the IC Chip Temperature with PCM Volume

The PCM volume content is varied inside the mini-channels by considering 100 and 50% volume for the LRBT case with a constant heat flux of 50 W/cm². The study

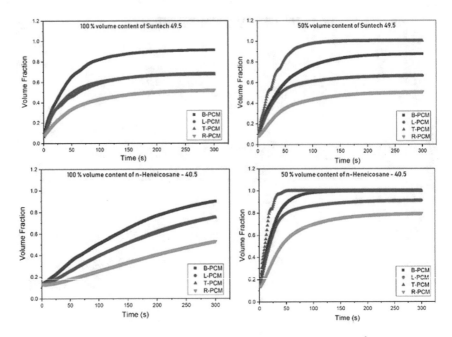

**Fig. 9** Variation of PCM volume content for the LRBT case with q = 50 W/cm²

is conducted for three different PCMs and the comparison of Suntech P116 and n-eicosane is shown in Fig. 9.

It is observed that the PCM with 50% volume content gets melted early in comparison with 100% volume content for all the PCMs for the LRBT case. The interesting fact to note is that despite the early melting of PCM, it has significantly impacted the temperature of the IC chips. The temperature of IC chips has increased by 2–10 °C for the 50% volume content of the PCM which is due to the less energy storage capacity during the sensible heating of PCM. It suggests that the 100% volume content of PCM is more effective in reducing the IC chip temperatures.

Figure 10 represents the temperature of the IC chips and the volume fraction of n-eicosane for the LRBT mini-channels. It is observed that the PCM in right and left channel has reached 70% melting and at the same time, the PCM in the top and bottom mini-channels has reached full melting. The IC chip $U_7$ and $U_4$ which have attained the maximum temperature are closer to the top mini-channels that have assisted in PCM melting through the conduction mode of heat transfer at an early stage.

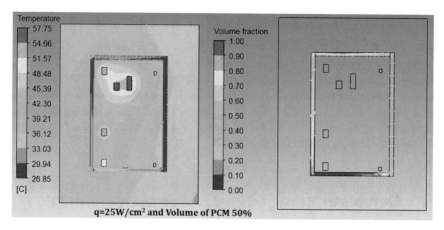

q=25W/cm² and Volume of PCM 50%

**Fig. 10** Temperature contour and volume fraction for n-eicosane with q = 25 W/cm²

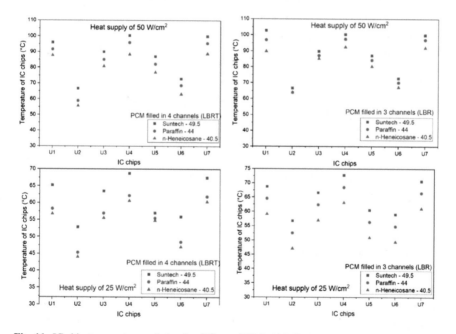

**Fig. 11** IC chip temperature variation for different PCMs with 50% volume content

### 4.4 Effect of IC Chip Temperature for the LRBT and LRB Case Using Different PCMs

Figure 11 shows the IC chip temperature variation of different PCMs with 50% volume content for the LRBT and LRB cases. It is evident that the IC chip temperatures have decreased by 3.25 °C–8.13 °C for the LRBT case in comparison to the LRB case for the heat flux of 50 W/cm$^2$ and 25 W/cm$^2$, respectively, which follows the same trend as explained under Sect. 4.2.

### 4.5 Correlation

A correlation is put forth to study the variation of the input parameters on the temperature of the IC chips. All the variables are made non-dimensional in which the independent variables are the IC chips input heat flux ($q_o = \frac{q*L*Tamb}{kf}$), PCM volume content ($v_o = \frac{vactual}{vmin}$), IC chips size ($\delta = \frac{t}{l}$), and the dependent variable is the non-dimensional temperature ($\theta = \frac{T_{sim}-T_\infty}{T_{max}-T_\infty}$) and is given in Eq. (1). The Eq. (1) is based on the transient simulations performed on all the cases as mentioned in Table 1. This equation has a coefficient of regression 0.80 and the root mean square value of 0.03811. The equation is valid for $1225009.800 \leq q_o \leq 9628577.029$, $1 \leq v_o \leq 2$, and $0.175 \leq \delta \leq 0.350$.

$$\theta = 0.045(1 + q_o)^{0.2}(1 + v_o)^{-0.003}(1 + \delta)^{-0.041} \tag{1}$$

Figure 12 shows the error plot between the $\theta$corr and $\theta$sim which suggests the arbitration between both the values with a 15% error.

## 5  Constrained Optimization Using Genetic Algorithm

The transient simulations are carried out for the 24 different cases as mentioned in Table 1 for three different PCMs Suntech P116, Paraffin wax, and n-eicosane having melting points of 49.5 °C, 44 °C, and 40.5 °C, respectively. The study focuses on maintaining the temperature of the IC chips below the critical temperature and also to study the effect of input heat flux, PCM volume content, and IC chips size on the temperature of the IC chips.

From the transient simulation and the result interpretation, it is noticed that the velocity of air cannot be increased beyond 13 m/s which is the threshold limit for the forced convection cooling using a fan and the heat flux supplied to the IC chips is 25 and 50 W/cm$^2$. The PCM-based mini-channels help to bring down the temperature of the IC chips under the forced convection as represented in Fig. 5 where the excess use of PCM is beneficial to decrease the temperature of the IC chips but it makes the

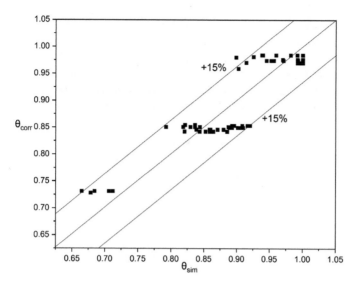

**Fig. 12** Parity plot between $\theta_{corr}$ and $\theta_{sim}$

system bulky due to the high density of the PCM. Hence an optimum PCM volume is also required in the system to keep the temperature of the IC chips below the critical temperature. Seven asymmetric IC chips are considered with the smallest size of the chip is $U_2$ (5 mm × 4 mm × 1.75 mm), reducing the size of the IC chip further will make it a flush-mounted chip which is critical from a thermal management perspective, therefore the need of optimum size of the IC chips arises. The input heat flux, PCM volume content, and IC chips size are the important constraint parameters for maintaining the IC chip temperature below the critical value.

The current study is further extended in evaluating the optimum values of these constrained parameters by conducting a nature-inspired optimization called Genetic Algorithm (GA). The Genetic Algorithm (GA) [8] is a search-based technique that works on the principle of genetics and natural selection. It rapidly searches for the most possible solutions to a given problem. Hence, this method can be handy to reach the global optimum by escaping from the local optima. Due to these merits, the GA is employed for the present study. The detail of the genetic algorithm is reported in Mathew and Hotta [18]. The objective function of GA is to minimize the non-dimensional IC chip temperature given in Eq. (2) and is subjected to the constraints $1225009.800 \leq q_o \leq 9628577.029$, $1 \leq v_o \leq 2$, and $0.175 \leq \delta \leq 0.350$.

$$\text{Minimize } \theta \tag{2}$$

The MATLAB toolbox is used for minimizing the Eq. (2) with $\theta$ as a function of $q_o$, $v_o$, and $\delta$ with lower bounds and upper bounds of the respective independent variable. The optimization is simulated until the convergence has been achieved which yields the minimum of the non-dimensional maximum temperature excess with the optimum values of the independent variables. The different parameters considered for simulations are, population size = 50, cross over function = 0.8 which gives the fitness value of θ for the optimal configuration as 0.7367, as represented in Fig. 13. The optimum values obtained for the independent variables are 1255009, 2, and 0.35 for the input heat flux, PCM volume content, and IC chips size, respectively. The generation independency study is also represented in Fig. 14 which signifies

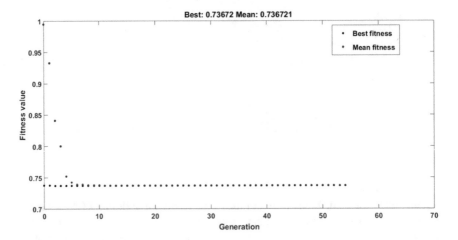

**Fig. 13**  Fitness value generated using GA

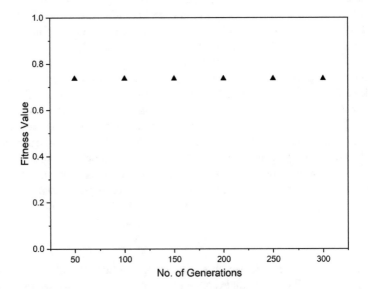

**Fig. 14**  Independency study on the number of generations

**Table 3** Sensitivity analysis of the constrained variable on $\theta$

| Variables | % Variation | Values | $\theta_{variation}$ | $\theta_{optimum}$ | % Error |
|---|---|---|---|---|---|
| $q_0 = 1225009$ | $-10\%$ | 1,102,508 | 0.7214 | 0.7367 | 2.084 |
| | $-20\%$ | 980,007 | 0.7046 | 0.7367 | 4.362 |
| $v_0 = 2$ | $-10\%$ | 1.8 | 0.7369 | 0.7367 | 0.0184 |
| | $-20\%$ | 1.6 | 0.7370 | 0.7367 | 0.0383 |
| $\delta = 0.35$ | $-10\%$ | 0.315 | 0.7375 | 0.7367 | 0.1081 |
| | $-20\%$ | 0.28 | 0.7383 | 0.7367 | 0.2193 |

that there is a negligible effect on the temperature of IC chips with the increase in generations.

The optimal temperature of the IC chip obtained is 85.36 °C when the $\theta$ is converted using the expression given under Sect. 4.5 and the temperature excess ($T_{GA} - T_\infty$) becomes 60.36 °C. The optimal non-dimensional values are then converted into the corresponding input heat flux, PCM volume content, and IC chips size and are found to be 25 W/cm², 100%, and 10.16 mm ($l$) × 3.55 mm ($t$) (U$_4$ chip), respectively.

## 5.1 Sensitivity Analysis of the Constrained Parameters on $\theta$

The optimum values obtained from the GA are further used to study the effect of constrained parameter variation on the objective function. The value of the single constrained independent variable ($q_0$) is varied by 10% and 20% within the given range as mentioned in Eq. (2), keeping the other two variables ($v_0$ and $\delta$) constant, as obtained from GA. The study is also carried for the other two variables ($v_0$ and $\delta$) as given in Table 3.

Table 3 shows the sensitivity analysis of the constrained variables on $\theta$. It depicts that the heat flux is a vital function of temperature, where there is a variation of 2–4.36% in the non-dimensional temperature due to the variation in the input heat flux keeping the other two variables constant.

## 6 Conclusions

The transient analysis is performed using the hybrid cooling technique (forced convection and PCM) on seven asymmetric IC chips for the LRBT and LRB case with three different PCMs; Suntech P116—49.5 °C, Paraffin wax—44 °C and n-eicosane—40.5 °C. The main objective is to maintain the temperature of the IC chips below the critical value (125 °C). The constrained optimization strategy is performed using the GA on the heat flux supplied to the IC chips, PCM volume content, and IC chips size. The following conclusions are summarized in the study.

- The hybrid cooling technique is more significant in keeping the temperature of IC chips below the critical value.
- The IC chip's temperature drops by 15.33%–22.2% and 8.62%–12.89% using the n-eicosane and Paraffin wax, respectively, in comparison to Suntech P116.
- The IC chips temperature has increased by 2–10 °C for the 50% PCM volume content which is due to the less energy storage capacity during the sensible heating of PCM.
- It suggests that 100% of PCM volume content is more effective in reducing the IC chip's temperature.
- A correlation is put forth in terms of the non-dimensional temperature ($\theta$) of the IC chips, IC chip input heat flux ($q_o$), PCM volume content ($v_o$), and IC chips size ($\delta$).
- The constrained optimization using the GA gives rise to the optimum value of the temperature as 85.36 °C and the optimal values of IC chips input heat flux, PCM volume content, and IC chips size as 25 W/cm$^2$, 100%, and 10.16 mm ($l$) × 3.55 mm ($t$) ($U_4$ chip), respectively.
- The sensitivity analysis for the parameters suggests that the IC chips temperature is a strong function of their input heat flux, size, and PCM volume content.

# References

1. Humphries, W.R., Griggs, E.I.: A design handbook for phase change thermal control and energy storage devices. NASA STI/recon Tech. Rep. N. 78 (1977)
2. Saha, S.K., Dutta, P.: Thermal management of electronics using PCM-based heat sink subjected to cyclic heat load. IEEE Trans. Compon. Packag. Manuf. Technol. 2, 464–473 (2012)
3. Sabbah, R., Farid, M.M., Al-Hallaj, S.: Micro-channel heat sink with slurry of water with micro-encapsulated phase change material: 3D-numerical study. Appl. Therm. Eng. 29, 445–454 (2009)
4. Baby, R., Balaji, C.: Thermal performance of a PCM heat sink under different heat loads: an experimental study. Int. J. Therm. Sci. 79, 240–249 (2014)
5. Faraji, M., El Qarnia, H.: others: Thermal analysis of a phase change material based heat sink for cooling protruding electronic chips. J. Therm. Sci. 18, 268–275 (2009)
6. Parlak, M., Etiz, U.: Thermal management of an electronic device using PCM embedded in a pin fin heat sink. In: 2010 12th IEEE Intersociety Conference on Thermal and Thermomechanical Phenomena in Electronic Systems, pp. 1–7 (2010)
7. Wang, Y., Yang, Y.: Three-dimensional transient cooling simulations of a portable electronic device using PCM (phase change materials) in multi-fin heat sink. Energy 36, 5214–5224 (2011)
8. Baby, R., Balaji, C.: A neural network-based optimization of thermal performance of phase change material-based finned heat sinks—an experimental study. Exp. Heat Transf. 26, 431–452 (2013)
9. Kozak, Y., Abramzon, B., Ziskind, G.: Experimental and numerical investigation of a hybrid PCM–air heat sink. Appl. Therm. Eng. 59, 142–152 (2013)

10. Ollier, E., Soupremanien, U., Remondière, V., Dijon, J., Le Poche, H., Seiler, A.L., Lefevre, F., Lips, S., Kinkelin, C., Rolland, N.: others: Thermal management of electronic devices by composite materials integrated in silicon. Microelectron. Eng. **127**, 28–33 (2014)
11. Sridharan, S., Srikanth, R., Balaji, C.: Multi-objective geometric optimization of phase change material based cylindrical heat sinks with internal stem and radial fins. Therm. Sci. Eng. Prog. **5**, 238–251 (2018)
12. Zaretabar, M., Asadian, H., Ganji, D.D.: Numerical simulation of heat sink cooling in the mainboard chip of a computer with temperature-dependent thermal conductivity. Appl. Therm. Eng. **130**, 1450–1459 (2018)
13. Loganathan, A., Mani, I.: A fuzzy-based hybrid multi-criteria decision-making methodology for phase change material selection in electronics cooling system. Ain Shams Eng. J. **9**, 2943–2950 (2018)
14. Mathew, V.K., Hotta, T.K.: Role of PCM based mini-channels for the cooling of multiple protruding IC chips on the SMPS board-A numerical study. J. Energy Storage. **26**, 100917 (2019)
15. Balaji, C., Mungara, P., Sharma, P.: Optimization of size and shape of composite heat sinks with phase change materials. Heat Mass Transf. **47**, 597–608 (2011)
16. Li, C., Zhang, B., Liu, Q.: N-eicosane/expanded graphite as composite phase change materials for electro-driven thermal energy storage. J. Energy Storage. **29**, 101339 (2020)
17. Arshad, A., Jabbal, M., Sardari, P.T., Bashir, M.A., Faraji, H., Yan, Y.: Transient simulation of finned heat sinks embedded with PCM for electronics cooling. Therm. Sci. Eng. Prog. 100520 (2020)
18. Mathew, V.K., Hotta, T.K.: Numerical investigation on optimal arrangement of IC chips mounted on a SMPS board cooled under mixed convection. Therm. Sci. Eng. Prog. **7**, 221–229 (2018)
19. Patil, N.G., Hotta, T.K.: A combined numerical simulation and optimization model for the cooling of IC chips under forced convection. Int. J. Mod. Phys. C. 2050081 (2020)
20. Zeng, J.L., Cao, Z., Yang, D.W., Sun, L.X., Zhang, L.: Thermal conductivity enhancement of Ag nanowires on an organic phase change material. J. Therm. Anal. Calorim. **101**, 385–389 (2010)
21. Usman, H., Ali, H.M., Arshad, A., Ashraf, M.J., Khushnood, S., Janjua, M.M., Kazi, S.N.: An experimental study of PCM based finned and un-finned heat sinks for passive cooling of electronics. Heat Mass Transf. **54**, 3587–3598 (2018)
22. Kannan, K.G., Kamatchi, R.: Augmented heat transfer by hybrid thermosyphon assisted thermal energy storage system for electronic cooling. J. Energy Storage. **27**, 101146 (2020)
23. Someshower Dutt Sharma, H.K., Hausotter, Zu W.: Fibromyalgie–ein entbehrlicher Krankheitsbegriff? Res. Rep. Fac. Eng. Mie Univ. **29**, 31–64 (2004)
24. Totten, G.E., Westbrook, S.R., Shah, R.J.: Fuels and Lubricants Handbook. ASTM Int. 593–594 (2003)

# Maximizing Downlink Channel Capacity of NOMA System Using Power Allocation Based on Channel Coefficients Using Particle Swarm Optimization and Back Propagation Neural Network

Shailendra Singh and E. S. Gopi

**Abstract** One of the methods among the pool of technologies for the 5G wireless communication is the NOMA (Non-orthogonal Multiple Access). Conventionally, the channel between the base station and the various mobile station users are shared based on the orthogonality principle. To increase the capacity of the channel, NOMA has been technologically enhanced and explored for 5G standards. In this case, the channel is shared between the users based on the difference in the transmitted power allocated to the individual users. The Successive Interference Cancellation (SIC) is adopted to detect the signal during the uplink and the downlink. In SIC, during the uplink scenario, the base station usually collects and detects the symbol in the decreasing order of the channel gain (between a particular user and the base station). In the same order, during the downlink scenario, the SIC uses the increasing order of the channel coefficients or gain. The power is allocated based on the corresponding channel gain. If the channel gain is larger, then less power is allotted and vice versa. In NOMA, the method used is Multiplexing in Power Domain and after allocation of Power to the users, it has been changed to the problem of Constraint Optimization. In this paper, an attempt is made to demonstrate the proposed methodology used for power allocation and handling the given constraints in NOMA downlink scenario using Particle Swarm Optimization (PSO) and Back propagation Neural Network (BPNN). The experimental results have shown the importance of the proposed technique for power allocation in the Downlink NOMA scenario.

**Keywords** PSO · ANN · NOMA · SIC · Power allocation · Channel capacity

## 1 Introduction

In wireless communication systems, various multiple access methods have been the key technologies ranging from the very first generation that is called (1G) to the advanced fourth generation (4G) that is called LTE-Advanced. The multiple access

S. Singh (✉) · E. S. Gopi
National Institute of Technology Tiruchirappalli, Chennai, Tamil Nadu, India
e-mail: Shailendrarana.medi@gmail.com

© The Author(s), under exclusive license to Springer Nature Singapore Pte Ltd. 2021    251
A. J. Kulkarni et al. (eds.), *Constraint Handling in Metaheuristics and Applications*,
https://doi.org/10.1007/978-981-33-6710-4_11

technologies include Frequency Division Multiple Access (FDMA), Time Division Multiple Access (TDMA), Code Division Multiple Access (CDMA) and Orthogonal Frequency Division Multiple Access (OFDMA). In all the above methods, access to the channel has been shared by using orthogonality in Frequency, Time and Codes, respectively, allocated to the individual users. In the fourth generation of mobile communication systems such as long-term evolution (LTE) and LTE-Advanced, OFDMA has been widely adopted and very much popular to achieve higher data rate. The near future demand for mobile traffic data volume could be expected to be 500–1,000 times larger than that in 2010. The technologies like mm wave transmission, Full duplex, Multicarrier transmission, Non-Orthogonal Multiple Access (NOMA) are being explored in 5G to increase the channel capacity. The working principle behind NOMA is Power Domain Multiplexing that means access to the channel is shared by using different Power levels allocated to the individual users, because of Power Domain Multiplexing in NOMA, it discards the Orthogonality principle and that helps in the improvement of the channel capacity, compared to other orthogonal multiple access (OMA) schemes.

## 2 Literature Survey

How fascinatingly things are changing, lets take the brief history of cellular phone over 50 years which itself is a milestone. The 1G based phones were big, heavy and analogue with heavy price. The 1990s saw the second generation i.e. 2G cell phones, embedded digitally with that we could make calls, send text messages, and a smiling face. In 2000, the third generation, i.e. 3G, cell phones were revolutionized and came with an Internet browser. MIMO technology was used in the 3G cell phones and for the data transmission packet switching was used. Now forth generation, i.e. 4G cellular system came in 2010. Methods that were used in 4G: long-term evolution (LTE), WiMAX, internet protocols, and packet switching were the key technologies. We could say it was a new computer or digital machine in our hands. Now it's time to move to a new generation, 2020 will be the fifth-generation era, i.e. 5G. It can be 100 times faster than the 4G. The downlink maximum throughput can offer a 10–20 Gbps, which means we can easily download 2–3 HD DVD movies in just 1s. Some technologies among others possibly used for 5G cellular networks are millimetre-wave for 5G 24–100 GHz is proposed, Massive MIMO, Beam forming and NOMA. It is highly anticipated that the connection density would become 106 connections for a square kilometre area in future. In the typical OFDM, we have the sub carriers, and different users are given different sets of sub carriers but in case of NOMA, a particular sub carrier or group of sub carriers can be given to more than one user. In [1–5], it has been observed and surveyed that all the NOMA schemes such as Single carrier, Multi carrier, Power Domain, Cognitive radio-NOMA and including Single Input Single Output (SISO) and Multiple Inputs Multiple Outputs (MIMO) don't use the multiple antennas for transmitting and receiving the particular sub carrier or group of sub carriers, it usually combine the signals and transmit it through the single transmitting antenna and receives by the single receiving antenna. While the MIMO system requires multiple antennas for transmitting and receiving the signals, this is the main difference between MIMO and NOMA. In case of OMA, if one user only needs to be served with low data rate, e.g. sensors, then OMA gives the more data rate than it's needed but NOMA provides a satisfactory solution for this problem faced by OMA. Because of the availability of new power dimensions, NOMA systems can be amalgamated with present multiple access (MA) models. NOMA consists of two types of Multiplexing. The Code Domain Multiplexing and other one is Power Domain multiplexing. In the NOMA the allocation of power can be initialized or performed by implementing different methods based on the channel conditions of users. Taking single input single output (SISO) system, the algorithm for power allocation has been determined by the parent source for maximizing the rate simply considering the Downlink NOMA [6]. However, most of the previously used methods are applied to only two users. To increase its domain and reach to more users, the proposed system is based on Pascal's triangle for the power allocation. The well-known French mathematician and philosopher Blaise pascal had proposed the Pascal's triangle method [7]. But the Power allocation in NOMA creates the problem for constrained optimization because the allocation must satisfy the distribution according to the channel conditions, and the distributed power among the users must be equal to the power at the base station. There are many classical methods that have been already developed by the researchers to solve the problems of constrained optimization. In the case of multiple input multiple output (MIMO) system, the powers are being allocated optimally to the 'n' number of communication channels for maximizing the sum rate by using Karush-Kuhn-Tucker (KKT) conditions subjected to the total power constrained and non negativity constrained. Similarly, the power allocation in NOMA to maximize the sum-rate is also the important task having same Power constrained like MIMO, but the power allocation is just opposite in NOMA for channel conditions as compared with MIMO. The same KKT conditions have proposed by the researchers to find the optimum power values by taking the weighted sum of received rates for the individual users after applying lagrangian Optimization [8]. The optimum power values for NOMA have also achieved by the researchers by using Jensen's inequality criteria to calculate Ergodic Capacity by applying closed form lower bound and after that a scheme for power allocation is applied to satisfy the ergodic capacity according to the requirements for all the users just by solving a problem of convex optimization [9]. In [10], a scheme for power allocation in NOMA called Proportional Fairness Scheduling (PFS) has discussed. Some other techniques that have proposed to allocate the power by satisfying the given constrained are dynamic power allocation and users scheduling [11, 12]. When the conventional methods or classical methods fail or not suitable for estimation then computational intelligence comes into the picture. It provides the solution to a complex problem by imitating the human behaviour. Recognition, classification and clustering can be done by using computational intelligence. The objective of this paper is to allocate the power for a novel power domain (PD) NOMA using machine learning techniques (PSO and ANN) according to the estimated channel conditions. PSO generate the optimum solution for any kind of optimization problem by minimizing or maximizing the problem [13, 14]. In this paper, we have allocated the power such that the total sum rate is maximized so that PSO is suitable for it. Back propagation neural network is used to make the results obtained from PSO more efficient. There are many applications like constraint satisfaction, associative storage, Optimization, planning control and classification, for the Neural Networks that have been designed. Because of having special characteristics like robustness, ability to learn and massive parallelism etc. Neural networks are being preferred [15].

The further discussion in this paper is as follows. Section 3 discusses about Conventional Power Domain NOMA in detail. Section 4 elaborates the Power allocation algorithm and optimization with constraints. The Experiments and the Results are presented in Sect. 5 followed by Conclusion.

## 3   Power Domain NOMA

In the Power Domain NOMA, different levels of power are allotted to the individual users based on the channel conditions, it means, the channel state information (CSI) must be required at the base station for the power allocation. If the channel is good between the base station and the receiver, then the receiver usually detects the signal of less strength, so the users having good channel conditions are supposed to allocate less power values and users having poorer channel conditions are supposed to allocate more power values. However, in case of conventional OMA, the water filling policy is used for the power allocation. The Superposition Coding (SC) and Successive Interference Cancellation (SIC) are the two key enabling technologies for NOMA, keeping the generality, SC is used at the base station that permits it to transmit the combined Superposition coded messages to all the users. SIC helps in efficiently managing the interference at the receiver end. It has been observed that using SIC at receiver end users rate can be improve up to Shannon limit. SIC technology mitigates the interference from the users with poorer channel conditions.

### 3.1   Downlink NOMA

Let $h_r$ be the Rayleigh channel coefficient between the base station and the $r$th user with $r = 1, 2, 3$. Figure 1a illustrates the typical power domain NOMA with three users with $|h_1| > |h_2| > |h_3|$ in downlink scenario. The channel link between the base station and the 3 receivers are represented as $|h_1|, |h_2|, |h_3|$, respectively, with $|h_1| > |h_2| > |h_3|$. Let the total power used to broadcast is given by P and let $a_i$ be the fraction of the power alloted for the $i$th user with $a_1 + a_2 + a_3 = 1$

$$u_i = h_i(\sqrt{(a_1)}X_1 + \sqrt{(a_2)}X_2 + \sqrt{(a_3)}X_3) + N \qquad (1)$$

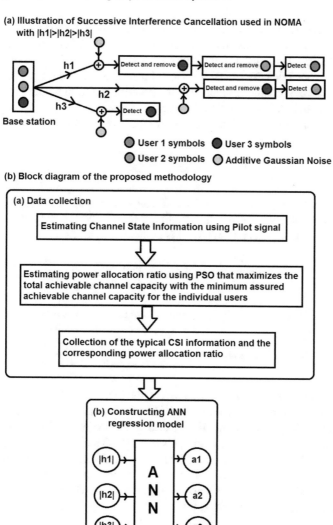

**Fig. 1** **a** Illustration of successive interference cancellation used in power domain NOMA, **b** proposed methodology for power allocation using PSO and ANN

where $N$ is the additive Gaussian noise with mean zero and variance $\sigma^2$. Let $X_i$ be the symbols associated with the $i$th user. The Successive Interference Cancellation (SIC) (refer Fig. 1a) is used in the receiver section as described below. As $|h_1| > |h_2| > |h_3|$, user 1 detects $X_3$ and is removed from the received signal. Further $X_2$ is detected from the remaining signal and is also removed from the remaining signal. Finally, detect $X_1$ from the remaining signal. Thus, the channel capacity attained between the base station and the user 1 is computed as follows:

$$C_3^d(1) = \log_2\left(1 + \frac{|h_1|^2 a_1 P}{\sigma^2}\right) \tag{2}$$

In the similar fashion, user 2 detect $X_3$ and removed from the received signal, which is followed by detecting $X_2$ from the remaining signal. The channel capacity attained between the base station and the User 2 is computed as follows:

$$C_2^d(2) = \log_2\left(1 + \frac{|h_2|^2 a_2 P}{|h_2|^2 a_1 P + \sigma^2}\right) \tag{3}$$

Finally, user 3 detect $X_3$ directly, and hence channel capacity is computed as follows:

$$C_1^d(3) = \log_2\left(1 + \frac{|h_3|^2 a_3 P}{|h_3|^2 a_1 P + |h_3|^2 a_2 P + \sigma^2}\right) \tag{4}$$

$c_r^k$ is the maximum achievable rate attained by the user $k$th user using SIC technique, $r$ is the order at which the data corresponding to the $k$th user is the detected. The Quality of service (QOS) is determined based on the demand of the data rate. Let the demand rate requirement of the User 1, User 2, User 3 are, respectively, represented as the $R_1$, $R_2$, $R_3$. We would like to obtain the optimal values for $a_1$, $a_2$ and $a_3$, such that $C_3^d(1) + C_2^d(2) + C_1^d(3)$ is maximized, satisfying the constraints $C_3^d(1) > R_1$, $C_2^d(2) > R_2$ and $C_1^d(3) > R_3$. It is noted that the order in which the data are detected is from weak signal to the strong signal (3, 2, 1), i.e. $|h_1| \geq |h_2| \geq |h_3|$.

## 3.2   Uplink NOMA

During the Uplink, the base station receives the signal as shown below.

$$s = h_1\sqrt{(a_1)}X_1 + h_2\sqrt{(a_2)}X_2 + h_3\sqrt{(a_3)}X_3 + N \tag{5}$$

The Successive Interference Cancellation (SIC) is used in the receiver section (base station) described below. In the case of uplink, the strongest signal is detected first, i.e. $X_1$ is detected first, followed by $X_2$ and $X_3$.

Thus, $X_1$ is detected first and is removed from the received signal. From the remaining signal, the signal $X_2$ is detected and is removed. Finally, the signal $X_3$ is detected from the remaining signal. The channel capacity attained between the base station and all the users during the uplink is computed as follows:

$$C_1^u(1) = \log_2 \left( 1 + \frac{|h_1|^2 a_1 P}{|h_2|^2 a_2 P + |h_3|^2 a_3 P + \sigma^2} \right) \tag{6}$$

$$C_2^u(2) = \log_2 \left( 1 + \frac{|h_2|^2 a_2 P}{|h_3|^2 a_3 P + \sigma^2} \right) \tag{7}$$

$$C_3^u(3) = \log_2 \left( 1 + \frac{|h_3|^2 a_3 P}{\sigma^2} \right) \tag{8}$$

# 4 Proposed Methodology

The block diagram illustrating the proposed methodology is given in the part (b) of Fig. 1 that is Fig. 1b. In the very first part, Particle Swarm Optimization is used to estimate the power allocation ratio corresponding to the given magnitude of the channel state information (CSI). Pilot signal is transmitted through the channel one after another to the individual users and the corresponding CSI (between the base station and the individual users are obtained). For the given CSI, power allocation ratio is estimated using PSO that maximizes the maximum achievable channel capacity (refer Sect. 3.1). It is also noted that the minimum achievable channel capacity of the individual users are incorporated while using PSO.

## 4.1 Constrained Optimization

A plethora of classical methods existed for constraint optimisation problems, basically depend on the nature of the constraints whether they are equality or inequality or together. Some of the methods among the pool are Lagrange's multiplier, Penalty Function method and augmented Lagrange method. Suitability of usage depends on the constraints; these mentioned methods has been useful for a problem with inequality constraints. Methods such as gradient projection and quadratic projection are very much useful for equality constraints differences existed between Constrained optimization and unconstrained optimization because of their approaches and because the local optima are not the intended goal. Generally, a subset of unconstrained optimization is useful for the Constrained optimization methods [16, 17]. In this paper, the proposed methods are Particle Swarm Optimization and Neural Network for solving the Constrained Optimization problem. Using Particle Swarm Optimization (PSO) algorithm, the constraint optimization has been achieved by designing the genuine Objective function and applying the required upper and lower bounds on each of the Particles and after that, the selection of the desired Global bests are considered as the Optimized Solutions [18]. In [19], Neural networks are extraordinarily intelligent and intuitive. To solve any problem, we need to train the network to perform the specific task, and hence to solve nonlinear non-convex optimization problems, we need to train the neural networks [20]. Different approaches and methods have been proposed by researchers. The first method which revolutionizes all analysis were classical methods. The most popular and prevalent algorithm to train the neural networks is error back propagation [21]. It has a specialty of minimizing an error function using the steepest descent algorithm. All algorithms have their own pros and cons and so as with the error back propagation, usually its implementation is quite easy but it comes with a price, i.e. convergence problem etc. It has all the disadvantages in optimization algorithms of Newtown based, which inherits slow convergence rate and trapping in local minima [20, 21]. Over the years Researchers have proposed different supervised learning methods such as the Step net. The tilling algorithms cascade-correlation algorithm and the scaled conjugate algorithm in order to mitigate deficiencies and to enhance its applicability, global optimization methods ate another available alternative for Newtown based methods and to learn the deepest of the structure of the neural network. Along with the learning techniques for ANN discussed in the above sentences, we have another widely used and flaw-

---

**Algorithm 1** Algorithm for PSO

Inputs: Generated channel coefficients ($h_1$, $h_2$, $h_3$), transmitted power (P) and noise power ($\sigma^2$)

1. **Costfunction** $= \frac{1}{C_3(1)} + \frac{1}{C_2(2)} + \frac{1}{C_1(3)}$, considering $a_1$, $a_2$, $a_3$ as variable or particle

2. Define parameters: number of dimension variables = n, number of iterations = it, number of particles = N, inertia coefficient = W, personal acceleration coefficient = $C_1$, global acceleration coefficient = $C_2$

3. Initialize: particle position(normalize), particle velocity, personal best position, personal best cost, global best cost

4. for j = 1:it

for k = 1:N

particle velocity = W × (particle velocity) + $C_1$ × (particle best position-particle position) + $C_2$ × (global best position-particle position)

particle position = particle position+particle velocity

particle cost = cost function (particle position)

if $a_1 < a_2 < a_3$

if particle cost < particle best cost

particle best position = particle position

particle best cost = particle cost

if particle best cost < global best cost

global best = particle best

end

end

end

end

Outputs: global best positions ($a_1$, $a_2$, $a_3$)

less methods to train the Neural networks with optimized structure, are the quotient gradient system (QGS), genetic algorithms and stimulated annealing [19, 22–26]. Results achieved through neural networks are in convergence with standard results and performance parameters using classical error back propagation algorithm for the defined constrained optimization problem. Hence, the training of neural networks used here is error back propagation.

## 4.2 Problem Formulation

The requirement is to obtain the optimal values for $a_1$, $a_2$, $a_3$ such that $C_1(1) + C_2(2) + C_3(3)$ is maximized, with constraints $C_3(1) > R_1$, $C_2(2) > R_2$, $C_1(3) > R_3$. Also $a_1 + a_2 + a_3 = 1$. It is noted that the order in which the data are detected is from strong signal to the weak signal (1, 2, 3), i.e. $|h_1| \geq |h_2| \geq |h_3|$. In this paper, we propose to use Particle Swarm Optimization (PSO) to optimize the power allocation ratio $a_1$, $a_2$ and $a_3$ such that it maximizes the total channel capacity in the downlink. PSO is the optimization algorithm inspired by the natural behaviour of the birds on identifying the path to the destination. The position of the bird is the possible solution that minimizes or converges the cost function and the distance of the position of the bird from the destination is the corresponding functional value. This is the analogy used in PSO algorithm. The steps involved in the PSO based optimization for given channel coefficients (refer algorithm in Sect. 4.1). Thus, for the given $|h_1|$, $|h_2|$, $|h_3|$, the corresponding values $a_1$, $a_2$ and $a_3$ are obtained using the proposed PSO based methodology. The experiments are repeated for various combinations of $h_1$, $h_2$, $h_3$ and the corresponding optimal fractional constants $a_1$, $a_2$ and $a_3$ obtained using PSO are collected. Further in the second part (refer Fig. 1b), Back propagation Network is used to construct the relationship between the $h_1$, $h_2$ and $h_3$ as the input and the corresponding values $a_1$ $a_2$ and $a_3$ as the target values.

## 5 Experiments and Results

Experiments are performed by generating 200 instances of channel coefficients $h_1$, $h_2$ and $h_3$ (with variances 0.9, 0.5, 0.1, respectively) and the corresponding optimal fractional weights $a_1$, $a_2$ and $a_3$ that maximize the total channel capacity, satisfying the constraints are obtained using Particle Swarm Optimization. Figures 2, 5 and 6 illustrate how the maximization of the total channel capacity is achieved and also showing the good convergence plot for defined objective function using Particle Swarm Optimization. 50% of the collected instances are used as the training data to construct the Back propagation Network to predict the optimal fractional constants $a_1$, $a_2$ and $a_3$. Figure 3 shows the designed back propagation neural network having

100 instances of each channel coefficient $h_1$, $h_2$ and $h_3$ are the 3 inputs and their corresponding optimal fractional weights $a_1$, $a_2$ and $a_3$ are the targets taken from PSO. Figure 4 describing the training performance of the constructed Network. Convergence plot concludes that it is fast at the training stage. Figure 7 is concluding the relationship between targets and actual Outputs and we got almost 99% regression values which means the actual outputs are completely converging to the targets. Figure 8a shows the magnitude plots of the channel coefficients corresponding to the three users. Figure 8b shows the optimal fractional constants $a_1$, $a_2$ and $a_3$ obtained using PSO and the values predicted using the trained constructed Neural Network. Also, the Table 1 shows the generated values of channel Coefficients and Tables 2 and 3 are the optimum values of Power Allocation Ratios by PSO and ANN respectively. Tables 4 and 5 are demonstrating the values of achieved Individual Rates and Sum Rates of the 3 users by PSO and ANN respectively. The results thus obtained act as the proof of concept and reveal the importance of proposed technique.

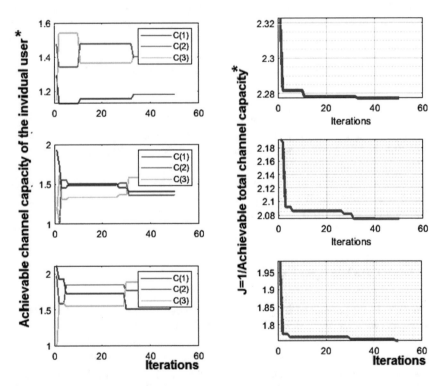

**\*Channel capacity is calculated in bits/Hz with the noise variance of 1 unit**

**Fig. 2** Illustrations on the performance of the PSO on achieving the maximum channel capacity for the individual users in the downlink scenario (Part 1)

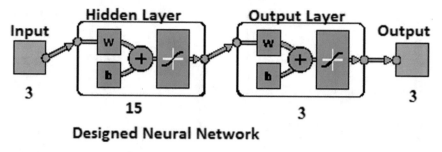

**Fig. 3** Designed neural network for channel coefficients as the inputs and Power allocation ratios as the targets and having 15 neurons in the hidden layer

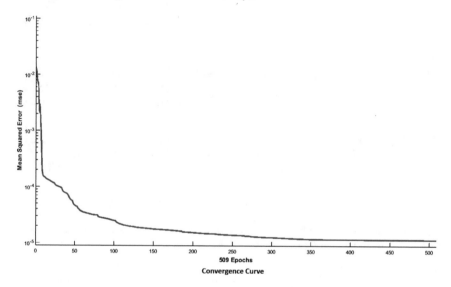

**Fig. 4** Performance plot of ANN showing convergence of MSE

**Table 1** Generated channel coefficient

| $h_1(0.9)$ | $h_2(0.5)$ | $h_3(0.1)$ |
|---|---|---|
| 1.65 | 0.56 | 0.03 |
| 0.60 | 0.20 | 0.05 |
| 1.11 | 0.69 | 0.09 |
| 1.47 | 0.21 | 0.001 |
| 0.78 | 0.20 | 0.06 |
| 0.35 | 0.28 | 0.06 |
| 1.32 | 0.48 | 0.19 |
| 0.94 | 0.09 | 0.06 |
| 2.24 | 0.85 | 0.09 |
| 1.42 | 0.25 | 0.14 |

For $P = 1000$ units and $\sigma^2 = 1$ units

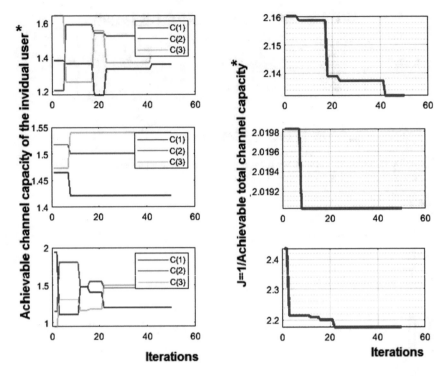

**\*Channel capacity is calculated in bits/Hz with the noise variance of 1 unit**

**Fig. 5** Illustrations on the performance of the PSO on achieving the maximum channel capacity for the individual users achieved in the downlink scenario (Part 2)

**Table 2** Optimum power allocation ratios by PSO

| $a_1$ | $a_2$ | $a_3$ |
|---|---|---|
| 0.0314 | 0.1731 | 0.7953 |
| 0.0464 | 0.2167 | 0.7368 |
| 0.0382 | 0.1734 | 0.7883 |
| 0.0140 | 0.0928 | 0.8930 |
| 0.0485 | 0.2208 | 0.7305 |
| 0.0616 | 0.2055 | 0.7328 |
| 0.0386 | 0.2167 | 0.7446 |
| 0.0224 | 0.1748 | 0.8006 |
| 0.0364 | 0.2370 | 0.7265 |
| 0.0430 | 0.2430 | 0.7139 |

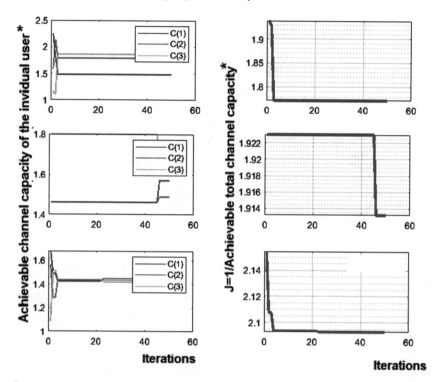

*Channel capacity is calculated in bits/Hz with the noise variance of 1 unit

**Fig. 6** Illustrations on the performance of the PSO on achieving the maximum channel capacity for the individual users achieved in the downlink scenario (Part 3)

**Table 3** Optimum power allocation ratios by ANN

| $a_1$ | $a_2$ | $a_3$ |
| --- | --- | --- |
| 0.0207 | 0.1428 | 0.8408 |
| 0.0518 | 0.2289 | 0.7188 |
| 0.0344 | 0.1810 | 0.7806 |
| 0.0195 | 0.1582 | 0.8327 |
| 0.0473 | 0.2366 | 0.7188 |
| 0.0683 | 0.2213 | 0.7101 |
| 0.0336 | 0.1825 | 0.7873 |
| 0.0195 | 0.1775 | 0.8021 |
| 0.0345 | 0.2310 | 0.7260 |
| 0.0451 | 0.2405 | 0.7105 |

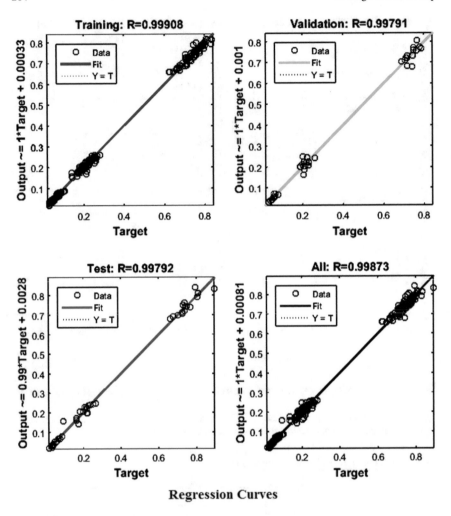

**Fig. 7** Plot for regression curves of ANN showing target and output relationship

## 6 Conclusion

The PSO based power allocation for the individual users of the NOMA downlink is demonstrated. Also, it is proposed to use the constructed Neural Network to obtain the power allocation as per one obtained using the proposed PSO based techniques. The experimental results reveal the importance of the proposed technique. The proposed technique can be extended to an uplink scenario as well as with various noise power and with an increasing number of users. In this paper, power allocation has been done for only 3 users, so the PSO algorithm having three dimensional (3D) search space has implemented for power allocation to the individual users such that the total

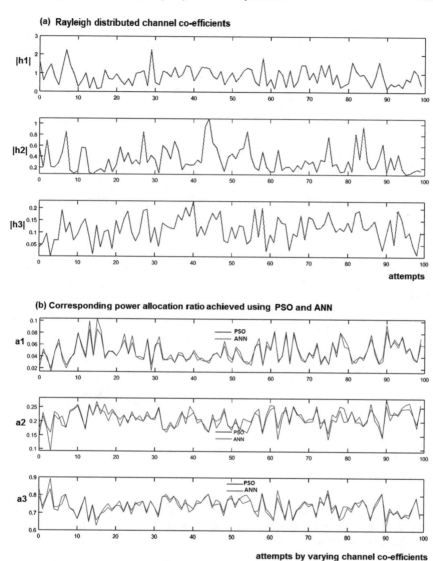

**Fig. 8** **a** Outcome of the Rayleigh distributed channel coefficients $h1$, $h2$ and $h3$ with variances 0.9, 0.5 and 0.1, respectively, **b** corresponding power allocation ratio achieved using PSO and ANN

**Table 4** Achievable rates using PSO

| $C_3(1)$ | $C_2(2)$ | $C_1(3)$ | Sum rate |
|---|---|---|---|
| 6.43 | 2.58 | 0.6821 | 9.69 |
| 4.156 | 2.0121 | 1.078 | 7.23 |
| 5.5994 | 2.40672 | 1.74533 | 9.73 |
| 4.9847 | 1.8197 | 1.00 | 7.79 |
| 4.93 | 2.00 | 1.22 | 8.15 |
| 3.113 | 1.912 | 1.22 | 6.24 |
| 6.0974 | 2.596 | 1.86040 | 10.54 |
| 6.9580 | 2.957 | 1.7988 | 11.69 |
| 6.224 | 2.465 | 1.6957 | 10.37 |
| 5.290 | 1.29 | 1.1823 | 7.76 |

**Table 5** Achievable rates using ANN

| $C_3(1)$ | $C_2(2)$ | $C_1(3)$ | Sum rate |
|---|---|---|---|
| 5.84 | 2.83 | 0.72 | 9.39 |
| 4.27 | 2.01 | 1.10 | 7.38 |
| 5.42 | 2.56 | 1.72 | 9.72 |
| 5.39 | 2.24 | 0.001199 | 7.63 |
| 4.88 | 2.09 | 1.18 | 8.15 |
| 3.22 | 1.89 | 1.17 | 6.28 |
| 5.87 | 2.54 | 2.08 | 10.49 |
| 6.56 | 3.17 | 1.81 | 11.56 |
| 6.14 | 2.47 | 1.71 | 10.32 |
| 5.44 | 1.27 | 1.17 | 7.88 |

sum rate will be maximized. And also, because of low dimensional error surface for the problem, the achieved results were in good agreement for Neural Network using error back propagation. But if we increase it for more number of users, for that we have to implement the PSO algorithm with higher dimensions that is according to the number of users and also increase in the number of users will give rise in dimensions of error surface for the problem hence for training the Neural Network, different alternatives methods can be used to achieve better results.

# References

1. Liu, Y., Qin, Z., Elkashlan, M., Ding, Z., Nallanathan, A., Hanzo, L.: Nonorthogonal multiple access for 5G and beyond. Proc. IEEE **105**, 2347–2381 (2017)
2. Ding, Z., Lei, X., Karagiannidis, G.K., Schober, R., Yuan, J., Bhargava, V.K.: A survey on non-orthogonal multiple access for 5G networks: research challenges and future trends. IEEE J. Sel. Areas Commun. **35**, 2181–2195 (2017)
3. Ye, N., Han, H., Zhao, L., Wang, A.: Uplink nonorthogonal multiple access technologies toward 5G: a survey. Wirel. Commun. Mob. Comput. 1–26 (2018). https://doi.org/10.1155/2018/6187580
4. Basharat, M., Ejaz, W., Naeem, M., Khattak, A.M., Anpalagan, A.: A survey and taxonomy on nonorthogonal multiple-access schemes for 5G networks. Trans. Emerg. Telecommun. Technol. **29**, e3202 (2018)
5. Dai, L., Wang, B., Ding, Z., Wang, Z., Chen, S., Hanzo, L.: A survey of non-orthogonal multiple access for 5G. IEEE Commun. Surv. Tutor. **20**, 2294–2323 (2018)
6. Yang, Z., Xu, W., Pan, C., Pan, Y., Chen, M.: On the optimality of power allocation for NOMA downlinks with individual QoS constraints. IEEE Commun. Lett. **21**, 1649–1652 (2017)
7. AbdelMoniem, M., Gasser, S.M., El-Mahallawy, M.S., Fakhr, M.W., Soliman, A.: Enhanced NOMA system using adaptive coding and modulation based on LSTM neural network channel estimation (2019)
8. Datta, S.N., Kalyanasundaram, S.: Optimal power allocation and user selection in non-orthogonal multiple access systems (2016)
9. Amin, S.H., Mehana, A.H., Soliman, S.S., Fahmy, Y.A.: Power allocation for maximum MIMO-NOMA system user-rate
10. Choi, J.: Power allocation for max-sum rate and max-min rate proportional fairness in NOMA. IEEE Commun. Lett. **20**(10) (2016)
11. Xie, S.: Power allocation scheme for downlink and uplink NOMA networks
12. Power allocation for downlink NOMA heterogeneous networks, Received April 1, 2018, Accepted May 1, 2018, date of publication May 11, 2018, date of current version June 5, 2018
13. Kennedy, J., Eberhart, R.C.: Particle swarm optimization. In: Proceedings of the IEEE International Conference on Neural Networks, vol. IV, pp. 1942–1948. IEEE Service Center, Piscataway, NJ (1995)
14. Shi, Y., Eberhart, R.C.: A modified particle swarm optimizer. In: Proceedings of the IEEE International Conference on Evolutionary Computation, pp. 69–73. IEEE Press, Piscataway, NJ (1998)
15. Kok, J.N., Marchiori, E., Marchiori, M., Rossi, C.: Evolutionary training of CLP-constrained neural networks
16. Dong, S.: Methods for Constrained Optimization. Spring (2006). https://www.researchgate.net/publication/255602767
17. Chong, E.K.P., Zak, S.H.: An Introduction to Optimization. Wiley, New York (1996)
18. Gopi, E.S.: Algorithm collections for digital signal processing applications using Matlab (2007)
19. Khodabandehlou, H., Sami Fadali, M.: Training recurrent neural networks as a constraint satisfaction problem
20. Livieris, I.E., Pintelas, P.: A survey on algorithms for training artificial neural networks. Technical report, Department of Math, University of Patras, Patras, Greece (2008)
21. Rumelhart, D.E., Hinton, G.E., Williams, R.J.: Learning internal representations by error propagation. In: Rumelhart, D., McClelland, J. (eds.) Parallel Distributed Processing: Explorations in the Microstructure of Cognition. MIT Press, Cambridge, MA (1986)
22. Gori, M., Tesi, A.: On the problem of local minima in backpropagation. IEEE Trans. Pattern Anal. Mach. Intell. **14**(1), 76–86 (1992)
23. Chen, C.L.P., Luo, J.: Instant learning for supervised learning neural networks: a rank-expansion algorithm. In: IEEE International Conference on Neural Networks (1994)

24. Werbos, P.J.: Backpropagation: past and future. In: Proceedings ICNN88, pp. 343–353. San Diego, CA, USA (1998)
25. Plagianakos, V.P., Sotiropoulos, D.G., Vrahatis, M.N.: Automatic adaptation of learning rate for backpropagation neural networks. In: Mastorakis, N.E. (ed.) Recent Advantages in Circuits and Systems, pp. 337–341 (1998)
26. Ribert, A., Stocker, E., Lecourtier, Y., Ennaji, A.: A survey on supervised learning by evolving multi-layer perceptrons. In: IEEE International Conference on Computational Intelligence and Multimedia Applications, pp. 122–126 (1999)

# Rank Reduction and Diagonalization of Sensing Matrix for Millimeter Wave Hybrid Precoding Using Particle Swarm Optimization

Mayank Lauwanshi and E. S. Gopi

**Abstract** Millimeter wave (mmwave) wireless communication systems is a promising technology which provides a high data rate (up to gigabits per second) due to the large bandwidth available at mmwave frequencies. But it is challenging to estimate the channel for mmwave wireless communication systems with hybrid precoding, since the number of radio frequency chains are much smaller as compared to a number of antennas. Due to limited scattering, the Beam space channel model using Dictionary matrices is proposed for mmwave channel model. There were many attempts made to design the precoder and decoder, along with the channel estimation for the mmwave channel model but it remains an unsolved problem. In this paper, we demonstrate the methodology of using Particle Swarm Optimization to design the precoder and decoder of the Beam space channel model with the prior knowledge of Angle of Arrival (AOA) and Angle of Departure (AOD). Particle swarm optimization is used to optimize the precoder and decoder such that the sensing matrix is diagonalized (diagonalization method) and is a reduced rank matrix (rank reduction method) and then the channel matrix is estimated. The results reveal the possible direction to explore the usage of computational intelligence technique in solving the mmwave channel model.

**Keywords** Compuational intelligence · Particle swarm optimization (PSO) · mmwave · Hybrid precoding · Diagonalization · Angle of arrival (AOA) · Angle of departure (AOD)

M. Lauwanshi (✉) · E. S. Gopi
Department of Electronics and Communication Engineering, National Institute of Technology, Tiruchirappalli, Tamil Nadu Chennai, India
e-mail: mayanklauwanshi@gmail.com

E. S. Gopi
e-mail: esgopi@nitt.edu

# 1 Introduction

The 5G technology demands 10 Gbps data rate, connecting 1 million devices per square km and 1ms round trip latency, requires 99.9999% availability and reduction in power consumption and improvement in efficiency. One way to achieve this is to use the unused high frequency mmwave band (6–300 GHz). The 57–64 GHz is considered as the oxygen absorption band and 164–200 GHz is considered as the water vapor absorption band. The remaining vast bandwidth of 252 GHz is available in the mmwave band. The Millimetre Wave MIMO technology [1, 2], is more suitable for Backhaul in urban environment in densely distributed small cells. This is also suitable for high data rate, low latency connectivity between vehicles. The conventional Sub 6 GHz MIMO assumes the model $\mathbf{y} = \mathbf{Hx+n}$, where $\mathbf{H}$ is the channel matrix, $\mathbf{x}$ is the transmitter symbol vector, $\mathbf{y}$ is the received symbol vector and $\mathbf{n}$ is the noise vector. Mostly, all signal processing action takes place in the baseband. There exists a separate RF chain for each antenna. The mmwave wireless propagation has higher propagation losses and reduced scattering. Hence the model adopted for Sub 6 GHz MIMO is not suitable for mmwave MIMO. Beam space channel model is more suitable for Millimeter wave. There were proposals made on the channel estimation and Hybrid precoding.

Digital baseband precoding with a large number of antennas is one of the baseband approaches used for the mmwave communication, where beamforming technique is used to increase spectral efficiency [3–5]. In digital baseband precoding, each antenna is driven with the RF chain and multiple streams of the data are transmitted simultaneously. Due to the large antenna, the energy consumption in the mmwave band is very high and also the hardware for digital precoder is complex and costly due to which it is not a suitable technique for channel estimation and precoding for mmwave. To overcome the above hardware limitation, analog beamforming solutions are proposed in [6–9]. In analog beamforming, the main idea is to vary the phase of the transmitted signal while keeping its magnitude fixed, i.e., analog beamformers are used as phase shifters. The analog beamformers have reduced system complexity because the antennas share only one RF chain. But as antennas share only one RF chain, only a single data stream is transmitted at a time due to which spectrum efficiency gets limited. The digital and analog beamforming techniques are not useful for mmwave communication individually, so the compromise is made between the spectral efficiency and hardware complexity and Hybrid beamforming (HBF) is proposed in which both analog and digital beamformers are used.

In [10], hybrid precoding algorithm was proposed in which phase shifters with quantized phase are required to minimize the mean-squared error of the received signals but the work in this paper does not account for mmwave characteristics. In [11] and [12], the hybrid precoding design problem was proposed such that the channel is partially known at the transmitter in the system. In [13] investigation of the hybrid precoding design is done for fully connected structure-based broadband mmWave multiuser systems with partial availability of Channel state information. Although the algorithms developed in [11–13] supports the transmission of multiple streams and the hardware limitations are also overcome to great extent but they are not as effective as compared to the digital precoding algorithm when it comes to system performance. In [14], hybrid precoding algorithm for mmWave communication system was proposed. In this algorithm, quantized beam steering direction is given importance. Also, multi-resolution codebook is designed for training the precoders and the codebook depends on hybrid (i.e., joint analog and digital) processing to generate different beamwidths beamforming vectors. This algorithm improves the system performance to some extent and also overcomes the hardware limitations but this algorithm is quite complex. So, there is a need to develop a less complex and more effective algorithm for the channel estimation and hybrid precoder and decoder design.

In this chapter, we propose a less complex and effective channel estimation and hybrid precoder and decoder designing algorithm for a mmWave system based on the computational intelligence algorithm, Particle swarm optimization. The main assumptions which we have considered on the mmwave hardware while developing the algorithm are (i) the analog phase shifters have constant magnitude and varying phases, and (ii) the number of RF chains are less than the number of antennas. Using particle swarm optimization, we are optimizing the baseband precoder and decoder such that the sensing matrix is a diagonal matrix (Diagonalization method) and reduced rank matrix (rank reduction method) and hence the channel matrix is estimated.

The rest of the chapter is organized as follows. Section 2 explains the particle swarm optimization algorithm. In Sect. 3, System model, problem formulation, and main assumptions considered in the chapter are discussed. Section 4 presents the methodology to solve the above problem. Here we discuss the pseudo code for the objective functions used in PSO for the diagonalization of the matrices and reduction of rank of the matrix and also the workflow to design the precoder and decoder and hence to estimate channel matrix. Section 5, demonstrates the simulation results obtained after performing the experiments, and finally the paper is concluded.

## 2  Particle Swarm Optimization

This section discusses the particle swarm optimization algorithm [15] which is used to diagonalize and reduce the rank of the matrix.

To understand PSO, let's consider the behavior of bird flocking. Suppose the birds are searching for the food in a particular area and they do not know about the exact location of the food. But they know how far the food is from them after each iteration. So what should birds do to find the exact location of the food? The effective way is to consider individual decisions along with the decisions taken by the neighbors to find the optimal path to be followed by the birds.

According to the PSO algorithm, Initialization with random particles (solutions) is done first and then optimum is searched by updating generations. In every iteration, each particle is updated by two values, personal best and global best. The personal best (pbest) value is the best solution achieved by the individual particle so far. And the global best (gbest) value is the common experience of all the particles in

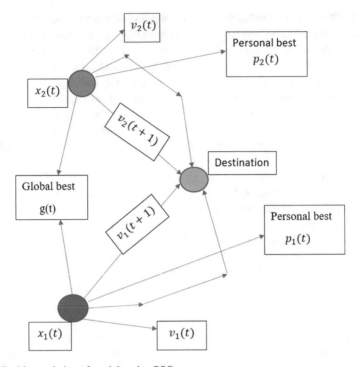

**Fig. 1** Position updation of particle using PSO

the population. It is the best value achieved so far by any of the particles in the population.

After finding the personal best and global best, the particle updates their position according to the following equations:

**next=present+C1× rand × (pbest-present)+ C2 × rand × (gbest-present)** (1)

Figure 1 shows how the position of particles ($x_1(t)$ and $x_2(t)$) is updated based on the value of the global best and individual personal best using the PSO algorithm.

## 3 Problem Formulation

### 3.1 Millimeter Wave System Model

The block diagram of the millimeter wave wireless communication system is shown in Fig. 2. It consists of baseband precoder and decoder, RF precoder and decoder, and RF chains as main blocks. From the block diagram, the baseband received signal **Y** can be modeled as follows:

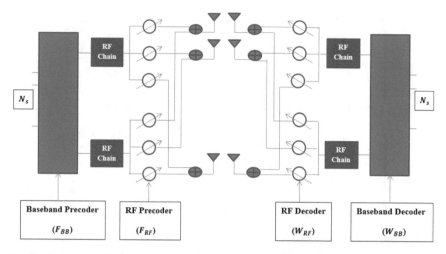

**Fig. 2** Illustration of the architecture of millimeter wave transceivers system

$$\mathbf{Y} = \sqrt{(\mathbf{P})}\mathbf{W}_{\mathbf{BB}}^{\mathbf{H}}\mathbf{W}_{\mathbf{RF}}^{\mathbf{H}}\mathbf{H}\mathbf{F}_{\mathbf{RF}}\mathbf{F}_{\mathbf{BB}}\mathbf{X} + \mathbf{N} \tag{2}$$

$$\mathbf{y} = \sqrt{(\mathbf{P})}(\mathbf{W}_{\mathbf{BB}}^{\mathbf{H}}\mathbf{W}_{\mathbf{RF}}^{\mathbf{H}} \otimes \mathbf{F}_{\mathbf{BB}}^{\mathbf{T}}\mathbf{F}_{\mathbf{RF}}^{\mathbf{T}})\mathbf{h} + \mathbf{n} \tag{3}$$

Equation (3) is the vector form of (2) which is obtained by considering input $\mathbf{X}$ as identity matrix. Here $\mathbf{H}$ is the channel matrix and $\mathbf{N}$ is the gaussian noise. $\mathbf{F}_{\mathbf{BB}}$ is the baseband precoder and $\mathbf{F}_{\mathbf{RF}}$ is the RF precoder. RF precoder is practically realized using phase shifters. Hence the elements of the matrix $\mathbf{F}_{\mathbf{RF}}$ are having the magnitude unity. Similarly, $\mathbf{W}_{\mathbf{BB}}$ is the baseband decoder and $\mathbf{W}_{\mathbf{RF}}$ is the RF decoder. $\mathbf{X}$ (with size $N_s \times 1$ ) is the symbol vector to be transmitted. $\mathbf{F}_{\mathbf{BB}}$ is of the size $N_{RF} \times N_s$. Also the size of the matrix $\mathbf{F}_{\mathbf{RF}}$ is $N_t \times N_{RF}$, where $N_t$ is the number of transmitter antennas and $N_{RF}$ is the number of RF blocks. The channel matrix is of size $N_r \times N_t$. The size of the matrix $\mathbf{W}_{\mathbf{RF}}^H$ is $N_{RF} \times N_r$ and the size of the matrix $\mathbf{W}_{\mathbf{BB}}^H$ is $N_s \times N_{RF}$.

The channel matrix $\mathbf{H}$ can further be modeled as following:

$$\mathbf{H} = \mathbf{A}_{\mathbf{R}}\mathbf{H}_{\mathbf{b}}\mathbf{A}_{\mathbf{T}}^{\mathbf{H}} \tag{4}$$

where $\mathbf{A}_{\mathbf{R}}$ is the dictionary matrix in the receiver array antenna and $\mathbf{A}_{\mathbf{T}}$ is the dictionary matrix in the transmitter array antenna as given below. The size of the matrices $\mathbf{A}_{\mathbf{R}}$ and $\mathbf{A}_{\mathbf{T}}$ are given as $N_r \times G$ and $N_t \times G$ respectively. In this the angle $\theta_i^r$ are the angle of arrivals of the receiving antenna (Mobile station) and $\theta_i^d$ are the angle of departures of the transmitting antenna (Base station).

$$A_R = \begin{pmatrix} 1 & 1 & 1 & \cdots \\ e^{-\frac{j2\pi}{\lambda}d_r\cos(\theta_1^r)} & e^{-\frac{j2\pi}{\lambda}d_r\cos(\theta_2^r)} & e^{-\frac{j2\pi}{\lambda}d_r\cos(\theta_3^r)} & \cdots \\ e^{-\frac{j2\pi}{\lambda}(2d_r)\cos(\theta_1^r)} & e^{-\frac{j2\pi}{\lambda}2d_r\cos(\theta_2^r)} & e^{-\frac{j2\pi}{\lambda}2d_r\cos(\theta_3^r)} & \cdots \\ e^{-\frac{j2\pi}{\lambda}(3d_r)\cos(\theta_1^r)} & e^{-\frac{j2\pi}{\lambda}3d_r\cos(\theta_2^r)} & e^{-\frac{j2\pi}{\lambda}3d_r\cos(\theta_3^r)} & \cdots \\ \cdots & \cdots & \cdots & \cdots \\ e^{-\frac{j2\pi}{\lambda}(N_r-1)d_r\cos(\theta_1^r)} & e^{-\frac{j2\pi}{\lambda}(N_r-1)d_r\cos(\theta_2^r)} & e^{-\frac{j2\pi}{\lambda}(N_r-1)d_r\cos(\theta_3^r)} & \cdots \end{pmatrix}$$

$$A_T^T = \begin{pmatrix} 1 & 1 & 1 & \cdots \\ e^{-\frac{j2\pi}{\lambda}d_t\cos(\theta_1^t)} & e^{-\frac{j2\pi}{\lambda}d_t\cos(\theta_2^t)} & e^{-\frac{j2\pi}{\lambda}d_t\cos(\theta_3^t)} & \cdots \\ e^{-\frac{j2\pi}{\lambda}(2d_t)\cos(\theta_1^t)} & e^{-\frac{j2\pi}{\lambda}2d_t\cos(\theta_2^t)} & e^{-\frac{j2\pi}{\lambda}2d_t\cos(\theta_3^t)} & \cdots \\ e^{-\frac{j2\pi}{\lambda}(3d_t)\cos(\theta_1^t)} & e^{-\frac{j2\pi}{\lambda}3d_t\cos(\theta_2^t)} & e^{-\frac{j2\pi}{\lambda}3d_t\cos(\theta_3^t)} & \cdots \\ \cdots & \cdots & \cdots & \cdots \\ e^{-\frac{j2\pi}{\lambda}(N_t-1)d_t\cos(\theta_1^t)} & e^{-\frac{j2\pi}{\lambda}(N_t-1)d_t\cos(\theta_2^t)} & e^{-\frac{j2\pi}{\lambda}(N_t-1)d_t\cos(\theta_3^t)} & \cdots \end{pmatrix}$$

The matrix $H_b$ is the matrix with elements filled up with complex numbers (with real and imaginary part as Gaussian distributed) and is describing the multipath channel coefficients. For each path, one particular angle of departure and the corresponding angle of arrival is activated, and hence the matrix $H_b$ needs to be sparse so that only few paths are active at a time. Substituting (4) in (3), we get the following:

$$y = \sqrt{(P)}(W_{BB}^H W_{RF}^H A_R \otimes F_{BB}^T F_{RF}^T A_T^*)h_b + n \qquad (5)$$

From (5), we conclude that the requirement is to design the precoder and the decoder such that the sensing matrix $[\sqrt{(P)}(W_{BB}^H W_{RF}^H A_R \otimes F_{BB}^T F_{RF}^T A_T^*)]$ is the diagonal matrix (diagonalization method) and the rank of the sensing matrix is to be minimized (rank reduction method) to estimate the sparse matrix ($H_b$) and hence channel matrix ($H$) from (4).

# 4 Proposed Methodology

## 4.1 Diagonalization Method

Initially, we start by initializing matrices $F_{RF}$ and $W_{RF}$ as a DFT matrix, in which only the phase of each element of the matrix is varied while magnitude is constant (unity) i.e they are acting as a phase shifters only. Matrices $A_R$ and $A_T$ are evaluated based on the specific value of $\theta$, $d_r$ and $d_t$. Matrices $F_{RF}$, $W_{RF}$, $A_R$ and $A_T$ are considered as fixed matrices based on the above constraints while implementing PSO. Matrices $W_{BB}$ and $F_{BB}$ are selected randomly and PSO algorithm is applied. Matrices $W_{BB}$ and $F_{BB}$ are updated after every iteration until sensing matrix $[\sqrt{(P)}(W_{BB}^H W_{RF}^H A_R \otimes F_{BB}^T F_{RF}^T A_T^*)]$ becomes diagonal matrix . Values of matrices $W_{BB}$ and $F_{BB}$ for which sensing matrix is the diagonal matrix are considered as the best value for baseband precoder ($F_{BB}$) and decoder ($W_{BB}$) matrices. Figure 3

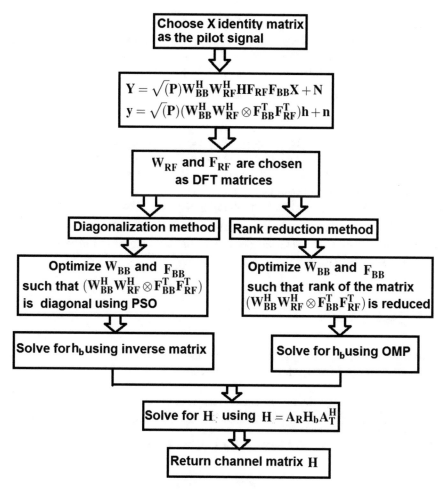

**Fig. 3** Flowchart to illustrating the methodology for obtaining channel matrix

shows the flowchart describing the abovediscussed process for the diagonalization of matrix.

Now substitute the obtained diagonalized matrix in the (5). Then take the inverse of the diagonalized matrix and multiple it with 'Y' so as to evaluate matrix $\mathbf{H_b}$ and hence the channel matrix ($\mathbf{H}$) using (1.4).

## 4.2 Rank Reduction Method

In this method, the first four steps are the same as that of diagonalization method. In the fifth step, minimization of the rank of the sensing matrix is done using PSO and corresponding values of matrices $\mathbf{W_{BB}}$ and $\mathbf{F_{BB}}$ are selected for which rank of the sensing matrix is minimized. Now, from this reduced rank sensing matrix, sparse matrix ($\mathbf{H_b}$) is recovered using orthogonal matching pursuit algorithm (OMP). The steps involved in the OMP algorithm are as follows:

- Consider the equation, $\mathbf{y} = \phi\mathbf{h_b}$, where $\phi$ is the sensing matrix.
- Now, the column of $\phi$ that has the largest correlation or projection with "y" is estimated.
- Then the best vector $\mathbf{h_b}$ is estimated using the maximum projection column estimated in the above step such that the least square norm is minimized.
- Then the residue or error is estimated and "y" is updated with the value of residue and is used in the next iteration.
- Above process is repeated until the stopping criterion is achieved.

Finally, from the estimated sparse matrix($\mathbf{H_b}$), channel matrix(H) is estimated using (4). Figure 3 shows the flowchart describing the abovediscussed process for the reduction of rank of a matrix.

## 4.3 Pseudo Code

This section explains the algorithm which we have followed to get the desired output (diagonalized matrix and reduced rank matrix).

---

**Algorithm 1** Algorithm for diagonalization of matrix

---
Input: $\mathbf{F_{RF}}$, $\mathbf{W_{RF}}$, $\mathbf{A_R}$ , $\mathbf{A_T}$ and $\mathbf{X}$ ;
Randomly generate the initial population for $\mathbf{F_{BB}}$ and $\mathbf{W_{BB}}$ for 'M' times;
for iteration= 1:N
for i= 1:M
Update the value of $\mathbf{F_{BB}}$ and $\mathbf{W_{BB}}$ using equation 1 for PSO;
($\mathbf{F_{BB}}new(i)$, $\mathbf{W_{BB}}new(i)$) ;
Update the cost function value;
Update the initial value of matrix $\mathbf{F_{BB}}$ and $\mathbf{W_{BB}}$;
$\mathbf{F_{BB}}initial(i) = F_{BB}new(i)$ ;
$\mathbf{W_{BB}}initial(i) = W_{BB}new(i)$ ;
end for
end for
Cost function;
residual ($\mathbf{F_{BB}}$ , $\mathbf{W_{BB}}$)
temp1=$\mathbf{W_{BB}} \times \mathbf{W_{RF}} \times \mathbf{A_R}$;
temp2=$\mathbf{F_{BB}} \times \mathbf{F_{RF}} \times \mathbf{A_T}$;;
res=**kron(temp1, temp2)**;
res1=matrix having only diagonal elements of 'res', rest of the elements are made zero;
res2= res;
res=ratio of absolute sum of elements of (res1) to (res2);
Output: Diagonalized matrix

---

| **Algorithm 2** Algorithm for rank reduction |
|---|
| Input: $F_{RF}$, $W_{RF}$, $A_R$ , $A_T$ and $X$ ; |
| Randomly generate the initial population for $F_{BB}$ and $W_{BB}$ for 'M' times; |
| for iteration= 1:N |
| for i= 1:M |
| Update the value of $F_{BB}$ and $W_{BB}$ using equation 1 for PSO; |
| $(F_{BB}new(i), W_{BB}new(i))$ ; |
| Update the cost function value; |
| Update the initial value of matrix $F_{BB}$ and $W_{BB}$; |
| $F_{BB}initial(i) = F_{BB}new(i)$ ; |
| $W_{BB}initial(i) = W_{BB}new(i)$ ; |
| end for |
| end for |
| Cost function; |
| residual $(F_{BB}$ , $W_{BB})$ |
| temp1=$W_{BB} \times W_{RF} \times A_R$; |
| temp2=$F_{BB} \times F_{RF} \times A_T$;; |
| res=**kron(temp1, temp2)**; |
| res1=Linear combination of the rows or column of res; |
| res2= Absolute sum of res1; |
| Output: Reduced rank matrix |

# 5   Experiment and Results

In this section, the simulation experiments are performed to demonstrate the proposed techniques (A) Diagonalization method and (B) Rank reduction method.

## 5.1   Diagonalization Method

RF precoder ($F_{RF}$) and decoder ($W_{RF}$) matrices are initialized as a DFT matrix, in which only the phase of each element of the matrix is changing while magnitude is constant (unity), i.e., they are acting as a phase shifters only. Dimensions of matrices ($F_{RF}$) and ($W_{RF}$) are considered as $32 \times 6$ and $64 \times 6$, respectively. Input $X$ is taken as a identity matrix with dimensions $4 \times 4$. Dictionary matrices $A_R$ and $A_T$ are evaluated based on the specific value of angle of departure ($\theta^t$), angle of arrival ($\theta^r$), spacing between antennas at receiver ($d_r$) and spacing between antennas at transmitter ($d_t$). While evaluating dictionary matrices resolution is taken as "4", i.e., only "4" different values of $\theta^t$ and $\theta^r$ are considered. Dimensions of $A_R$ and $A_T$ are taken as $64 \times 4$ and $32 \times 4$. Matrices $F_{RF}$, $W_{RF}$, $A_R$ and $A_T$ are considered as fixed matrices based on the above constraints while implementing PSO. Now the baseband decoder

**Table 1** Simulation prameters

| Parameters | Name | Dimension | Range |
|---|---|---|---|
| $\theta^t$ | Angle of Departure | – | $0\text{-}\pi$ |
| $\theta^r$ | Angle of arrival | – | $0\text{-}\pi$ |
| H | Channel Matrix | $64 \times 32$ | – |
| $H_b$ | Sparse matrix | $4 \times 4$ | – |
| $W_{BB}$ | Baseband Decoder | $6 \times 4$ | – |
| $F_{BB}$ | Baseband Precoder | $6 \times 4$ | – |
| $W_{RF}$ | RF Decoder | $64 \times 6$ | – |
| $F_{RF}$ | RF Precoder | $32 \times 6$ | – |
| $A_R$ | Dictionary Matrix | $64 \times 4$ | – |
| $A_T$ | Dictionary Matrix | $32 \times 4$ | – |

($\mathbf{W_{BB}}$) and precoder ($\mathbf{F_{BB}}$) matrices are selected randomly with dimensions $6 \times 4$ and $6 \times 4$ respectively and PSO algorithm is applied. Matrices $\mathbf{W_{BB}}$ and $\mathbf{F_{BB}}$ are updated after every iteration until matrix $\sqrt{(P)}(\mathbf{W_{BB}^H W_{RF}^H A_R} \otimes \mathbf{F_{BB}^T F_{RF}^T A_T^*})$ becomes diagonal matrix. Table 1 shows the simulation parameters which are considered while performing the experiment.

Now the obtained diagonalized matrix is substituted in (5) and its inverse is evaluated. Then it is multiplied by Y, so that sparse matrix $\mathbf{H_b}$ is evaluated and hence channel matrix $\mathbf{H}$ from (4).

After performing the experiment the obtained results are as follows:

Figure 4 shows the convergence of the PSO algorithm, it shows the minimization of the best cost as the number of iteration is increasing. Figure 5a and b shows the diagonal sensing matrix which we get after applying the PSO algorithm. In this, the diagonal elements (non-zero elements) are represented by the brighter color and the off diagonal element (approximately zero values) are represented by the darker color. Figure 5c and d shows the matrix $\mathbf{H_b}$ which is a sparse matrix, in which only a few elements are non-zero and the rest of the elements are nearly zero. Figure 5e and f shows the magnitude and phase of the 64*32 channel matrix ($\mathbf{H}$ ). Both magnitude and phase are varying for different elements of the matrix. Figure 6a and b shows the magnitude and phase of the baseband combiner matrix ($\mathbf{W_{BB}}$ ). Figure 6c and d shows that the magnitude of each element of RF Combiner matrix ($\mathbf{W_{RF}}$) is unity and only phase is changing. Figure 6e and f shows the magnitude and phase of the baseband precoder matrix($\mathbf{F_{BB}}$). Figure 6g and h shows that the magnitude of elements of RF Precoder matrix ($\mathbf{F_{RF}}$) is unity and only phase is changing.

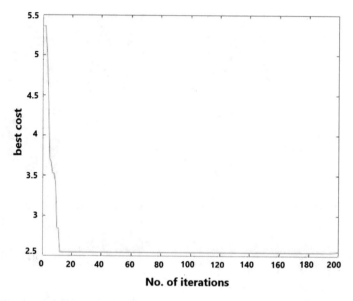

**Fig. 4** Convergence of PSO Algorithm for diagonalization of matrix

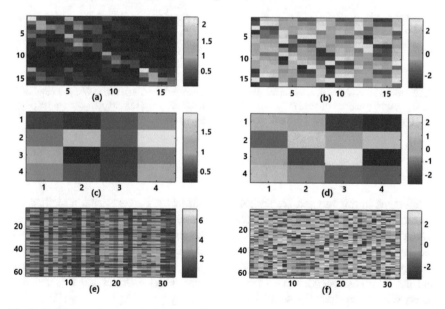

**Fig. 5** Illustration of matrices in the form of images, Diagonal sensing matrix obtained using PSO **a** magnitude and **b** phase, sparse matrix ($H_b$) **c** magnitude and **d** phase, channel matrix (H) **e** magnitude and **f** phase

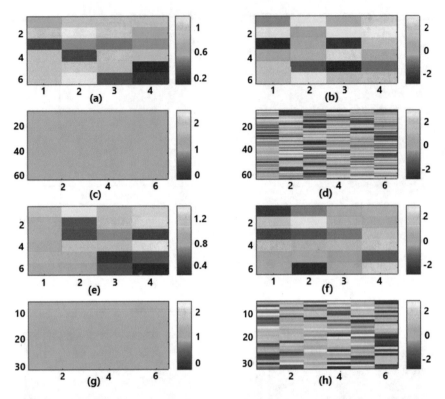

**Fig. 6** Illustration of matrices in the form of images, Baseband decoder matrix ($W_{BB}$) **a** magnitude and **b** phase, RF decoder matrix ($W_{RF}$) **c** magnitude and **d** phase, baseband precoder matrix ($F_{BB}$) **e** magnitude and **f** phase, RF precoder matrix ($F_{RF}$) **g** magnitude and **h** phase

## 5.2 Rank Reduction Method

Initial steps in this method are the same as that of the diagonalization method but the dimensions of some of the matrices are changed here. The dimension for sparse matrix is changed to 16 × 16, and the dimension of the dictionary matrices $A_R$ and $A_T$ are changed to 64 × 16 and 32 × 16 respectively. Now, same as in previous method, matrices $F_{RF}$, $W_{RF}$, $A_R$ and $A_T$ are considered as fixed matrices and the baseband decoder ($W_{BB}$) and precoder ($F_{BB}$) matrices are selected randomly with dimensions 6 × 4 and 6 × 4 respectively while implementing PSO. Now matrices $W_{BB}$ and $F_{BB}$ are updated after every iteration until the rank of the matrix $\sqrt{(P)}(W_{BB}^H W_{RF}^H A_R \otimes F_{BB}^T F_{RF}^T A_T^*)$ is minimized. Then the matrix $H_b$ is estimated using the OMP algorithm as explained in the Sect. 4.2 and hence the channel matrix (H) is estimated using (4). Results obtained from this experiment are as follows:

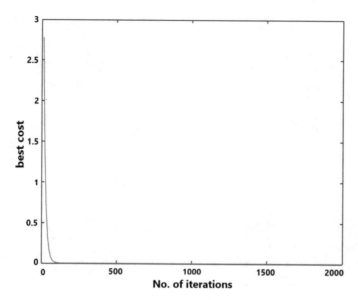

**Fig. 7** Convergence of PSO Algorithm for minimization of rank of matrix

Figure 7 shows the convergence of the PSO algorithm for the minimization of the rank of matrix, it shows the minimization of the best cost as the number of iteration is increasing. Figure 8a and b shows the initial sensing matrix with rank 16 before applying the PSO algorithm. Figure 8c and d shows reduced rank sensing matrix with rank 3 obtained using PSO. Figure 8e and f shows the comparison of the sparse matrix ($H_b$) obtained from both the methods and from the figure it is clear that the matrix obtained from the rank reduction method is of higher order and is more sparse as compared to the matrix obtained from the diagonalization method. So we can say that the rank reduction method is better than diagonalization method for estimating sparse matrix($H_b$). Figure 8g and h shows the magnitude and phase of the 64*32 channel matrix (**H**). Both magnitude and phase are varying for different elements of the matrix. Figure 9a and b shows the magnitude and phase of the baseband combiner matrix ($W_{BB}$). Figure 9c and d shows that the magnitude of each element of RF Combiner matrix ($W_{RF}$) is unity and only phase is changing. Figure 9e and f shows the magnitude and phase of the baseband precoder matrix($F_{BB}$). Figure 9g and h shows that the magnitude of elements of RF Precoder matrix ($F_{RF}$) is unity and only phase is changing.

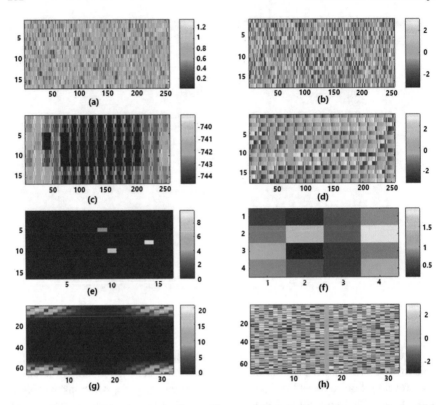

**Fig. 8** Illustration of matrices in the form of images, Initial sensing matrix **a** magnitude and **b** phase, reduced rank sensing matrix obtained using PSO **c** magnitude and **d** phase, sparse matrix ($H_b$) **e** obtained from rank reduction method and **f** obtained from diagonalization method, channel matrix (H) **g** magnitude and **h** phase

## 6   Conclusion

This chapter demonstrates the proposed methodology to design precoder and decoder for a given pilot signal x and corresponding y by rank reduction and diagonalization of sensing matrix using particle swarm optimization. This helps to estimate the sparse matrix $H_b$ and hence channel matrix $H$. The precoder and decoder are designed by considering large number of transmitter and receiver antennas (typically for $N_t=32$ and $N_r=64$). The results are obtained with high accuracy and speed, for a large number of transmitter and receiver antennas. The rank reduction method is considered as more suitable as compared to the diagonalization method on the basis of the sparse matrix $H_b$ estimated in both the methods. The complexity of the above proposed methodology increases if the number of transmitter and receiver antennas are increased further. So for the future work, it would be interesting to explore other constraint optimization techniques to overcome the above problem.

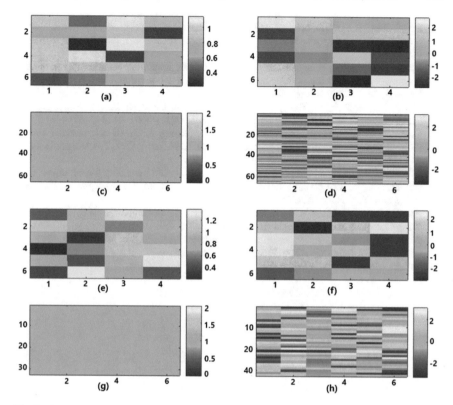

**Fig. 9** Illustration of matrices in the form of images, Baseband decoder matrix ($W_{BB}$) **a** magnitude and **b** phase, RF decoder matrix ($W_{RF}$) **c** magnitude and **d**, baseband precoder matrix ($F_{BB}$) **e** magnitude and **f** phase, RF precoder matrix ($F_{RF}$) **g** magnitude and **h** phase phase

# References

1. Pi, Z., Khan, F.: An introduction to millimeter-wave mobile broadband systems. IEEE Commun. Mag. **49**(6), 101–107 (2011)
2. Rappaport, T., Sun, S., Mayzus, R., Zhao, H., Azar, Y., Wang, K., Wong, G., Schulz, J., Samimi, M., Gutierrez, F.: Millimeter wave mobile communications for 5G cellular: it will work!. IEEE Access **1**, 335–349 (2013)
3. Xiao, M., Mumtaz, S., Huang, Y., Dai, L., Li, Y., Matthaiou, M., Karagiannidis, G.K., Bjornson, E., Yang, K., C. L. I, Ghosh, A.: Millimeter wave communications for future mobile networks. IEEE J. Sel. Areas Commun. **35**(9), 1909–1935 (2017)
4. Xiao, Z., Xia, P., Xia, X.-G.: Full-duplex millimeter-wave communication. IEEE Wirel. Commun. Mag. **16** (2017)
5. Andrews, J.G., Bai, T., Kulkarni, M.N., Alkhateeb, A., Gupta, A.K., Heath, R.W.: Modeling and analyzing millimeter wave cellular systems. IEEE Trans. Commun. **65**(1), 403–430 (2017)
6. Wang, J., et al.: Beam codebook based beamforming protocol formulti-Gbps millimeter-waveWPAN systems. IEEE J. Sel. Areas Commun. **27**(8), 1390–1399 (2009)
7. Chen, L., Yang, Y., Chen, X., Wang, W.: Multi-stage beamforming codebook for 60GHz WPAN. In Proc. 6th Int. ICST Conf. Commun. Network. China, China, pp. 361–365 (2011)

8. Hur, S., Kim, T., Love, D., Krogmeier, J., Thomas, T., Ghosh, A.: Millimeter wave beamforming for wireless backhaul and access in small cell networks. IEEE Trans. Commun. **61**(10), 4391–4403 (2013)

9. Tsang, Y., Poon, A., Addepalli, S.: Coding the beams: Improving beamforming training in mmwave communication system. In: Proc. IEEE Global Telecomm. Conf. (GLOBECOM), Houston, TX, USA, 2011, pp. 1–6

10. Zhang, X., Molisch, A., Kung, S.: Variable-phase-shift-based RF-baseband code sign for MIMO antenna selection. IEEE Trans. Sign. Process. **53**(11), 4091–4103 (2005)

11. Venkateswaran, V., van der Veen, A.: Analog beamforming in MIMO communications with phase shift networks and online channel estimation. IEEE Trans. Sign. Process. **58**(8), 4131–4143 (2010)

12. Alkhateeb, A., El Ayach, O., Leus, G., Heath, R.W.: Hybrid precoding for millimeter wave cellular systems with partial channel knowledge. In: Proc. Inf. Theory Applicat. Workshop (ITA), San Diego, CA, USA, Feb. 2013, pp. 1–5

13. Xu, C., Yongming Huang, R.Y., He, S., Zhang, C.: Hybrid Precoding for Broadband Millimeter-Wave Communication Systems With Partial CSI, vol. 8. IEEE (2018)

14. Ahmed, A., El Ayach, O.: Channel estimation and hybrid precoding for millimeter wave cellular systems. IEEE J. Sel. Top. Sign. Proces. **8**(5) (2014)

15. Pattern Recognition and Computational Intelligence book. Springer

# Comparative Analysis of Constraint Handling Techniques Based on Taguchi Design of Experiments

**Misael Lopez-Sanchez, M. A. Cosío-León, and Anabel Martínez-Vargas**

**Abstract** In this chapter, we analyze the effect that constraint-handling techniques such as penalty functions, repair methods, and decoders have on a steady-state genetic algorithm running on a smartphone. We examine these techniques on one particular problem: the tourist trip design problem. This problem selects a set of points of interest that matches tourist preferences to bring a personalized trip plan. Our points of interest are focused on Mexico City. In order to test the differences among the constraint-handling techniques, we apply the Taguchi design of experiments. The results support the decision that a random decoder is the best choice to handle constraints in the context of the tourist trip design problem.

## 1  Introduction

Optimization problems exist in different fields of science and engineering. They minimize or maximize some objective(s). Most of the real-world problems have restrictions on their variables and they must be considered to obtain a feasible solution. These problems correspond to constrained optimization.

Metaheuristics that deal with constrained optimization may create solutions that fall outside the feasible region; therefore a viable constraint-handling strategy is needed.

This chapter is devoted to a detailed analysis of constraint-handling techniques. We aim to analyze their effects on the quality of solutions provided by a steady-state genetic algorithm (SGA), solving the tourist trip design problem (TTDP) [1],

M. Lopez-Sanchez · M. A. Cosío-León · A. Martínez-Vargas (✉)
Universidad Politécnica de Pachuca, Zempoala, Hidalgo, Mexico
e-mail: anabel.martinez@upp.edu.mx

M. Lopez-Sanchez
e-mail: misael.lopez@micorreo.upp.edu.mx

M. A. Cosío-León
e-mail: ma.cosio.leon@upp.edu.mx

© The Author(s), under exclusive license to Springer Nature Singapore Pte Ltd. 2021
A. J. Kulkarni et al. (eds.), *Constraint Handling in Metaheuristics and Applications*,
https://doi.org/10.1007/978-981-33-6710-4_14

**Table 1** Constraint-handling techniques [6]

| Category | Variants |
| --- | --- |
| Penalty functions | Logarithmic |
| | Linear |
| | Quadratic |
| Repair methods | Random repair method |
| | Greedy repair method |
| Decoders | Random decoder |
| | Greedy decoder |

an extension of the orienteering problem (OP). The TTDP selects a set of points of interest (PoI) that matches tourist preferences to bring a personalized trip plan.

The selected methodology driving this study is the *Taguchi design of experiments* (Taguchi DOE) procedure also called orthogonal array design [2]. Robust design methods were developed by Genichi Taguchi [3] to improve the quality of manufactured goods. Nowadays, it is a statistical approach that helps guide data analysis.

An additional result from this analysis is an optimal parameter vector of $P^*$. This vector is an SGA configuration to achieve its best performance in solving TTDP, considering a set of real-world instances. To do so, we take as guides works in [4, 5]. Both studies develop a description of the best practices in selecting $P^*$. They also provide results on computational experiments with metaheuristics. In concordance with results shown in [6], we propose the following hypothesis: there is an effect or a difference (significant differences between constraint-handling techniques) on the SGA.

The categories of constraints-handling techniques and variants analyzed in this chapter are shown in Table 1.

The approaches listed above are briefly described

- *Penalty functions* are the most common strategies to handle constraints. They transform a constrained-optimization problem into an unconstrained one. This is done by adding (or subtracting) a value (penalty) to the objective function. The penalty depends on the amount of constraint violations in a solution [7].
- *Repair methods* are heuristics that move infeasible solutions to the feasible space. However, this can lead to a loss of population diversity. Methods that place the infeasible solutions randomly inside the feasible space result in a loss of useful information gathered by this solution. Therefore, a balanced approach that utilizes the useful information from the solutions and brings them back into the search space in a meaningful way is desired [8]. The repair heuristics are specific to the optimization problem at hand.
- *Decoders* use special representation mappings. They guarantee the generation of a feasible solution at the cost of computation time. Furthermore, not all constraints can be easily implemented by decoders. The resulting algorithm is specific to the particular application [6].

There are some early attempts to handle constraints on metaheuristics. For example, authors in [8] explore constraint-handling strategies and propose two new single parameter constraint-handling methodologies based on parent-centric and inverse parabolic probability (IP) distribution. Their proposed constraint-handling methods exhibit robustness in terms of performance and also success in search spaces comprising up to 500 variables, while locating the optimum value within an error of $10^{-10}$. The work in [9] describes the process executed by metaheuristics resolving constrained-optimization problems (COP). The lack of constrained management algorithms discards the infeasible individuals in the process of evolution. Consequently, this leads to the loss of potential information. Authors conclude that a COP can be better solved by effectively using the potential information present in the infeasible individuals. The work in [10] shows a mechanism to deal with the constraints. Such a mechanism is applied in evolutionary algorithms (EAs) and swarm intelligence algorithms (SIAs) to solve a constrained non-linear optimization problem (CNOP). A case study of a modified artificial bee colony algorithm to solve constrained numerical optimization problems is used by the authors. The results showed an improvement in performance compared with classical algorithms. A different approach for dealing with CNOP is demonstrated in [11]. Support vector machines convert constraint problems into unconstrained ones by space mapping. Based on the above techniques, the authors propose this new decoder approach that constructs a mapping between the unit hypercube and the feasible region using the learned support vector model. Recently, the work in [12] proposes the use of the $\epsilon$ constraint method in the differential evolution algorithm. There, the $\epsilon$ level is defined based on the current violation in the population. For every constraint, an individual $\epsilon$ level is kept. Then by combining fitness and constraint violation, the infeasible solutions are determined.

The chapter is organized as follows: Sect. 2 shows an overview of TTDP and the knapsack problem. Section 3 describes how SGA generates a personalized trip plan. Section 4 shows computational experiments, these include instance design process, instance generation process, and Taguchi's DOE. Section 5 is devoted to results and discussion. Finally, Sect. 6 concludes the chapter and provides direction for future research.

# 2 Overview

## 2.1 Case Study: TTDP

TTDP selects a set of PoIs that matches with the tourist's preferences and geographical current (or selected) location. This maximizes tourist satisfaction while considering constraints (e.g., distances among PoIs, a time required to visit each PoI, open periods of PoIs days/hours, opening time/closing time, entrance fees, weather conditions, and public health policies).

**Fig. 1** A solution to the optimization model

A solution for the TTDP is decoded as *a route-planning for tourists* interested in visiting multiple PoIs, so-called optimal scheduling trip plan (OSTP). In this work, a solution for TTDP is decoded *as a star type itinerary* (see Fig. 1). The star type itinerary is a solution to the knapsack problem (or rucksack problem) [13]. The aforementioned combinatorial optimization problem deals with a set of items having a weight and a profit. Then, a solution is a subset of those items that maximize the sum of the profits, while keeping the summed weight within a certain capacity $C$.

The above is the kernel-philosophy of our mobile application called *Turisteando Ando* (TA) which has embedded an SGA to maximize the tourist satisfaction when visiting a place. TA gathers a set of PoIs from *Places* API by Google. As shown in Fig. 1, those PoIs are filtered in a 30 km circle around the current tourist position (or selected start point). TA takes into account people's best recommendations (ratings), reducing fear of missing out on trending places. Then it delivers a subset of PoIs that maximize the tourist's experience while keeping the stay time at each place. It allows travel decision flexibility for the tourist; i.e., a tourist selects the PoIs sequence since it does not affect the stay time from the proposed trip plan. If a tourist wants to spend more time in a PoI, then the PoIs from the proposed trip plan can be removed. Most of the Apps to solve the TTDP cannot offer travel decision flexibility because the solution is a route. Then, the tourist must follow that route in the order it was designed. Table 2 summarizes characteristics of current commercial Apps including TA.

**Table 2** A comparative of TA versus relevant commercial Apps

| App | Internet | PTP | Guides | Opinion | iOS | Ads | Georef. | Time opt. |
|---|---|---|---|---|---|---|---|---|
| TA | ✓ | ✓ | ✗ | ✗ | ✗ | ✗ | ✓ | ✓ |
| Google trips [14] | ✗ | ✓ | ✗ | ✓ | ✓ | ✓ | ✓ | ✗ |
| Minube [15] | ✗ | ✓ | ✓ | ✓ | ✓ | ✓ | ✓ | ✗ |
| TripAdvisor [16] | ✗ | ✗ | ✓ | ✓ | ✓ | ✓ | ✓ | ✗ |
| Triposo [17] | ✗ | ✗ | ✓ | ✓ | ✓ | ✓ | ✓ | ✗ |

PTP: personalized trip plan; Time opt: time optimization; Georef: geo-referential information

## 2.2 Knapsack Problem

The optimization model is as follows: having a set of weights $W[i]$, profits $P[i]$, and capacity $C$, find a binary vector $x = \{x[1], \ldots, x[n]\}$. The knapsack problem (KP) [18] is an optimization problem as follows:

$$\text{Maximize} \quad P(x) = \sum_{i=1}^{n} x[i] \cdot P[i] \tag{1}$$

Considering the following restrictions:

$$\sum_{i=1}^{n} x[i] \cdot W[i] \leq C \tag{2}$$

$$x \in \{0, 1\}, i = 1, 2, \ldots, n \tag{3}$$

where $x[i] = 1$, if the $i$-th item is included in the knapsack, otherwise $x[i] = 0$.

We map the knapsack problem to solve the TTDP as follows:

- *Travel time* (knapsack capacity). This parameter is set by the tourist. It is defined in days or hours. The mobile application translates periods to seconds.
- *Solution size* (*n*). It is the cardinally of the set of PoIs around the current tourist location.
- *Rating* (profit). This parameter is given by Google *Places* API metadata. It is the average of ratings, determined by people's experience on each PoI.
- *Time by place* (weight). It is time in seconds. This data is gathered using API Directions by MapBox. Time by place is computed as follows: the time to move from tourist location to the PoI by two (go and back) plus stay time as Eq. (4) shows.

$$T = (t \cdot 2) + t_p \tag{4}$$

where:

- $T$ = Time by place.
- $t$ = Transportation time.
- $t_p$ = Stay time.

## 3 Personalized Trip Plan Algorithm

Personalized Trip Plan Algorithm (PTPA) embedded in the mobile app includes a SGA. Unlike a GA with a generational model, a SGA generates one or two new descendants. After that, it chooses survivors that are inserted into the population. That means that there are cycles instead of generations [19].

In PTPA, each individual has a subset of PoIs that represent a solution (personalized trip plan). A solution is coded in three vectors (see Fig. 2):

- The first vector (the individual) keeps a selection of PoIs, a choice coded in binary. That is, each gene indicates if a PoI is included (a bit with value 1) or not (a bit with value 0) in the personalized trip plan.
- The second vector (the profits) holds a set of ratings of the selected PoIs. This vector allows PTPA to evaluate each individual in the objective function described in (1).
- The third vector (the weights) stores the time by place. That is, the time that the tourist should spend in the round-trip to the PoI from the current location plus the stay time. This vector is used to compute the constraint of (2).

Constraint-handling techniques are applied to deal with individuals that violate the constraint. In this work, we apply the constraint-handling techniques described

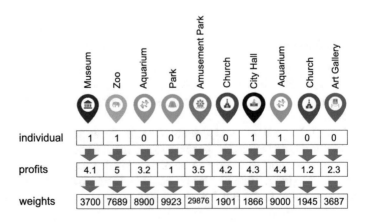

**Fig. 2** An individual in the PTPA

in [6] as penalty functions ($A_p[i]$, where $i$ is the index of a particular algorithm in this class), repair algorithms ($A_r[i]$), and decoders ($A_d[i]$). The first technique penalizes individuals by decreasing the "goodness" of the objective function. The second technique corrects any infeasible individual through special repair algorithms. The third technique uses integer representation where an individual is interpreted as a strategy that incorporates places into the solution.

We consider logarithmic, linear, and quadratic cases as penalty functions based in the study in [6]:

$$A_p[1] : Pen(x) = \log_2 \left( 1 + \rho \cdot \left( \sum_{i=1}^{n} x[i] \cdot W[i] - C \right) \right) \tag{5}$$

$$A_p[2] : Pen(x) = \rho \cdot \left( \sum_{i=1}^{n} x[i] \cdot W[i] - C \right) \tag{6}$$

$$A_p[3] : Pen(x) = \left( \rho \cdot \left( \sum_{i=1}^{n} x[i] \cdot W[i] - C \right) \right)^2 \tag{7}$$

In all three cases, $\rho = \max_{i=1...n}\{P[i]/W[i]\}$.

For repair algorithms, we apply the next two approaches described in [6] and based on Algorithm 1. They only differ in the selection procedure, i.e., the way that a PoI is removed from the personalized trip plan

- $A_r[1]$ (random repair). It selects a random element from the personalized trip plan.
- $A_r[2]$ (greedy repair). All PoIs in the personalized trip plan are sorted in the decreasing order from their profit to weight ratios. The strategy is then to always choose the last PoI (from the list of available PoIs) for deletion.

---

**Algorithm 1:** Repair procedure [6]

---

**Data:** Individual vector ($x$), profits vector ($P$), and weights vector ($W$).
**Result:** repaired individual ($x$).
1 $knapsack\_overfilled = False$;
2 $x' = x$;
3 **if** $\sum_{i=1}^{n} x'[i] \cdot W[i] > C$ **then**
4 $\quad$ $knapsack\_overfilled = True$;
5 **end**
6 **while** $knapsack\_overfilled$ **do**
7 $\quad$ $i$ = select an item from the knapsack;
8 $\quad$ $x'[i] = 0$;
9 $\quad$ **if** $\sum_{i=1}^{n} x'[i] \cdot W[i] \leq C$ **then**
10 $\quad\quad$ $knapsack\_overfilled = Flase$;
11 $\quad$
12 **end**
13 End;

---

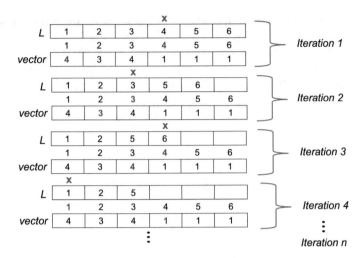

**Fig. 3** The mapping between a vector (individual) and a list $L$ in a decoder

The equation to determine the number of individuals that can be repaired in each run is given by:

$$nSolutions = (nCycles \cdot (nParents + nChildren)) \cdot \%repair \qquad (8)$$

where $nSolutions$ is the number of infeasible solutions that can be repaired by a run. $nCycles$ is the number of cycles of the SGA. $nParents$ is the number of parents selected to crossover. $nChildren$ is the number of children created at the crossover. $\%repair$ is the percentage of infeasible solutions that can be repaired.

In the case of decoders, we use the ones mentioned in [6]. Instead of binary representation, the decoders use integer representation (ordinal representation). Each individual is a vector of $n$ integers, then the $i$-th component of the vector is an integer in the range from 1 to $n - i + 1$. The ordinal representation references a list $L$ of PoIs. Take, for example, the list of PoIs $L = (1, 2, 3, 4, 5, 6)$ and the vector $\langle 4, 3, 4, 1, 1, 1 \rangle$ shown in Fig. 3. There, the vector holds the $j$-th position in the list $L$. Then, the $j$-th PoI is removed from the list $L$. If current total weight does not exceed the travel time by including that $j$-th PoI, then that $j$-th PoI is part of the solution (the personalized trip plan). As Fig. 3 suggests, the aforementioned procedure is performed until no more PoIs can be included in the solution due to travel time. Assuming that when mapping the vector and list $L$ in Fig. 3, the travel time has not been exceeded, then the following sequence of places is the solution: 4, 3, 6, 1, 2, 5.

Specifically, we apply the two decoders proposed in [6]. They are based on Algorithm 2. They differ only in the way that the list $L$ built

- $A_d[1]$ (random decoding). The list $L$ of PoIs is created randomly.
- $A_d[2]$ (greedy decoding). The list $L$ of PoIs is created in the decreasing order of their profit to weight ratios.

---

**Algorithm 2:** Decoder procedure [6]

---

**Data**: Individual vector ($x$), profits vector ($P$), and weights vector ($W$).
1  build a list $L$ of items;
2  $i = 1$;
3  $WeightSum = 0$;
4  $ProfitSum = 0$;
5  **while** $i \leq n$ **do**
6     $j = x[i]$;
7     remove the $j$-th item from the list $L$;
8     **if** $WeightSum + W[j] \leq C$ **then**
9         $WeightSum = WeightSum + W[j]$;
10        $ProfitSum = ProfitSum + P[j]$;
11     **end**
12     $i = i + 1$;
13 **end**
14 End;

---

Then PTPA performs parent selection. Tournament selection is applied in all the types of constraint-handling techniques mentioned above. It picks $k$ individuals randomly from the population and selects the best individual from this group. The best individual is the parent [20]. That procedure is repeated twice.

PTPA uses a two-point crossover operator [20] for the binary representation while applying the one-point crossover operator for the ordinal representation as suggested in [6]. Consequently, PTFA applies a bit-flip mutation operator for the binary representation. For ordinal representation, it undergoes a similar mutation like bit-flip but unlike changing bits, it takes a random value (uniform distribution) from the range $[1 \ldots n - i + 1]$ [6].

In an SGA, parents and offspring compete for survival. In this work, we apply family competition replacement methods [19]: elitist recombination and correlative family-based selection. For ordinal representation, PTFA performs elitist recombination. In this method, the best two of parents and offspring, go to the next generation. On the other hand, for binary representation, PTFA undergoes correlative family-based selection. This strategy chooses the best of the four individuals (two parents and two offspring) as the first survivor. Then, from the remaining three individuals, the one with the highest distance from the best becomes the second survivor [21].

When a predefined number of cycles is reached, PTFA returns the best personalized trip plan as a solution. Algorithm 3 summarizes the procedures described above to find the best-personalized trip plan.

---

**Algorithm 3:** Personalized trip plan

---

**Data**: Population size, crossover probability, mutation probability, and number of iterations.
Travel time, profits vector (rating), and weights vector (time by place).
**Result**: The best-personalized trip plan.
1 Generate population randomly;
2 **repeat**
3     Select two parents using a tournament selection;
4     **if** $random\_number <= crossover\_probability$ **then**
5         Perform two-point crossover for binary representation. Otherwise, perform one-point crossover for ordinal representation;
6     **else**
7         $offspring\_1 = Parent\_1$;
8         $offspring\_2 = Parent\_2$;
9     **end**
10    Mutate the two-resulting offspring according to the type of representation;
11    Perform replacement strategy according to the type of representation;
12 **until** $number\_of\_cycles > total\_number\_of\_cycles$;
13 Select the best solution from population;
14 End;

---

## 4 Computational Experiments

Design of experiments (DOE) is a procedure used to determine the influence of one or more independent variables (factors) on a response variable (solution). In the 1920s, Fisher first proposed the DOE with multiple factors known as Factorial Design of Experiments. In the full factorial design, all possible combinations of a set of factors are executed. In 1950, Taguchi proposes a different flavor on DOE, his approach reduces costs in time and effort by evaluating several factors with the least of experiments. This work uses a type of Taguchi DOE also called orthogonal arrays (OA), to establish the combinations of the values that each factor can take to run the algorithms under analysis.

The experiments were carried out to gather data to analyze the effect of constrained handling techniques on the quality of solutions provided by Algorithm 3. To do so, we select a Taguchi DOE, different configurations of the Algorithm 3 are considered. Those configurations take into account different constraint-handling mechanisms and different values for the $k$ value, which controls the number of individuals who face each other in the selection tournament. We included this genetic operator in the Taguchi DOE, since it too has a strong impact on the search process. Selection guides the search towards promising regions of the search space. It exploits the information represented within the population [22].

The Algorithm 3 was implemented for mobile application in Android operating system. It was coded in Java with Android Studio (version 3.5.3 and build #A-191.8026.42.35.6010548) integrated development environment (IDE). It was run by a *2.0 GHz Octa-core Qualcomm Snapdragon 625* CPU with 3 GB of RAM and *MIUI 9.5 based on Android 7.1.2 Nougat* as the operating system.

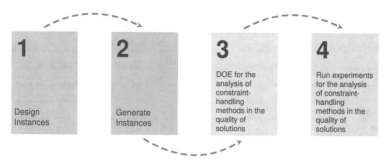

**Fig. 4** Pipeline of computational experiments

To carry out computational experiments, the procedures shown in the block diagram in Fig. 4 were executed.

## 4.1 Instance Design Process

TTDP instances were designed to represent different scenarios to be analyzed. Three sets of instances were considered, each group with defined characteristics.

A total of 90 instances were set up for the current study. These instances were categorized into three groups ($A$, $B$, and $C$), each with 30 elements.

Datasets used for the KP mapped to the TTDP were collected from internet data sources (*Places* API by Google, and Directions API by MapBox) that provide information about PoIs.

The search for PoIs information is based on the *tourist-location* geo-referential information. These coordinates will be used to search for the PoIs within a 30 km radius. Each group has 30 different origin coordinates. They are located in the territorial area of Mexico City (Mexico) and its surroundings.

Subsequently, the types of PoIs to be included in each instance are established. PoIs are grouped into four categories: Museum ($T_1$), Park ($T_2$), Zoo ($T_3$) and Historical Place ($T_4$), as well as their stay time (See Table 3).

Let

$$M = \{x \,|\, x \text{ is a museum less than 30 km radius from the } tourist\text{-}location\}$$
$$G = \{x \,|\, x \text{ is an art gallery less than 30 km radius from the } tourist\text{-}location\}$$
$$Z = \{x \,|\, x \text{ is a zoo less than 30 km radius from the } tourist\text{-}location\}$$
$$Q = \{x \,|\, x \text{ is an aquarium less than 30 km radius from the } tourist\text{-}location\}$$
$$R = \{x \,|\, x \text{ is a park less than 30 km radius from the } tourist\text{-}location\} \tag{9}$$
$$K = \{x \,|\, x \text{ is an amusement park less than 30 km radius from the } tourist\text{-}location\}$$
$$H = \{x \,|\, x \text{ is a church less than 30 km radius from the } tourist\text{-}location\}$$
$$I = \{x \,|\, x \text{ is a city hall less than 30 km radius from the } tourist\text{-}location\}$$
$$S = \{x \,|\, x \text{ is a tourist attraction less than 30 km radius from the } tourist\text{-}location\}$$

**Table 3** Stay time by type of place [23–26]

| Place type | Stay time in seconds |
|---|---|
| Museum | 3600 |
| Zoo | 7200 |
| Park | 9900 |
| Amusement Park | 28,800 |
| Historic place | 1800 |

Group A characteristics

$$T_1 = M \cup G$$
$$T_2 = Z \cup Q$$
$$T_3 = R \cup K$$
$$T_4 = I \cup H$$

(10)

Each one of the 30 instances $A_{1,2,...,30}$ has the following characteristics:

$$A_{1,2,...,30} = \bigcup_{i=1}^{4} T_i$$

(11)

Group B characteristics

$$T_1 = (M \cap S) \cup (G \cap S)$$
$$T_2 = (R \cap S) \cup (K \cap S)$$
$$T_3 = (Z \cap S) \cup (Q \cap S)$$
$$T_4 = (I \cap S) \cup (H \cap S)$$

(12)

Each one of the 30 instances $B_{1,2,...,30}$ has the following characteristics:

$$B_{1,2,...,30} = \bigcup_{i=1}^{4} T_i$$

(13)

Group C characteristics:

$$T_1 = (M \cap S)$$
$$T_2 = (R \cap S)$$
$$T_3 = (Z \cap S)$$
$$T_4 = (H \cap S)$$

(14)

Each one of the 30 instances $C_{1,2,...,30}$ has following characteristics:

$$C_{1,2,\dots,30} = \bigcup_{i=1}^{4} T_i \tag{15}$$

For the instance design process, we consider worst-case scenarios for the TA app. Therefore, we chose a city with a high number of PoIs, a reduced travel time (a 12-hour day), and include all types $(T_1, T_2, T_3, T_4)$ of proposed PoIs for the instances.

## 4.2 Instance Generation Process

*Real-world Instances* of TTDP were generated by the process of search around from the mobile app *Turisteando Ando*, gathering data and metadata from *Places API by Google* and *Directions API by MapBox*, the sequence of actions to build an instance are as follow:

1. On a 30 km radius around the *current tourist-location*, a request is sent to *Places API*.
2. The last action on *Places API* will return the set of places (PoIs) along with information describing each PoI.
3. Finally, travel time from the *tourist-location* to the POIs is calculated on the *Directions API by MapBox*.

Each instance has four key data which are processed by the algorithm to offer an OSTP:

- *Size of instance* $(n)$: Every instance has $n$ elements. Index $(i)$ is a correlative number that indicates the position of each PoI of the instance. For this study, the index represents the order in which the information on each PoI was received from Google data source.
- *Profits* $(P_i)$: These values are assigned by tourists once they visited a PoI. $(P_i)$ is part of data gathered from *Places* API. It is obtained for each item (PoI) in the knapsack. In all items, $P_i \in \mathbb{R} : 0 \le P_i \le 5$.
- *Weights* $(W_i)$: Weights of knapsack are calculated by the Eq. (4). A request is sent to *Directions API* with the intention to gather data to calculate the travel time from the *tourist-location* to each PoI returned from *Places API*. In all items $W_i \in \mathbb{R} : W_i > 0$.
- *Knapsack capacity* $C$: It affixes to travel time (see Sect. 2.2). For every instance, independently of the group to which it corresponds, the knapsack capacity value is set to 43,200 s, the equivalent of a 12-hour day.

Table 4 shows the number of PoIs $(n)$ per instance in the groups.

Strong relation exists between correlation and difficulty of problems, the latter is greatly affected by the correlation between profits and weights. A higher correlation implies a smaller value of the difference, and also means hardness to be resolved. So instances in Table 9 column *correlation*, show a low complexity.

**Table 4** Size of instances per groups

| Group | Number of PoIs per instance |
| --- | --- |
| A | $n \in \mathbb{Z} : 400 \leq n \leq 450$ |
| B | $n \in \mathbb{Z} : 230 \leq n \leq 240$ |
| C | $n \in \mathbb{Z} : 180 \leq n \leq 185$ |

### 4.3  Taguchi DOE for the Analysis of Constraint-Handling Methods in the Quality of Solutions

The steps followed to design our experiments using the Taguchi DOE method are detailed below.

**Step 1: Identify Control Factors Influencing a Quality Solution**

The control factors are those that affect the solution of the problem. The control factors for the different experiments are listed below:

- Penalty Functions Experiments.

  – $k$: Quantitative parameter.
  – *Penalty Function*: Qualitative parameter.

- Repair Methods Experiments.

  – *Repair method*: Qualitative parameter.
  – $k$: Quantitative parameter.
  – *Repair percentage*: Qualitative parameter.

- Decoders Experiments.

  – *Decoder type*: Qualitative parameter.
  – $k$: Quantitative parameter.

**Step 2: Determine the Number of Levels and Their Values for All Factors**

Experiments are run with different factor values, called levels. Their values remain static once they have been chosen and determine the solution. The factor levels for this study were selected according to the guidelines below. The collection of factors for the experiments, with their levels, is found in Table 5.

- **Penalty functions experiments**

  – **Factor 1** $k$: It refers to the number of individuals that compete in the parent selection tournament.

    Level 1: 2 individuals.
    Level 2: 5 individuals.
    Level 3: 10 individuals.

**Table 5** Factors and levels determined

| Factor | Level 1 | Level 2 | Level 3 |
|---|---|---|---|
| *Penalty functions experiments* | | | |
| $k$ | 2 | 5 | 10 |
| $A_p[i]$ | $A_p[1]$ | $A_p[2]$ | $A_p[3]$ |
| *Repair methods experiments* | | | |
| $A_r[i]$ | $A_r[1]$ | $A_r[2]$ | |
| $k$ | 2 | 5 | 10 |
| *% repair* | 5 | 50 | 100 |
| *Decoders experiments* | | | |
| $A_d[i]$ | $A_d[1]$ | $A_d[2]$ | |
| $k$ | 2 | 5 | 10 |

– **Factor 2** Penalty functions ($A_p[i]$): They are the strategies used to estimate the penalty value for infeasible solutions. In this case, we considered three levels that describe the growth of the penalization function on the degree of violation

Level 1: Logarithmic ($A_p[1]$). See Eq. (5).
Level 2: Linear ($A_p[2]$). See Eq. (6).
Level 3: Quadratic ($A_p[3]$). See Eq. (7).

- **Repair methods experiments**

  – **Factor 1** Repair method ($A_r[i]$): It is the approach employed to repair infeasible solutions. We examine two levels that vary in the process that selects the items (PoIs) to remove from the knapsack.

  Level 1: Random repair method ($A_r[1]$).
  Level 2: Greedy repair method ($A_r[2]$).

  – **Factor 2** $k$: It refers to the number of individuals that compete in the parent selection tournament.

  Level 1: 2 individuals.
  Level 2: 5 individuals.
  Level 3: 10 individuals.

  – **Factor 3** Repair percentage (*% repair*): It corresponds to the percentage of infeasible solutions that can be repaired.

  Level 1: 5%.
  Level 2: 50%.
  Level 3: 100%.

- **Decoders experiments**

    - **Factor 1** Decoder ($A_d[i]$): This indicates the strategy used to decode solutions. We consider two levels that vary in the order of the PoIs list that the decoder receives.

        Level 1: Random decoder ($A_d[1]$).
        Level 2: Greedy decoder ($A_d[2]$).

    - **Factor 2** $k$: It refers to the number of individuals that compete in the parent selection tournament.

        Level 1: 2 individuals.
        Level 2: 5 individuals.
        Level 3: 10 individuals.

## Step 3: Select the Orthogonal Array

Taguchi suggested a method based on orthogonal arrays. Taguchi developed a series of arrays that he called, $L_a(b^c)$. There, $L$ is a Latin Square. $a$ is the number of rows in an array that represent the number of tests or experiments that must be performed. $b$ is the number of levels at which each factor will be evaluated. $c$ is the number of columns (levels) in the array. For penalty functions experiments, we consider the $L_9(3^2)$, i.e., three-level series OA. In the case of repair methods experiments, we select the $L_{18}(2^1, 3^2)$, i.e., mixed-level series OA. Finally, for decoders experiments, we chose the $L_6(2^1, 3^1)$, i.e., mixed-level series OA. Details of selected configurations are in the next step 4.

## Step 4: Assign Factors and Interactions to the Columns of OA

Factors and their levels selected in previous steps are assigned to OA. Resulting arrays have configurations to execute different experiments using the SGA algorithm. Tables 6, 7, and 8 present the factors and levels for Taguchi DOE for the experiments (penalty functions, repair methods, and decoders).

## Step 5: Conduct the Experiment

The total number of instances to analyze the effect of constrained handling techniques was 18. This is due to the rule of the Pareto principle (20/80) which states that 20% of cases explain 80% of cases, considering the correlation metric (see Table 9). By using uniform distribution to select instances, six instances were chosen from each group up to 18 instances (20% of the total instances). The selection procedure uses a random number $r$ where $r \in \mathbb{Z} : 1 \leq r \leq 30$. Selected instances for experiments and their characteristics are described in Table 9. Section 5 shows the results obtained from the experiments. It reports solution fitness value ($f$), the number of PoIs in the itinerary ($n$), and time in milliseconds that the SGA needed to find a solution ($t$).

**Table 6** Taguchi DOE application for penalty functions experiments

| Run | $k$ | $A_p[i]$ |
|-----|-----|----------|
| 1 | 2 | $A_p[1]$ |
| 2 | 2 | $A_p[2]$ |
| 3 | 2 | $A_p[3]$ |
| 4 | 5 | $A_p[1]$ |
| 5 | 5 | $A_p[2]$ |
| 6 | 5 | $A_p[3]$ |
| 7 | 10 | $A_p[1]$ |
| 8 | 10 | $A_p[2]$ |
| 9 | 10 | $A_p[3]$ |

**Table 7** Taguchi DOE application for repair methods experiments

| Run | $A_r[i]$ | $k$ | % repair |
|-----|----------|-----|----------|
| 1 | $A_r[1]$ | 2 | 5 |
| 2 | $A_r[1]$ | 2 | 50 |
| 3 | $A_r[1]$ | 2 | 100 |
| 4 | $A_r[1]$ | 5 | 5 |
| 5 | $A_r[1]$ | 5 | 50 |
| 6 | $A_r[1]$ | 5 | 100 |
| 7 | $A_r[1]$ | 10 | 5 |
| 8 | $A_r[1]$ | 10 | 50 |
| 9 | $A_r[1]$ | 10 | 100 |
| 10 | $A_r[2]$ | 2 | 5 |
| 11 | $A_r[2]$ | 2 | 50 |
| 12 | $A_r[2]$ | 2 | 100 |
| 13 | $A_r[2]$ | 5 | 5 |
| 14 | $A_r[2]$ | 5 | 50 |
| 15 | $A_r[2]$ | 5 | 100 |
| 16 | $A_r[2]$ | 10 | 5 |
| 17 | $A_r[2]$ | 10 | 50 |
| 18 | $A_r[2]$ | 10 | 100 |

## 5 Results and Discussion

This section includes steps 6 and 7 of Taguchi DOE. Step 6 comprises the analysis of resulting data, to make conclusions from them in step 7.

In those steps, Taguchi DOE uses a logarithmic quality loss function to measure the performance characteristics, defined as signal-to-noise ratio (SNR). It is inversely proportional to variance, so the chosen factors should produce the maximum value of

**Table 8** Taguchi DOE application for decoder methods experiments

| Run | $A_d[i]$ | $k$ |
|-----|----------|-----|
| 1 | $A_d[1]$ | 2 |
| 2 | $A_d[1]$ | 5 |
| 3 | $A_d[1]$ | 10 |
| 4 | $A_d[2]$ | 2 |
| 5 | $A_d[2]$ | 5 |
| 6 | $A_d[2]$ | 10 |

**Table 9** Selected instances for the experiments

| No | User-location | | PoIs per instance | | | | | Correlation |
|----|---------------|---|-------------------|---|---|---|---|-------------|
| | Latitude | Longitude | Museums ($T_1$) | Parks ($T_2$) | Zoos ($T_3$) | Historical ($T_4$) | Total | |
| *Group A instances* | | | | | | | | |
| 29 | 19.497038 | −99.139048 | 120 | 120 | 77 | 120 | 437 | 0.002193 |
| 2 | 19.432751 | −99.133114 | 120 | 120 | 77 | 120 | 437 | 0.002204 |
| 7 | 19.303104 | −99.150506 | 120 | 120 | 75 | 120 | 435 | 0.001597 |
| 22 | 19.318541 | −99.032161 | 120 | 120 | 74 | 118 | 432 | 0.001126 |
| 5 | 19.546176 | −99.147456 | 120 | 120 | 77 | 120 | 437 | 0.001902 |
| 1 | 19.435726 | −99.143893 | 120 | 120 | 77 | 120 | 437 | 0.001747 |
| *Group B instances* | | | | | | | | |
| 8 | 19.427061 | −99.167514 | 102 | 71 | 5 | 60 | 238 | 0.001938 |
| 5 | 19.546176 | −99.147456 | 100 | 71 | 5 | 60 | 236 | 0.001306 |
| 24 | 19.402627 | −99.208645 | 99 | 71 | 5 | 60 | 235 | 0.001579 |
| 9 | 19.44645 | −99.150297 | 100 | 72 | 5 | 60 | 237 | 0.002148 |
| 28 | 19.352401 | −99.14289 | 96 | 71 | 5 | 60 | 232 | 0.001658 |
| 16 | 19.360845 | −99.164755 | 102 | 71 | 5 | 60 | 236 | 0.001726 |
| *Group C instances* | | | | | | | | |
| 16 | 19.360845 | −99.164755 | 60 | 60 | 4 | 60 | 184 | 0.001497 |
| 10 | 19.452487 | −99.137261 | 60 | 60 | 4 | 60 | 184 | 0.001380 |
| 14 | 19.405579 | −99.09965 | 60 | 60 | 5 | 60 | 185 | 0.001390 |
| 12 | 19.254841 | −99.089113 | 60 | 60 | 4 | 60 | 184 | 0.000689 |
| 22 | 19.318541 | −99.032161 | 60 | 60 | 4 | 60 | 184 | 0.000917 |
| 15 | 19.385868 | −99.226543 | 60 | 60 | 4 | 60 | 184 | 0.001196 |

SNR to achieve minimum variability. The reference of performance characteristics is available in three categories to determine the SNR ratio [27]. For this work, we selected *Larger-the-Best* due to the objective function shown in the Eq. (1).

**Step 6: Analyze the Data**

Next subsections describe resulting data from Taguchi DOE experimental configurations organized in summary tables.

## 5.1  Summary of Results on Penalty Functions Experiments

A common challenge in employing penalty functions is in selecting an appropriate *penalty parameter* ($P_p$). Usually, this parameter is selected by the trial-and-error method. Work in [28] proposes a mechanism to select $P_p$. So implementing a mechanism to select $P_p$ is a key factor to improve the effect on solutions, our experimental results of the SGA using penalty functions techniques could not find feasible solutions to solve problem like authors in [6], concluding that it is necessary to explore the use of the abovementioned mechanism, which was not considered in this work.

## 5.2  Summary of Results on Repair Techniques Experiments

Table 10 shows the results of using repair techniques. The running time increases due to repair methods using time to become feasible solutions. There is an upward trend in groups for the running time parameter.

Table 11 reports the mean values and the SNR of mean values from Table 10. Configuration 15 shows the highest values on SNR. The *coefficient of determination* $R^2$ is the proportion of the variance in the dependent variable that is predictable from the independent variable. The linear model produced has an $R^2$ equal to 96.91%, while the same coefficient considering the number of variables in the so-called model $R\,adjusted$ is at 95.63%. The factor with biggest effect in $f$ was *repair method* with a delta value of 3.59, while the second one was $\%\,repair$ with 1.01, and the third one $k$ at 0.45 (see Table 12). On Taguchi DOE, it means that the most influential factor on $f$ value is the repair method, and the less influential factor is $k$.

Figures 5 and 6 have three sections. The first one includes the repair method graph. It has the highest slope compared to $k$-graph and $\%\,repair$. The second higher slope is $\%\,repair$, and finally, the third one is $k$.

The reference of performance is Larger-the-best (Table 13), and considering information in Fig. 5 the optimal vector $P_1^*$ is built by taking the highest value by graph: {greedy repair, 10, 100} = {$A_r$[2], $k = 10$, $\%\,repair = 100\%$}. Similar behavior was found in Fig. 6, resulting in the same $P_1^*$.

$p$-value or probability value is, for a given statistical model, the probability that, when the null hypothesis is true, the statistical summary would be greater than or

**Table 10**  Summary of results on repair methods including groups $A$, $B$, $C$

| Run | Group A | | | Group B | | | Group C | | |
|---|---|---|---|---|---|---|---|---|---|
| | $f$ | $n$ | $t$ | $f$ | $n$ | $t$ | $f$ | $n$ | $t$ |
| 1 | 23.93 | 5.33 | 23795.00 | 32.38 | 7.17 | 6100.50 | 27.12 | 6.00 | 3812.67 |
| 2 | 29.72 | 6.50 | 128335.00 | 37.97 | 8.33 | 24378.50 | 32.97 | 7.17 | 13373.17 |
| 3 | 29.87 | 6.67 | 131252.83 | 37.42 | 8.17 | 24207.67 | 32.60 | 7.17 | 13354.50 |
| 4 | 25.70 | 5.83 | 27947.83 | 31.80 | 7.00 | 7274.17 | 28.93 | 6.33 | 4666.00 |
| 5 | 30.20 | 6.67 | 127324.00 | 40.63 | 8.83 | 24577.17 | 35.80 | 7.83 | 13423.17 |
| 6 | 32.50 | 7.17 | 133610.17 | 41.37 | 9.00 | 24600.67 | 36.75 | 8.00 | 13336.83 |
| 7 | 27.08 | 6.33 | 30206.83 | 35.18 | 7.67 | 7605.00 | 29.77 | 6.50 | 4676.67 |
| 8 | 26.53 | 6.50 | 127708.67 | 43.88 | 9.50 | 25029.00 | 36.88 | 8.00 | 13541.83 |
| 9 | 33.85 | 7.67 | 130876.17 | 43.23 | 9.33 | 24283.17 | 36.78 | 8.00 | 13668.83 |
| 10 | 48.38 | 10.33 | 10659.17 | 54.87 | 12.00 | 3319.33 | 44.05 | 9.67 | 2192.83 |
| 11 | 50.15 | 10.67 | 81782.67 | 55.10 | 12.00 | 20523.83 | 45.72 | 10.00 | 11504.83 |
| 12 | 50.30 | 10.67 | 87429.33 | 55.10 | 12.00 | 20145.17 | 45.55 | 10.00 | 11326.83 |
| 13 | 49.40 | 10.50 | 10855.83 | 54.87 | 12.00 | 3416.67 | 45.57 | 10.00 | 2275.67 |
| 14 | 49.73 | 10.50 | 82172.83 | 55.33 | 12.00 | 20928.67 | 45.62 | 10.00 | 11689.83 |
| 15 | 50.97 | 10.83 | 87812.33 | **55.83** | 12.17 | 20630.33 | 45.73 | 10.00 | 11565.17 |
| 16 | 50.17 | 10.67 | 10872.33 | 55.00 | 12.00 | 3529.17 | 44.90 | 9.83 | 2323.33 |
| 17 | 48.55 | 10.33 | 82162.83 | 55.33 | 12.00 | 20984.83 | 45.77 | 10.00 | 11788.17 |
| 18 | **51.12** | 10.83 | 87538.33 | 55.35 | 12.00 | 20740.83 | **45.80** | 10.00 | 11616.33 |

**Table 11**  SNR in repair methods in groups $A$, $B$, $C$ mean

| Run | Repair method | $k$ | %$repair$ | $f$ | SNR |
|---|---|---|---|---|---|
| 1 | Random repair | 2 | 5 | 27.811 | 28.8843 |
| 2 | Random repair | 2 | 50 | 33.550 | 30.5139 |
| 3 | Random repair | 2 | 100 | 33.294 | 30.4473 |
| 4 | Random repair | 5 | 5 | 28.811 | 29.1912 |
| 5 | Random repair | 5 | 50 | 35.544 | 31.0153 |
| 6 | Random repair | 5 | 100 | 36.872 | 31.3339 |
| 7 | Random repair | 10 | 5 | 30.678 | 29.7365 |
| 8 | Random repair | 10 | 50 | 35.767 | 31.0697 |
| 9 | Random repair | 10 | 100 | 37.956 | 31.5856 |
| 10 | Greedy repair | 2 | 5 | 49.100 | 33.8216 |
| 11 | Greedy repair | 2 | 50 | 50.322 | 34.0352 |
| 12 | Greedy repair | 2 | 100 | 50.317 | 34.0343 |
| 13 | Greedy repair | 5 | 5 | 49.944 | 33.9697 |
| 14 | Greedy repair | 5 | 50 | 50.228 | 34.0189 |
| 15 | Greedy repair | 5 | 100 | 50.844 | 34.1248 |
| 16 | Greedy repair | 10 | 5 | 50.022 | 33.9832 |
| 17 | Greedy repair | 10 | 50 | 49.883 | 33.9591 |
| 18 | Greedy repair | 10 | 100 | 50.756 | 34.1097 |

**Table 12** Repair methods response (Larger-the-best)

| Level | For SNR values | | | For means | | |
|---|---|---|---|---|---|---|
| | Repair method | $k$ | % *repair* | Repair method | $k$ | % *repair* |
| 1 | 30.42 | 31.96 | 31.60 | 33.36 | 40.73 | 39.39 |
| 2 | 34.01 | 32.28 | 32.44 | 50.16 | 42.09 | 42.55 |
| 3 | | 32.41 | 32.61 | | 42.51 | 43.34 |
| Delta | 3.59 | 0.45 | 1.01 | 16.79 | 1.78 | 3.95 |
| Rank | 1 | 3 | 2 | 1 | 3 | 2 |

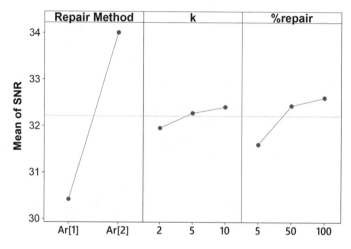

**Fig. 5** Response plot of repair methods effects for SNR on SGA results

equal to the actual observed results. The smallest the $p$-value the strongest the evidence on the data.

Table 14 in the first row, column seven (Repair method) the $p$-value is zero, denoting high statistical evidence to support information on *repair method values* facing the hypothesis proposed, similar result is found in the third row, column seven, (% *repair*) the $p$-value is 0.012. In the other case, for $k$ there is not enough statistical evidence to support results.

The effect of repair methods in solutions is clear, (see $f$ in Table 10), which has two groups, the second group has better results on $f$ than the first one. As % *repair* increase, the quality of solutions does too, this pattern is present in the rest of subsets with size three in the above mentioned table. Another pattern is: the best solutions are related to the percentage of repair being equal to 100%.

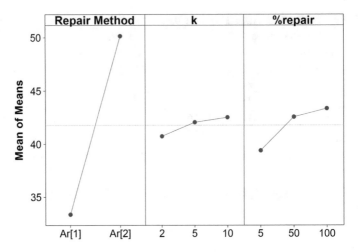

**Fig. 6** Response plot of repair methods effects for means on SGA results

**Table 13** Types of problems and signal-to-ratio function [27]

| Reference of performance | SNR formulas | Goal is to… |
|---|---|---|
| Smaller-the-best | $SNR = -10 \log[\frac{1}{n} \sum_{i=1}^{n} y_i^2]$ | Minimize the response |
| Nominal-the-best | $SNR = \log \frac{\mu^2}{\sigma^2}$ | Target the response and you want to base the S/N ratio on means and standard deviations |
| Larger-the-best | $SNR = -10 \log[\frac{1}{n} \sum_{i=1}^{n} \frac{1}{y_i^2}]$ | Maximize the response |

**Table 14** Analysis of variance for SNR values

| Source | DF | Seq SS | Adj SS | Adj MS | F | P |
|---|---|---|---|---|---|---|
| Repair method | 1 | 57.8843 | 57.8843 | 57.8843 | 214.39 | 0.000 |
| $k$ | 2 | 0.6461 | 0.6461 | 0.3230 | 1.20 | 0.336 |
| $\% repair$ | 2 | 3.4942 | 3.4942 | 1.7471 | 6.47 | 0.012 |
| Residual error | 12 | 3.2399 | 3.2399 | 0.2700 | | |
| Total | 17 | 65.2644 | | | | |

## 5.3 Summary of Results in Decoders Techniques Experiments

Decoders map from a solution space to another one in which it becomes simpler to find a feasible solution for the SGA. Comparing decoders $f$ values with the ones from repair techniques, as Table 15 shows, decoders have the worst performance to solve the TTDP.

**Table 15** Summary of results in decoders in groups $A$, $B$, $C$

| Run | Group $A$ | | | Group $B$ | | | Group $C$ | | |
|-----|-----------|---|---|-----------|---|---|-----------|---|---|
| | $f$ | $n$ | $t$ | $f$ | $n$ | $t$ | $f$ | $n$ | $t$ |
| 1 | **31.43** | 6.83 | 1927.17 | 37.02 | 8.17 | 597.00 | **35.57** | 7.83 | 413.67 |
| 2 | 31.40 | 7.17 | 1497.83 | 39.58 | 8.67 | 592.17 | 33.40 | 7.33 | 415.67 |
| 3 | 30.92 | 7.00 | 1497.67 | **39.98** | 8.83 | 596.00 | 32.05 | 7.00 | 416.83 |
| 4 | 30.17 | 6.67 | 1497.83 | 38.65 | 8.50 | 593.67 | 34.52 | 7.67 | 417.67 |
| 5 | 29.82 | 7.00 | 1498.33 | 36.48 | 8.00 | 597.50 | 33.48 | 7.33 | 416.00 |
| 6 | 31.22 | 7.00 | 1484.83 | 38.15 | 8.33 | 598.50 | 33.35 | 7.33 | 418.00 |

**Table 16** SNR in decoder methods in groups $A$, $B$, $C$ mean

| Run | Decoder method | $k$ | $f$ | SNR |
|-----|----------------|-----|-----|-----|
| 1 | Random decoder | 2 | 34.6722 | 30.7996 |
| 2 | Random decoder | 5 | 34.7944 | 30.8302 |
| 3 | Random decoder | 10 | 34.3167 | 30.7101 |
| 4 | Greedy decoder | 2 | 33.2611 | 30.4387 |
| 5 | Greedy decoder | 5 | 34.4444 | 30.7424 |
| 6 | Greedy decoder | 10 | 34.2389 | 30.6904 |

**Table 17** Analysis of Variance for SNR values

| Source | DF | Seq SS | Adj SS | Adj MS | F | P |
|--------|----|--------|--------|--------|---|---|
| Decoder method | 1 | 0.03657 | 0.03657 | 0.03657 | 2.24 | 0.273 |
| $k$ | 2 | 0.02793 | 0.02793 | 0.01397 | 0.86 | 0.539 |
| Residual error | 2 | 0.03260 | 0.03260 | 0.01630 | | |
| Total | 5 | 0.09710 | | | | |

The best solution of decoders corresponds to Taguchi's DOE design three, having $k = 5$ and a random decoder ($A_d[1]$) (see Table 16). Table 16 shows mean $f$ and SNR values. Configuration one is the second best of Taguchi's DOE design. From the linear model analysis, $R^2$ value and $R$ $adjusted$ are both 100%. The largest coefficient of the model was *decoding method* with 0.3065 while the strongest interaction was $k$-decoder having 0.4602. On Taguchi DOE, this means that the most influential factor on $f$ is the decoding method, and the less influential factor is $k$. For decoders, solutions are not good enough to exceed the solution values of repair methods.

Analyzing Table 17 along with Table 18, we found that ranking of factors have not enough statistical evidence on data about the hypothesis, due to the short delta showed by them. Same information we found on Table 17, analyzing column $F$ and $P$ (See Fig. 7 and Fig. 8).

**Table 18** Decoders response (Larger-the-best)

| Level | For SNR values | | For means | |
|---|---|---|---|---|
| | Decoder method | $k$ | Decoder method | $k$ |
| 1 | 30.78 | 30.62 | 34.59 | 33.97 |
| 2 | 30.62 | 30.79 | 33.98 | 34.62 |
| 3 | | 30.70 | | 34.28 |
| Delta | 0.16 | 0.17 | 0.61 | 0.65 |
| Rank | 2 | 1 | 2 | 1 |

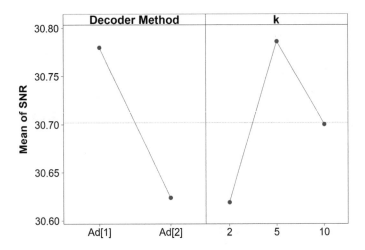

**Fig. 7** Response plot of decoders effects for SNR on SGA results

**Step 7: Interpret Results**

After analyzing the three families of constraint-handling techniques, we found that the best results are provided by repair methods. However, the time to provide them is high compared to decoder methods; thus decoders are good options in mobile devices since they have constrained resources.

## 5.4 Optimal Scheduling Trip Plans Offered by SGA

Results on Table 19 were obtained using the $P_1^*$ according to the experiments defined in Sect. 4.2. Information on $x$ about the sum of the weights of items in the knapsack is included once the optimization process ended. Results show that the quality of solutions from the SGA having a greedy repair method considerably outperforms the other constraint-handling mechanisms analyzed here.

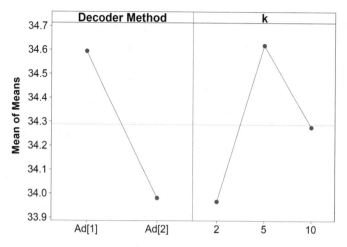

**Fig. 8** Response plot of decoders effects for means on SGA results

Through the $x$-values, we observed a recurring pattern in all experiments. In 23 out of the 24 experiments per group, solutions found by the algorithm are near the optimal solution; an optimal solution exists when the knapsack is full. This is based on the sum of the weights of the items included in the knapsack $(x)$. For example, Group $A$, in 23 out of the 24 experiments, $x \in Z : 41641.2 \leq x \leq 43176.0$. considering that, for any $x$-value, theoretically, their smallest weight of elements to be included in the knapsack is 1800; and the knapsack capacity is $C = 43200$. The 23rd instance is an atypical case that does not present such a property. However, its results have the lowest fitness values. The before-mentioned behavior is repeated in the instance with the same identifier by all groups. It occurs in the two remaining groups of instances.

Regarding the fitness of the solutions, in all groups, the algorithm obtains good solutions. The mean of the fitness per group is 50.44, 49.53, and 49.11, respectively. All groups presented atypical data. The 23rd instance presented the worst fitness for Groups A and C. In the case of Group B, the 18th and 23rd instances performed the worst fitness values, considering them to be atypical data for the group.

Regarding the number of PoIs (items) included in the knapsack, the mean per group was 10.7, 10.8, and 10.7, respectively. The pattern of atypical data (worse solutions) is the same as the case for the fitness value for groups A and B. But, Group C does not present any atypical data in the places included in the knapsack.

Inspection of Table 19 reveals that the performance of the SGA using the greedy repair method is directly related to the dimensionality of the instances: the smaller the instance (according to group classification) lengths the shorter the execution time. In fact, the size of instances did not have profound implications in the quality of solutions. The maximum time for one of 30 instances of Group A is about 90 s; and for any of 30 instances of groups B and C, the maximum time is around 22 and 12 s, respectively.

**Table 19** Optimal scheduling trip plans offered by SGA using the $P_1^*$, (Repair Techniques)

| Instance | Group A | | | | Group B | | | | Group C | | | |
|---|---|---|---|---|---|---|---|---|---|---|---|---|
| | $f$ | $n$ | $x$ | $t$ | $f$ | $n$ | $x$ | $t$ | $f$ | $n$ | $x$ | $t$ |
| 1 | | | | | 67.4 | 15 | 42729.6 | 20947 | 67.6 | 15 | 42843.2 | 11287 |
| 2 | | | | | 68.6 | 15 | 43066.6 | 22172 | 68 | 15 | 41765.2 | 11728 |
| 3 | 62.4 | 13 | 42883.2 | 87155 | 55 | 12 | 42370.6 | 21664 | 54.9 | 12 | 42013.2 | 13379 |
| 4 | 53.3 | 11 | 43172.8 | 98691 | 50.3 | 11 | 43080.8 | 20747 | 50.3 | 11 | 43080.8 | 13549 |
| 5 | | | | | | | | | 36.8 | 8 | 42694.8 | 13729 |
| 6 | 53.1 | 11 | 41880.3 | 90021 | 55.7 | 12 | 41735.6 | 23190 | 55.7 | 12 | 42175.6 | 12469 |
| 7 | | | | | 45.6 | 10 | 43177.2 | 21697 | 41.9 | 9 | 42288.6 | 13397 |
| 8 | 59.5 | 13 | 42994.2 | 89825 | | | | | 64.8 | 14 | 42502.8 | 11577 |
| 9 | 61.6 | 13 | 42176.8 | 88862 | | | | | 68.2 | 15 | 43012.8 | 11179 |
| 10 | 56.6 | 12 | 43156.2 | 90555 | 58.5 | 13 | 42944.4 | 24891 | | | | |
| 11 | 61.5 | 13 | 42886.2 | 87066 | 68.6 | 15 | 42828.4 | 22397 | 68.5 | 15 | 43180.2 | 12125 |
| 12 | 36 | 8 | 41084.6 | 84237 | 28.2 | 6 | 42709.2 | 22216 | | | | |
| 13 | 48.2 | 10 | 42742 | 90206 | 51 | 11 | 41811 | 21299 | 51.1 | 11 | 42770.4 | 12242 |
| 14 | 56.9 | 12 | 42796.6 | 86915 | 54.7 | 12 | 42596.6 | 22171 | | | | |
| 15 | 45.1 | 10 | 42903.8 | 91029 | 46.1 | 10 | 42127.2 | 21037 | | | | |
| 16 | 56.9 | 12 | 41641 | 91313 | | | | | | | | |
| 17 | 50.3 | 11 | 43150.8 | 94287 | 46.1 | 10 | 43164.4 | 22168 | 42 | 9 | 41944.8 | 13093 |
| 18 | 40.4 | 9 | 43163 | 78932 | 21.8 | 5 | 43036.4 | 22001 | 18.5 | 4 | 42840.8 | 12951 |
| 19 | 36 | 8 | 42400 | 88979 | 27.7 | 6 | 42301.6 | 24040 | 27.6 | 6 | 42222.4 | 12414 |
| 20 | 57.5 | 12 | 43096.8 | 91439 | 65.1 | 14 | 42859.6 | 21069 | 64.7 | 14 | 43120.6 | 11679 |
| 21 | 52 | 11 | 43176 | 90616 | 45.9 | 10 | 42733 | 22452 | 45.9 | 10 | 43174.8 | 13617 |
| 22 | | | | | 31.6 | 7 | 43093.8 | 23513 | | | | |
| 23 | 19.6 | 4 | 42428 | 85565 | 14.2 | 3 | 40272.4 | 25165 | 14.1 | 3 | 37423.4 | 13069 |
| 24 | 50.2 | 11 | 42893 | 98290 | | | | | 51.3 | 11 | 42963 | 11369 |
| 25 | 45.6 | 10 | 42491.6 | 92782 | 46.2 | 10 | 42391 | 22072 | 46.2 | 10 | 42646.6 | 11787 |
| 26 | 53 | 11 | 42783.2 | 94260 | 49.1 | 11 | 43085.2 | 25022 | 49.3 | 11 | 42950.4 | 13113 |
| 27 | 51.4 | 11 | 43030.6 | 91980 | 50.9 | 11 | 42872.2 | 22641 | 50.6 | 11 | 42900.4 | 12999 |
| 28 | 56.3 | 12 | 42860.8 | 94779 | | | | | 55.2 | 12 | 42508.2 | 12175 |
| 29 | | | | | 50.3 | 11 | 42462.8 | 23763 | 50.4 | 11 | 42379.2 | 12202 |
| 30 | 47.8 | 10 | 42760.2 | 96331 | 51.3 | 11 | 42614 | 20981 | 51.2 | 11 | 41720.2 | 12067 |

Figure 9 shows that $f P_1^*$ of repair methods and $f P_2^*$ of decoders have the same order of magnitude; as well as $n P_1^*$ and $n P_2^*$. $t P_1^*$ is almost five orders of magnitude $t P_2^*$ is near to three orders of magnitude. Concluding that $t P_1^*$ is two orders of magnitude larger than $t P_2^*$, while the quality of solution is the same magnitude.

$x P_1^*$ and $x P_2^*$ are values that correspond to the sum of the objects' weights that have been included in the knapsack. Both are almost five orders of magnitude values. If we compare OSTP by instance, the PoIs are different, because decoding handling techniques move to SGA on non-equal space of solutions compared to the space

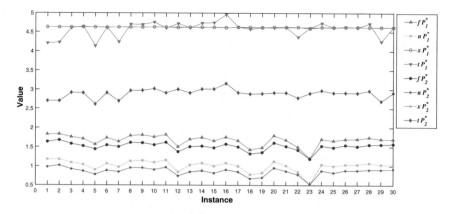

**Fig. 9** A logarithmic comparison of the constraint-handling techniques analyzed

explode-explored by SGA using repair methods. However OSTP, considering the knapsack size, is full using repair methods or decoding procedures.

Information gathered from Fig. 9 comparison allows to build around which is the technique that has greater effect on the quality of solutions. An answer is: the TA works on mobile devices, repair, and decoding methods; $tP_2^*$ is two orders of magnitude lower than $tP_1^*$. Therefore, decoders are an excellent option to be implemented in these before-mentioned devices (see Table 20). In this study, we did not find enough statistical evidence to select one of the analyzed methods.

We observe a hint of bias in OSTP given by the SGA using configuration vector $P_1^*$. This bias occurs due to the way that the greedy repair method works. Table 21 shows an OSTP example. There, the highest values of the $P/W$ value were selected by the greedy repair method. In contrast, the configuration vector $P_2^*$ for the random decoder technique has greater diversity in PoIs types than the ones from the configuration vector $P_1^*$ for the greedy repair. This is due to historical places having high P/W values (Profit ($P$) and Weight ($W$)), while random decoder does not use P/W values, so bias related to P/W does not exist.

## 6 Conclusions and Future Work

Based on the results shown in this chapter, the following conclusions are drawn:

There is enough statistical evidence on data to support that the greedy repair method is the best in terms of the quality of solutions. The greedy repair method offers solutions that outperform the solutions given by the other methods. However, it has a trend towards one type of PoIs on a personalized trip plan generated (see

**Table 20** Optimal scheduling trip plans offered by SGA using the $P_2^*$, (Decoders)

| Instance | Group A | | | | Group B | | | | Group C | | | |
|---|---|---|---|---|---|---|---|---|---|---|---|---|
| | $f$ | $n$ | $x$ | $t$ | $f$ | $n$ | $x$ | $t$ | $f$ | $n$ | $x$ | $t$ |
| 1 | | | | | 41.0 | 9 | 42376.2 | 593 | 45.3 | 10 | 42979.2 | 411 |
| 2 | | | | | 45.2 | 10 | 42267.4 | 592 | 50.2 | 11 | 43177.8 | 411 |
| 3 | 36.5 | 8 | 42701.0 | 1480 | 41.4 | 9 | 42507.0 | 581 | 36.5 | 8 | 42705.8 | 428 |
| 4 | 31.5 | 7 | 40617.4 | 1453 | 32.2 | 7 | 42831.2 | 587 | 36.6 | 8 | 42976.4 | 409 |
| 5 | | | | | | | | | 27.7 | 6 | 42941.0 | 409 |
| 6 | 32.4 | 7 | 42220.2 | 1477 | 36.3 | 8 | 42946.6 | 584 | 37.1 | 8 | 42141.6 | 407 |
| 7 | | | | | 31.9 | 7 | 42445.6 | 589 | 32.0 | 7 | 42635.8 | 407 |
| 8 | 36.6 | 8 | 41256.8 | 1463 | | | | | 45.7 | 10 | 43147.8 | 407 |
| 9 | 35.4 | 8 | 43180.6 | 1484 | | | | | 45.4 | 10 | 42633.2 | 414 |
| 10 | 30.8 | 7 | 42581.2 | 1461 | 40.8 | 9 | 43162.2 | 647 | | | | |
| 11 | 34.7 | 8 | 42975.4 | 1463 | 40.6 | 9 | 42901.2 | 594 | 50.0 | 11 | 42974.8 | 409 |
| 12 | 24.4 | 6 | 42976.6 | 1423 | 23.2 | 5 | 43136.4 | 616 | | | | |
| 13 | 27.9 | 6 | 41789.8 | 1469 | 32.4 | 7 | 43175.2 | 566 | 36.1 | 8 | 42528.2 | 415 |
| 14 | 30.6 | 7 | 40600.0 | 1467 | 36.0 | 8 | 43170.2 | 595 | | | | |
| 15 | 27.6 | 6 | 43164.6 | 1474 | 32.4 | 7 | 42921.0 | 597 | | | | |
| 16 | 37.6 | 8 | 42054.2 | 1445 | | | | | | | | |
| 17 | 31.5 | 7 | 42407.4 | 1468 | 31.7 | 7 | 42277.8 | 624 | 32.2 | 7 | 42020.0 | 410 |
| 18 | 26.9 | 6 | 41501.0 | 1314 | 18.1 | 4 | 40845.4 | 594 | 18.3 | 4 | 43198.6 | 410 |
| 19 | 22.7 | 5 | 43148.0 | 1361 | 23.0 | 5 | 41740.8 | 599 | 23.1 | 5 | 42530.4 | 411 |
| 20 | 35.3 | 8 | 43122.6 | 1489 | 45.8 | 10 | 41736.4 | 597 | 41.0 | 9 | 42941.2 | 408 |
| 21 | 30.0 | 7 | 40930.4 | 1485 | 36.5 | 8 | 42530.2 | 593 | 32.4 | 7 | 42121.2 | 415 |
| 22 | | | | | 26.5 | 6 | 42795.8 | 612 | | | | |
| 23 | 18.2 | 4 | 42293.0 | 1397 | 14.4 | 3 | 42293.0 | 608 | 14.1 | 3 | 42501.2 | 403 |
| 24 | 31.4 | 7 | 42153.0 | 1484 | | | | | 36.1 | 8 | 42820.2 | 411 |
| 25 | 28.0 | 6 | 43049.2 | 1471 | 32.1 | 7 | 39690.4 | 596 | 32.0 | 7 | 42666.0 | 405 |
| 26 | 32.5 | 7 | 43063.0 | 1472 | 37.0 | 8 | 42745.6 | 598 | 36.5 | 8 | 42462.6 | 414 |
| 27 | 30.1 | 8 | 42217.2 | 1453 | 36.1 | 8 | 42821.2 | 588 | 32.1 | 7 | 42931.0 | 409 |
| 28 | 32.2 | 7 | 42709.0 | 1457 | | | | | 40.6 | 9 | 42776.4 | 411 |
| 29 | | | | | 36.9 | 8 | 42941.6 | 576 | 36.7 | 8 | 42888.8 | 417 |
| 30 | 32.3 | 7 | 43170.8 | 1481 | 40.2 | 9 | 42748.6 | 575 | 40.6 | 9 | 43091.6 | 410 |

Table 21). In data analyzed, the $P^*$ vectors proposed by the Taguchi DOE, do not confirm an effect or a significant difference between decoding techniques on the SGA.

Although the greedy repair method offers solutions that outperform the solution given by the other methods, considering the running time factor to offer a solution, decoder techniques outperform greedy repair technique. The running time of the last one is two orders of magnitude greater than decoder techniques.

**Table 21** Example of a personalized trip plan: PoI characteristics (Instance 1, Group $B$)

| Place name | Profit ($P$) | Weight ($W$) | $P/W$ |
|---|---|---|---|
| *Personalized trip plan using $P_1^*$* | | | |
| San Juan de Dios Church | 4.3 | 2032.6 | 0.002116 |
| Parroquia de San Fernando | 4.2 | 2315.8 | 0.001814 |
| Rectory Our Lady of Guadalupe Queen of Peace | 4.7 | 2639.2 | 0.001781 |
| Church of San Hipolito | 4.6 | 2632.8 | 0.001747 |
| Church of San Francisco | 4.7 | 2739.8 | 0.001715 |
| Parroquia del Inmaculado Corazón de María | 4.6 | 2832.2 | 0.001624 |
| Regina Coeli parish | 4.4 | 2713.8 | 0.001621 |
| Church of San Bernardo | 4.4 | 2787.2 | 0.001579 |
| Nuestra Señora de Loreto Church | 4.3 | 2800.4 | 0.001535 |
| Iglesia San Rafael Arcángel y San Benito Abad | 4.6 | 3037.6 | 0.001514 |
| Santuario Parroquial Nuestra Señora de los Angeles | 4.2 | 2803.2 | 0.001498 |
| Parroquia de la Sagrada Familia | 4.7 | 3153.2 | 0.001491 |
| Parroquia Merced de las Huertas | 4.4 | 3185.4 | 0.001381 |
| Parroquia de la Santísima Trinidad y Nuestra Señora del Refugio | 4.6 | 3434.6 | 0.001399 |
| Parroquia Coronación de Santa María de Guadalupe | 4.7 | 3621.8 | 0.001298 |
| *Personalized trip plan using $P_2^*$* | | | |
| Parroquia De La Resurrección Del Señor | 4.4 | 4657.8 | 0.000945 |
| Church of San Francisco | 4.7 | 2739.8 | 0.001715 |
| Museo Anahuacalli | 4.7 | 7428.0 | 0.000633 |
| Parroquia Francesa–Cristo Resucitado y Nuestra Señora de Lourdes | 4.7 | 4526.2 | 0.001038 |
| Rectory Our Lady of Guadalupe Queen of Peace | 4.7 | 2639.2 | 0.001781 |
| Museum of Torture | 4.2 | 4201.0 | 0.001000 |
| Parroquia de la Santa Cruz de la Herradura | 4.6 | 5892.6 | 0.000781 |
| Museo Jumex | 4.6 | 5837.0 | 0.000788 |
| INBA Galería José María Velasco | 4.4 | 4454.6 | 0.000988 |

The random decoder will be implemented in TA because it offers more diverse personalized trip plans with more diversified types of PoIs. In the solution proposed by the random decoder, the knapsack is almost full. This has an impact on the tourist experience since he can visit the recommended places without losing a lot of travel time.

For future work, we propose two topics:

Taguchi DOE, as a methodology for analysis, can gather relevant information. However, Step 1: *Identify control factors influencing quality solution* has challenges achieving its goal. Open questions in this work considering results on penalty functions and decoding methods are:

- Hybrid methods on tuning could be used with the intention to know: Which are penalty parameter $P_p$ methods effective for penalty functions into a SGA resolving the TTDP?
- Random decoder method by itself presents a good performance. Therefore, if a hybrid decoder mechanism is included: Could this reduce the gap between repair method and decoder mechanism on $f$-value?

Another future work is related to the real-world instances here proposed, since they have low complexity. However, instance 23 has characteristics that lead to the poor performance of the SGA. Hence, it is necessary to analyze which are the challenges imposed by this instance to the SGA.

**Acknowledgements** Misael Lopez-Sanchez gratefully acknowledges the scholarship from CONA-CyT to pursue his graduate studies.

# References

1. Gavalas, D., Kasapakis, V., Konstantopoulos, C., Pantziou, G., Vathis, N.: Scenic route planning for tourists. Pers. Ubiquitous Comput. **21**(1), 137–155 (2017)
2. Krishnaiah, K., Shahabudeen, P.: Applied Design of Experiments and Taguchi Methods. PHI Learning Pvt. Ltd., New Delhi (2012)
3. Taguchi, G., Chowdhury, S., Wu, Y., Taguchi, S., Yano, H.: Quality Engineering: The Taguchi Method, Chapt. 4, pp. 56–123. Wiley (2007)
4. Coy, S.P., Golden, B.L., Runger, G., Wasil, E.A.: Using experimental design to find effective parameter settings for heuristics. J. Heuristics **7**(1), 77–97 (2001)
5. Barr, R.S., Golden, B.L., Kelly, J.P., Resende, M.G.C., Stewart, W.R.: Designing and reporting on computational experiments with heuristic methods. J. Heuristics **1**(1), 9–32 (1995)
6. Michalewicz, Z.: Genetic Algorithms + Data Structures = Evolution Programs, 3rd edn. Springer-Verlag, Berlin, Heidelberg (1996)
7. Coello Coello, C.A.: Theoretical and numerical constraint-handling techniques used with evolutionary algorithms: a survey of the state of the art. Comput. Methods Appl. Mech. Eng. **191**(11), 1245–1287 (2002)
8. Padhye, N., Mittal, P., Deb, K.: Feasibility preserving constraint-handling strategies for real parameter evolutionary optimization. Comput. Optim. Appl. **62**(3), 851–890 (2015)
9. Mallipeddi, R., Jeyadevi, S., Suganthan, P.N., Baskar, S.: Efficient constraint handling for optimal reactive power dispatch problems. Swarm Evol. Comput. **5**, 28–36 (2012)
10. Mezura-Montes, E., Cetina-Domínguez, O.: Empirical analysis of a modified artificial bee colony for constrained numerical optimization. Appl. Math. Comput. **218**(22), 10943–10973 (2012)
11. Bremer, J., Sonnenschein, M.: Constraint-handling with support vector decoders. In: International Conference on Agents and Artificial Intelligence, pp. 228–244. Springer (2013)
12. Stanovov, V., Akhmedova, S., Semenkin, E.: Combined fitness-violation epsilon constraint handling for differential evolution. Soft. Comput. **24**, 7063–7079 (2020)
13. Kellerer, H., Pferschy, U., Pisinger, D.: Knapsack Problems. Springer, Berlin Heidelberg (2013)

14. Google LLC: Google trips: planificador de viajes (2019)
15. minube: minube - mis viajes (2019)
16. TripAdvisor. TripAdvisor: hoteles, restaurantes, vuelos (2019)
17. Triposo: World Travel Guide by Triposo (2019)
18. Martello, S., Toth, P.: Knapsack Problems: Algorithms and Computer Implementations. Wiley, USA (1990)
19. Lozano, M., Herrera, F., Cano, J.R.: Replacement strategies to preserve useful diversity in steady-state genetic algorithms. Inf. Sci. **178**(23), 4421–4433 (2008). Including Special Section: Genetic and Evolutionary Computing
20. Eiben, A.E., Smith, J.E.: Introduction to Evolutionary Computing, 2nd ed. Springer Publishing Company, Incorporated (2015)
21. Matsui, K.: New selection method to improve the population diversity in genetic algorithms. In: IEEE SMC'99 Conference Proceedings. 1999 IEEE International Conference on Systems, Man, and Cybernetics (Cat. No. 99CH37028), vol. 1, pp. 625–630 (1999)
22. Back, T.: Selective pressure in evolutionary algorithms: a characterization of selection mechanisms. In: Proceedings of the First IEEE Conference on Evolutionary Computation. IEEE World Congress on Computational Intelligence, vol. 1, pp. 57–62 (1994)
23. Kitayama, D., Ozu, K., Nakajima, S., Sumiya, K.: A Route Recommender System Based on the User's Visit Duration at Sightseeing Locations, pp. 177–190. Springer International Publishing, Cham (2015)
24. Huang, H., Gartner, G.: Using Context-Aware Collaborative Filtering for POI Recommendations in Mobile Guides, pp. 131–147. Springer Berlin Heidelberg, Berlin, Heidelberg (2012)
25. Alejandro Acebal Fernández Antonio Aledo Tur.: AnÁlisis de las caracterÍsticas de los visitantes de los parques naturales de la comunidad valenciana: Parque natural serra d'irta. Technical report, Generalite de Valencia (2011)
26. Instituto Nacional de Estadística y Geografía. Museos de méxico y sus visitantes 2017. Technical report, Instituto Nacional de Estadística y Geografía (2017)
27. Davis, R., John, P.: Application of Taguchi-based design of experiments for industrial chemical processes. In: Statistical Approaches With Emphasis on Design of Experiments Applied to Chemical Processes, p. 137 (2018)
28. Deb, K.: An efficient constraint handling method for genetic algorithms. Comput. Methods Appl. Mech. Eng. **186**(2), 311–338 (2000)

Printed in the United States
by Baker & Taylor Publisher Services